AF713708

Visualized Basic Mechanical Drawing

Second Edition

Paul Wallach
Joseph J. Almon

Glencoe Publishing Company
Encino, California

The author wishes to acknowledge and thank the following individuals for their help: Fred Leon, Steven Magnasco, Brian Booker, and Jamie Romero.

Printed in the United States of America

Glencoe Publishing Company
17337 Ventura Boulevard
Encino, California 91316

Library of Congress Catalog Card Number: 80-84979

ISBN 0-02-829670-2

4 5 6 7 8 9 90 89 88 87 86

Contents

TO THE STUDENT

Like the earlier edition of this text, the second edition of *Visualized Basic Mechanical Drawing* is a basic book. It is an introduction to mechanical drawing, a book for beginners. It can be used in class or as a self-teaching sourcebook for anyone wanting to learn mechanical drawing.

The art in each lesson in this new edition has been completely redrawn and redimensioned. Teachers and students alike should find the art to be clear, understandable, and well suited for learning the basic techniques of draftsmanship. It is recommended to students that they study the drawings carefully and in detail.

There are many new end-of-lesson assignments. These new problems should be used to test your grasp of the content of the lesson. It is also recommended that, on your own or under your teacher's supervision, you practice drawing the examples of mechanical illustration that are shown in the lessons themselves.

As before, the visual approach, with a minimum of reading matter, has been chosen as the teaching method. This is because:

1. You learn mechanical drawing by looking at and drawing pictorial representations of objects. By carefully studying these pictures, you can see how the drafter solves drafting problems. And then you can use the same methods to learn the skills of drafting.
2. By studying drawings, you learn that each line and shape you draw is a surface, or an edge of an actual object. This teaches you to *see* that object in your drawing. It also teaches you the habit of visualizing an object before putting it on paper.

This approach has been successfully used by students over the last twenty years. With the proper application and practice, you will quickly acquire the basic skills of mechanical drawing.

UNIT 1

AN INTRODUCTION TO DRAFTING

This book uses the *visual* approach to drafting to teach you how to draw and read mechanical drawings. You learn drafting by studying the materials and visualizing how a drafter would solve the problem and complete the drawings.

Industry requires many millions of drawings every year. Each product made must have a drawing of each of its parts. These drawings communicate ideas to skilled people who use the drawings to make products of all kinds. To manufacture the airplane shown on this page, more than 100,000 drawings were required.

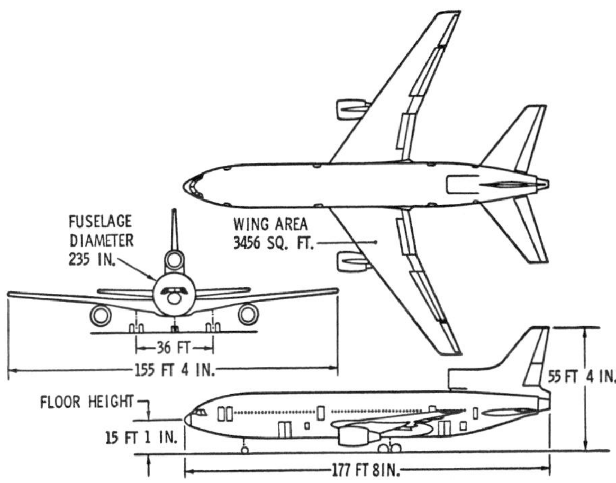

McDONNELL DOUGLAS DC-10
Credit: Western Airlines

The drawings and drawing problems in this book are done in dimensioning systems used in industry. These systems are the:

1. Decimal System (70 percent of the drawings)
2. Fractional System (20 percent)
3. Metric System (10 percent)

The people who make these drawings are called drafters. There are seven different working levels within the drafting profession:

1. A tracer makes simple drawings and does office errands.
2. A detailer makes changes on existing drawings.
3. A senior detailer makes the difficult and more complicated drawings.
4. A checker looks for errors on other people's drawings.
5. A designer helps to create and draw designs with the engineers and scientists.
6. A technical Illustrator does pictorial and artistic drawings, usually in ink.
7. An architect designs and draws structures.

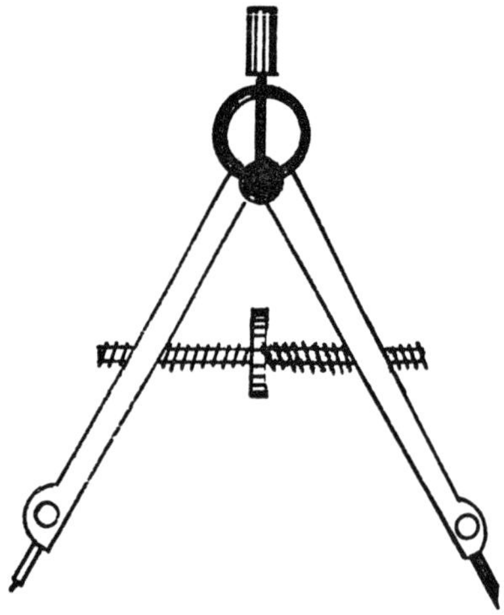

UNIT 2

DRAFTING EQUIPMENT

The basic tools of mechanical drafting are shown in the illustrations below. These are the tools that the beginner must have. For advanced drawing techniques, additional equipment will be necessary.

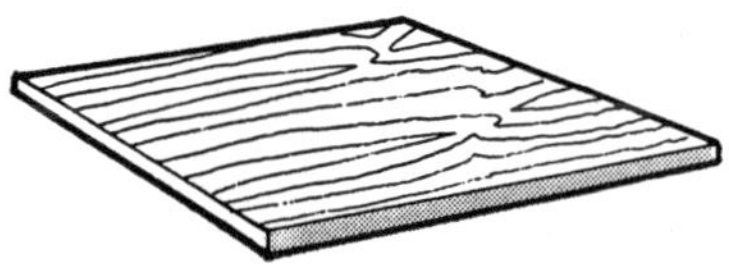

20"x26" DRAWING BOARD

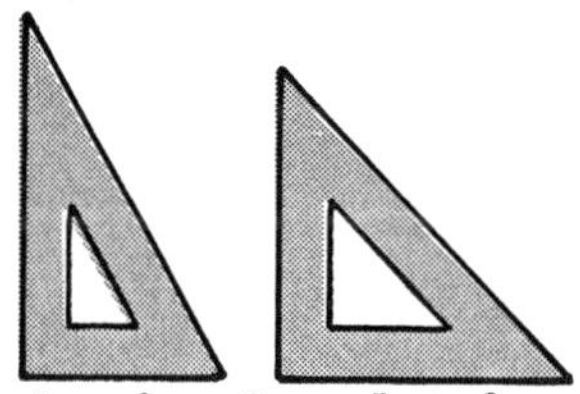

10" 30°-60° 8" 45°

TRIANGLES

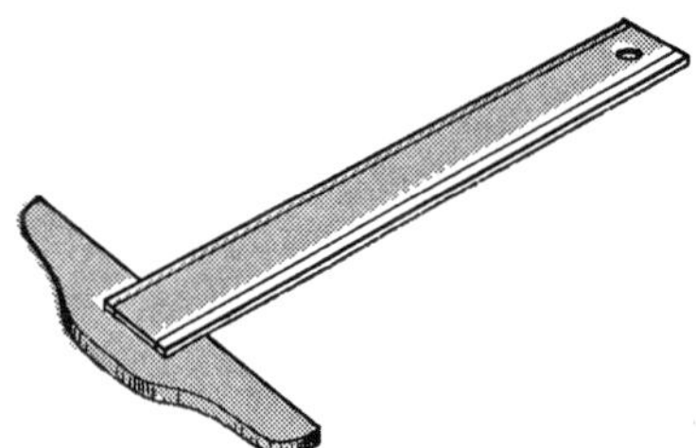

24" T SQUARE

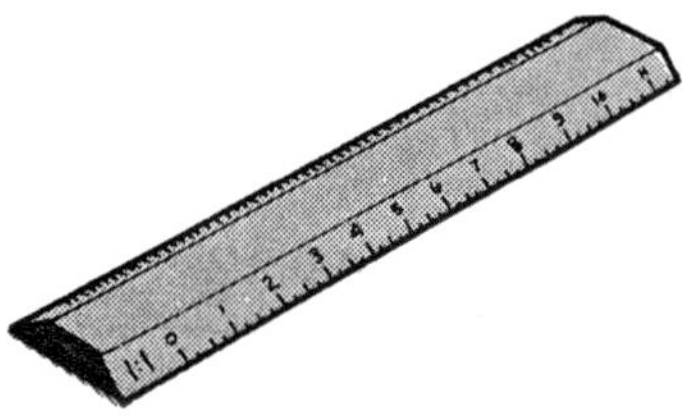

12" SCALE

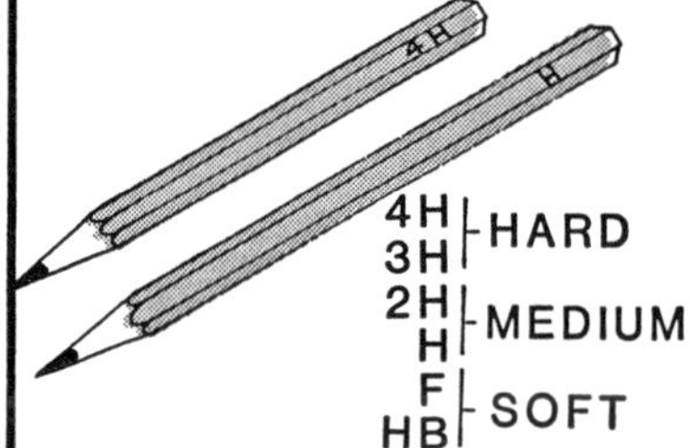

DRAFTING PENCILS

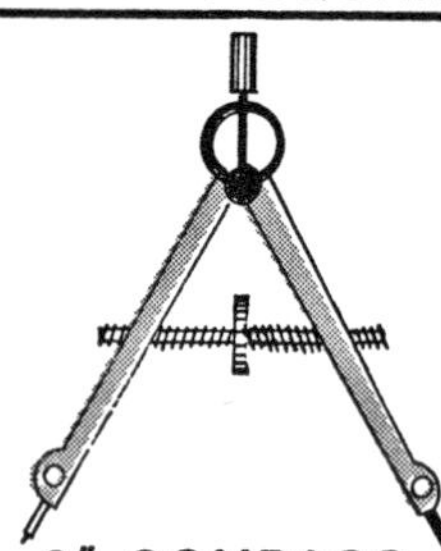

6" COMPASS

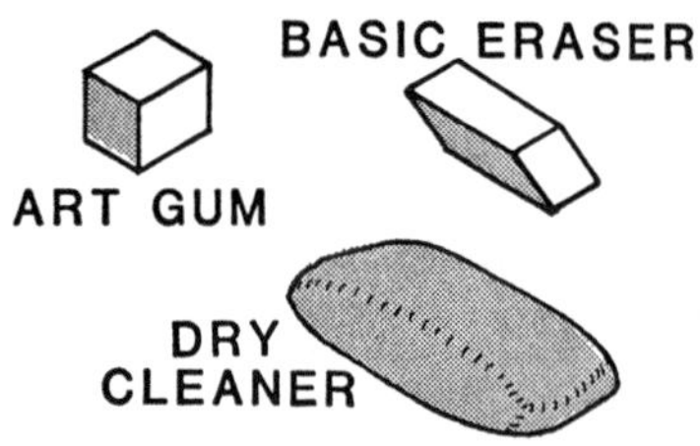

ERASERS/CLEANERS

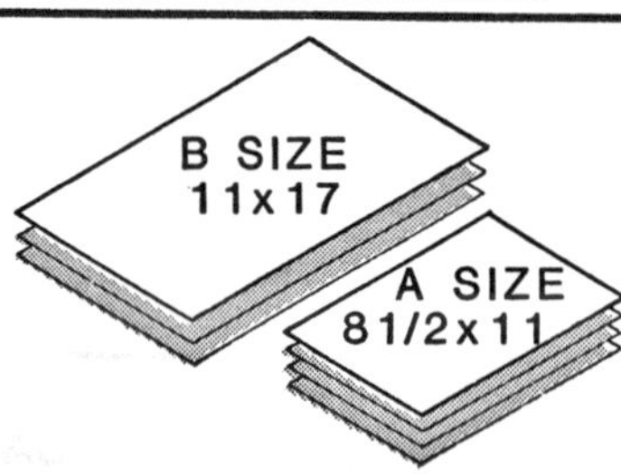

DRAFTING PAPER

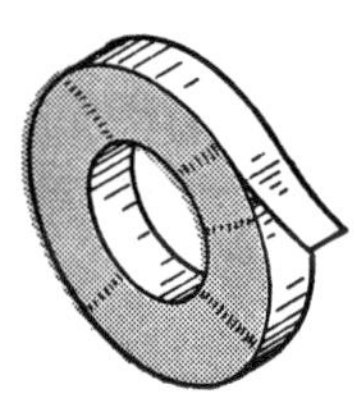

DRAFTING TAPE

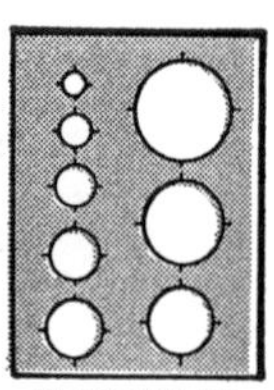

CIRCLE

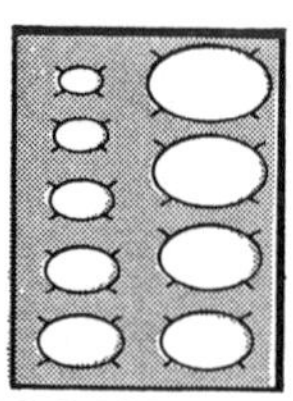

ISOMETRIC

TEMPLATES

PENCIL SHARPENER

12x19 ENVELOPE

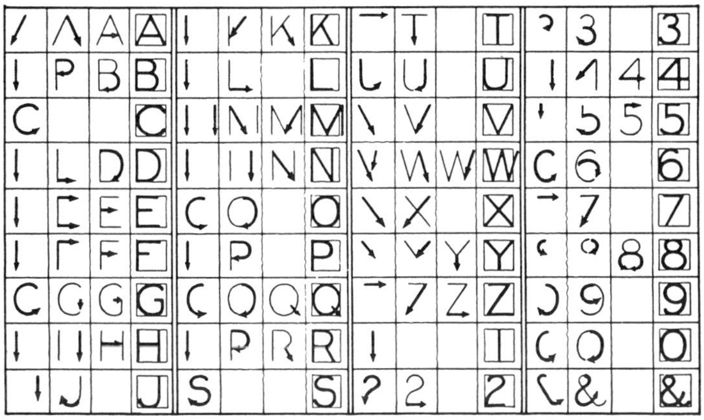

UNIT 3

LETTERING

Skill in lettering is as important as the ability to make good drawings.

A PLAN FOR BEGINNERS

1. It would be best to practice only one style of alphabet during your work throughout the entire book. Use the vertical or slant uppercase gothic alphabet shown below.
2. It would be best to practice all of your lettering using letters 1/8" high. This will help you to develop a uniform stroke and enable you to "get the feel" of lettering in the shortest possible time.
3. Microfont lettering is used when drawings are to be microfilmed.

Vertical single-stroke gothic letters and numbers.

ABCDEFGHIJKLMNOPQRSTUVWXYZ 1234567890

Slanted single-stroke gothic letters and numbers.

ABCDEFGHIJKLMNOPQRSTUVWXYZ 1234567890

Microfont letters and numbers.

ABCDEFGHIJKLMNOPQRSTUVWXYZ 1234567890

A STUDY OF THE VERTICAL UPPERCASE GOTHIC ALPHABET

A close study of the gothic alphabet will show that the letters are not all of the same width. However, they are all the same height.

LETTERS THAT COMPLETELY FILL A SQUARE

A C O Q T V X Y Z 4

LETTERS THAT DO NOT COMPLETELY FILL A SQUARE

B D E F G H J K L N 9 0 I

P R S U 2 3 5 6 7 8

LETTERS THAT ARE WIDER THAN A SQUARE

M W

FORMING 1/8″ VERTICAL UPPERCASE GOTHIC LETTERS

Important Note: The alphabet shown below was made larger than 1/8″. This was done in order that you may clearly see the direction of each stroke used to make each letter. However, these strokes are to be used only on 1/8″ letters.

When you have to make letters larger than 1/8″, you will find that a slightly different set of strokes is necessary when forming the letters "B," "C," "P," "S," and "U."

STEPS IN FORMING VERTICAL GOTHIC LETTERS (for 1/8″ letters only)

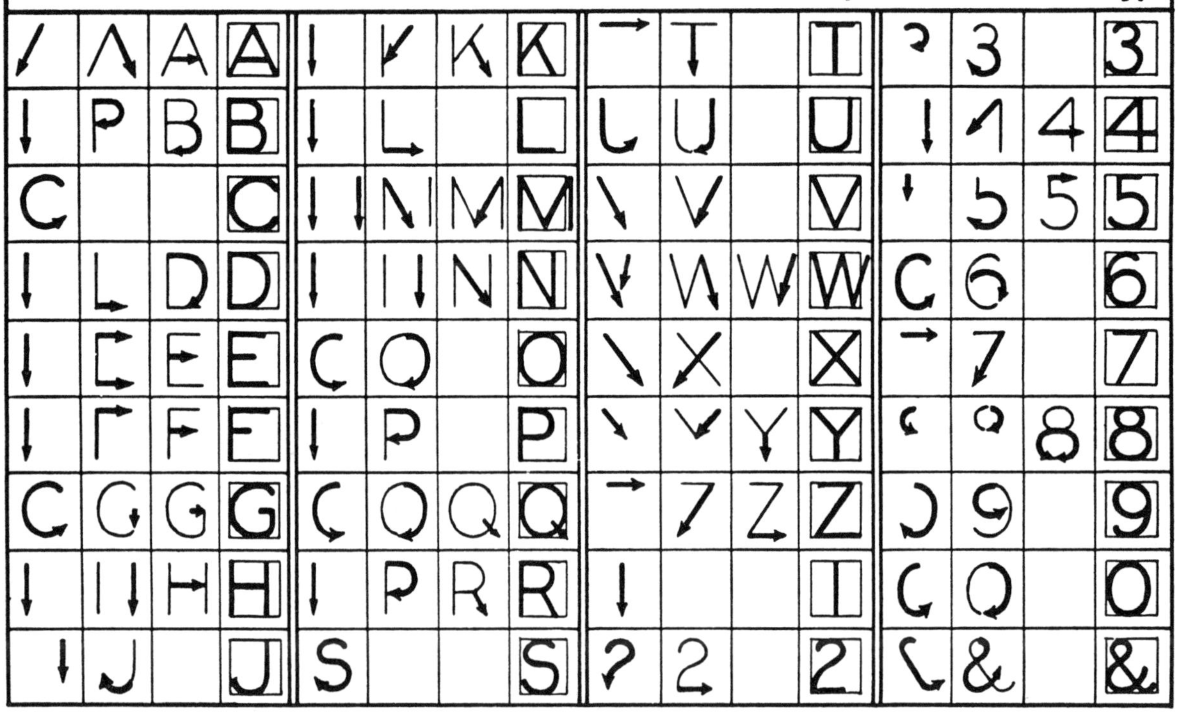

BASIC LETTERING STROKES (for 1/8″ letters)

Letters are formed by vertical downstrokes, by oblique downstrokes, by curved downstrokes, and by strokes to the right.

NEVER USE AN UPSTROKE WHEN LETTERING.

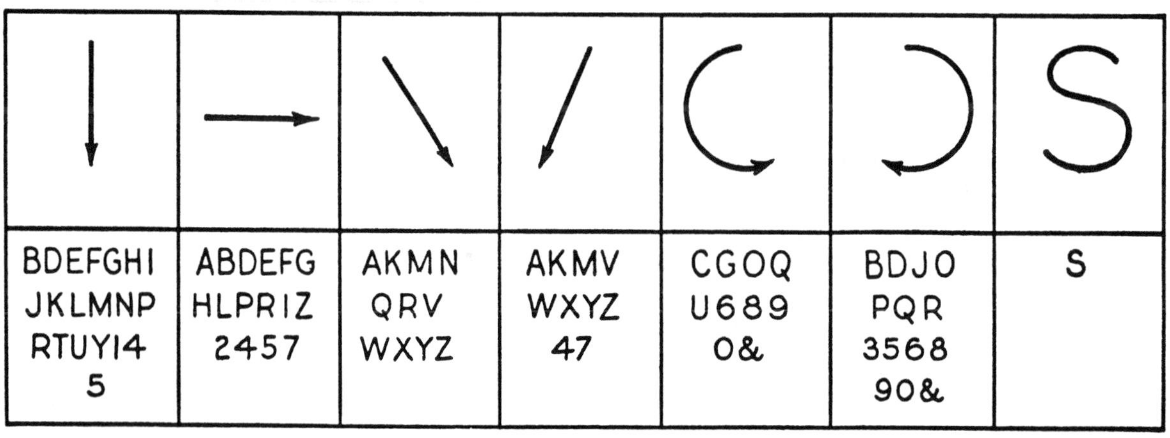

BDEFGHI JKLMNP RTUYI4 5	ABDEFG HLPRIZ 2457	AKMN QRV WXYZ	AKMV WXYZ 47	CGOQ U689 O&	BDJO PQR 3568 9O&	S

LETTER FAMILIES

EFHILT	AKMNVWXYZ	COGJQU BDPRS

The alphabet may be divided into letter families as shown above. Learn to recognize the basic characteristics of a letter and to associate that letter with other letters.

THE HEIGHT OF LETTERS (Use extremely light guidelines)

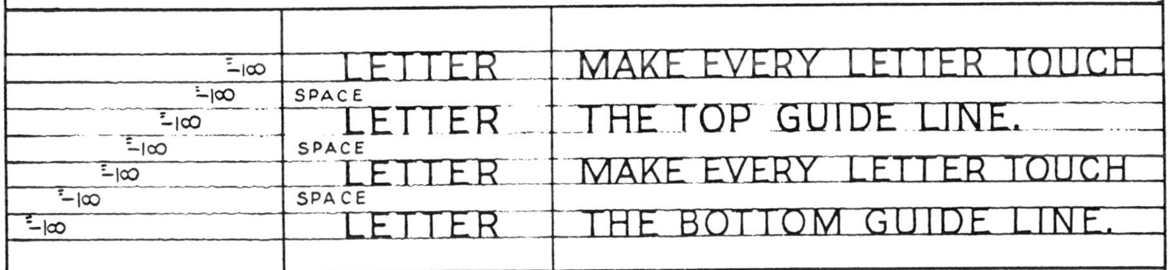

(1) Guidelines must always be drawn. Accurate measuring is essential.	(2) Leave a space equal to the height of a letter between lines of lettering.	(3) Every letter must touch the top and bottom guidelines. Draw the guidelines so lightly that they do not have to be erased.

CONTROLLING THE PHYSICAL SIDE WHEN LETTERING

1. Sit comfortably. Sit up! Do not lean over your lettering.
2. Learn to use your middle vision when lettering. Sit back in your chair and letter from there.
3. Keep both arms on the drawing board when lettering.
4. Move your drawing board and paper to the left at least twice as you letter a line which is 6″ long. Sit directly in front of your lettering area.
5. Keep your pencil sharp at all times. Revolve your pencil one sixth of a turn after every word.

CORRECT SPACING IS THE KEY TO LEGIBLE LETTERING

There is much more to good lettering than the correct formation of the individual letters. The key to good lettering lies in the proper spacing of the individual letters to form words.

Do not spend too much time practicing the alphabet. Study the alphabet carefully until you have learned the proper formation of each letter. Then, devote all of your practice to learning how to letter words, phrases, and sentences.

The next page contains the key to spacing letters to form words.

CONTROLLING THE SPACING BETWEEN LETTERS

The amount of white space that appears between the first two letters should appear **between each of the letters that follow.**

① The moment that you have lettered the "E" following the "L" (above) you have set the pattern for the amount of white space that **must** appear between the letters that follow.

② The white area (shaded) between the "T" and the "E" must appear to be the same size as the white area between the first two letters.

③ The letter "I" and the first stroke of the "N" are parallel lines. When parallel lines occur, an optical illusion takes place. The two parallel lines appear to be closer together than they really are. To overcome this, increase the space between parallel lines until the white area between these letters appears the same as the white area between other letters.

④ In the word "DRAW," the "A" and the "W" which follow it create parallel lines. Move the "W" away in order to create an open white area equal to that found between the "D" and the "R," and between the "R" and the "A."

Parallel lines occur in the words "VAN" and "HIT."

AN EXPERIENCE IN CREATING EQUAL AREAS BETWEEN LETTERS		
MECHANICAL DRAWING	MECHANICAL DRAWING	MECHANICAL DRAWING
① Make the spaces between the letters appear to be equal.	② Increase the space (slightly) between two letters that have elements parallel to each other.	③ Equalize the open space between certain letters by moving them closer together.

The **space between words** should be equal to the height of a letter.

The **space between lines of lettering** should be equal to the height of a letter.

The **speed of your lettering** should be a little slower than the speed at which you do your best lettering. Do not hurry.

FRACTIONS for 1/8" lettering will be between 3/16" to 1/4" in height. Each number in a fraction is approximately 3/32" in height. The numbers must clear the horizontal fraction bar. The fraction will be difficult to read if the numbers touch the fraction bar.

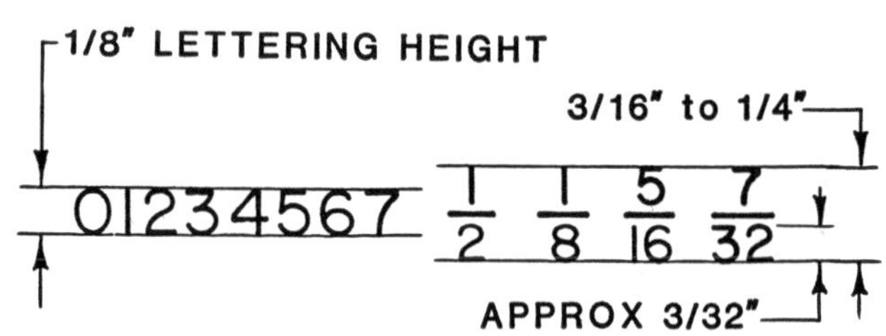

USING 1/8" GRAPH PAPER FOR LETTERING PRACTICE

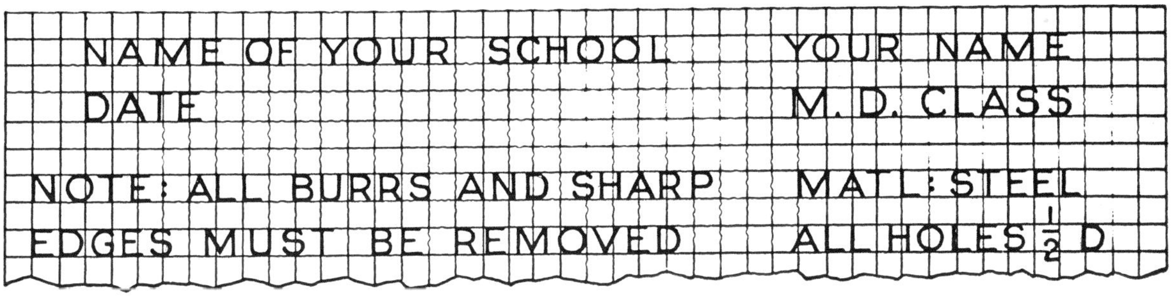

While it is absolutely necessary that you learn to draw 1/8" guidelines lightly and accurately, it is also convenient to practice lettering outside of class by using graph paper.

1. Use graph paper with 1/8" squares.
2. Use the horizontal lines as guidelines (see above).
3. **Important:** Use the vertical lines only as a guide in making vertical strokes. **Under no circumstances should you attempt to letter by drawing a letter in each square.** Control the spacing between letters by eye.
4. Practice with a fountain pen will convince you not to use "up" strokes.
5. Always use "copy" directly related to drafting (see above).

COMMON ERRORS AND CORRECTIONS		
DRAW I N G POOR!	MECHANICAL POOR!	MEcHANlCAL DRAWNG POOR!
① The "ING" is too widely spaced.	② The letters must be vertical through the entire line.	③ Guidelines were not used. The lettering was done too rapidly. "DRAWING" has been misspelled.

ASSIGNMENT (on 1/8" graph paper)

The following material and all other notes found throughout this book will make good copy for lettering practice.

1. NOTE: ALL BURRS & SHARP EDGES MUST BE REMOVED
2. MAT'L: SHEET BRASS
3. ALL ROUNDS & FILLETS 1/4"R EXCEPT WHERE NOTED
4. F. A. O.
5. TWO REQUIRED
6. SECTION A–A
7. NOTE: TO BE HEAT-TREATED TO RELIEVE WINDING STRESSES AND TO BE BLACK ENAMELED REQ.: 2 PER CAR

NAME________________ PERIOD___

ABCDEFGHIJKLMNOPQRSTUVWXYZ 1234567890 $\frac{1}{2}$ $\frac{3}{4}$ $\frac{7}{8}$

ABCDEFGHIJKLMNOPQRSTUVWXYZ 1234567890 $\frac{1}{2}$ $\frac{3}{4}$ $\frac{7}{8}$

LETTERING EXERCISE (Copy this page for the lettering exercise.)

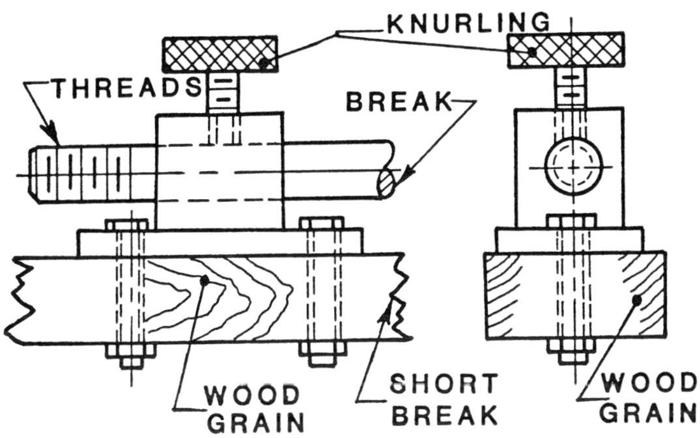

UNIT 4

LINE CONVENTIONS

TYPES OF LINES AND THEIR USES IN DRAFTING

An understanding of the purposes of the various line conventions is very important for reading and drawing mechanical drawings. Each line is standardized, has a definite meaning, and is drawn the same way in all mechanical drawings.

The width of these lines must be consistent on all drawings so errors will not be made when reading them. Each line must be sharp, clean, and crisp. This will make interpretation easier and enable better reproduction copies to be made from the original drawing.

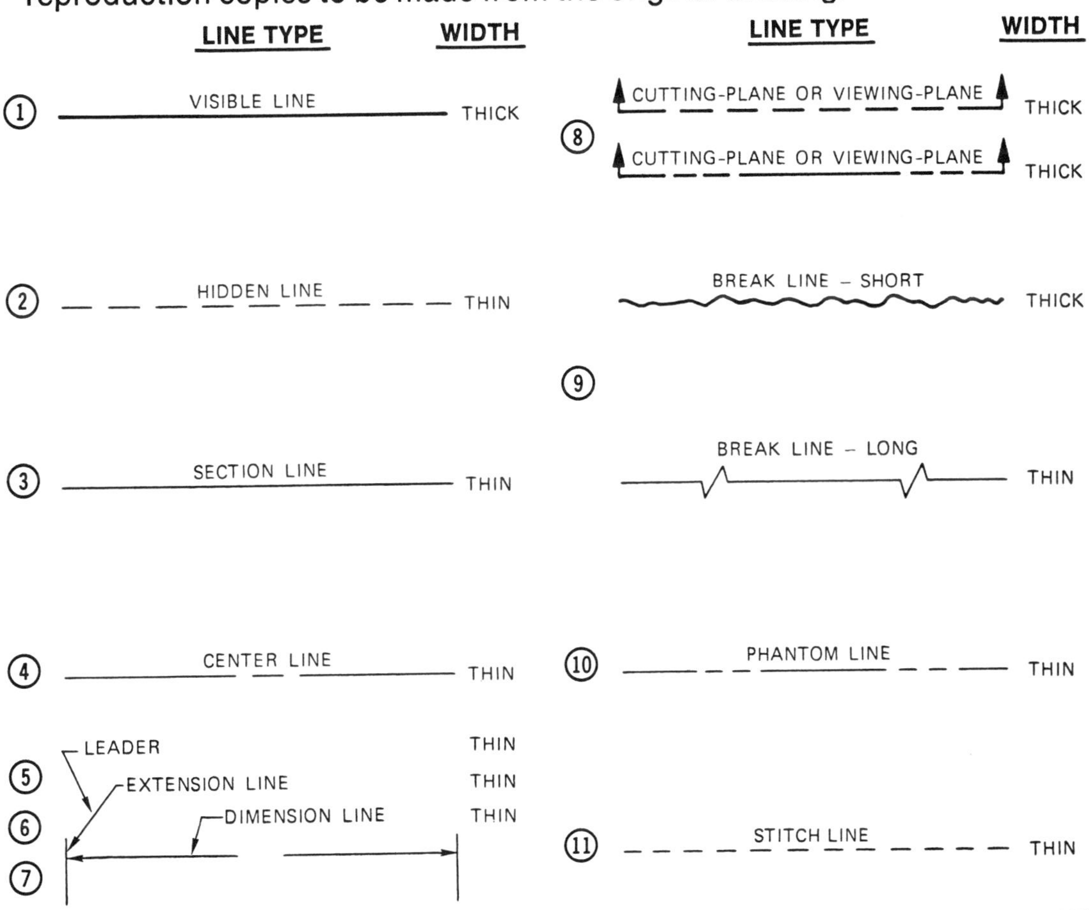

LINE CONVENTIONS

NAME	LINE	DESCRIPTION	SIZES
1. BORDER LINE		A VERY HEAVY LINE THAT FRAMES THE DRAWING.	A VERY WIDE SOLID LINE
2. CUTTING PLANE LINE		A VERY HEAVY LINE THAT SHOWS WHERE A LINE CUTS THROUGH AN OBJECT TO SHOW HOW THE INSIDE LOOKS.	A VERY HEAVY BROKEN LINE. LONG DASH IS 1" SHORT DASH IS 1/4", SPACES 1/8"
3. VISIBLE LINE		A HEAVY LINE THAT SHOWS THE FORM OF AN OBJECT. IT IS ALSO CALLED AN "OBJECT LINE".	A WIDE SOLID LINE
4. HIDDEN LINE		A THIN LINE THAT HAS BROKEN DASHES. IT SHOWS A HIDDEN EDGE. ALSO CALLED AN "INVISIBLE LINE".	DASHES ARE 1/8" LONG WITH 1/32" SPACES.
5. CENTER LINE		A THIN LINE WITH LONG AND SHORT DASHES. IT SHOWS THE CENTER OF OBJECTS.	DASHES ARE 1" LONG AND THE SHORT DASH 1/8" WITH 1/16" SPACES.
6. SECTION LINE		A THIN LINE THAT SHOWS A SOLID SURFACE WHEN A CUTTING PLANE LINE PASSES THROUGH IT. ALSO CALLED CROSS-HATCHING LINES.	THIN LINE NORMALLY DRAWN AT A 45° ANGLE.
7. LEADER		A THIN LINE WITH AN ARROWHEAD ON ONE END. ON THE OTHER END IS A DIMENSION OR NOTE.	A THIN LINE USED TO SHOW THE PARTS RELATED TO THE NOTE OR DIMENSION.
8. EXTENSION LINE		A THIN LINE USED TO EXTEND THE SIZE OF THE OBJECT BEING DRAWN SO THAT DIMENSION CAN BE PLACED NEAR THE VIEW.	A THIN LINE THAT PROJECTS FROM AN OBJECT.
9. DIMENSION LINE		A THIN LINE WITH ARROWHEADS ON EACH END THAT TOUCH THE EXTENSION LINE. IT IS OPENED FOR THE DIMENSION.	A THIN LINE THE LENGTH OF EACH DIMENSION.
10. LONG BREAK LINE		A THIN LINE WITH SEVERAL FREEHAND ZIG-ZAGS. IT CUTS OFF A WIDE OBJECT.	A THIN LINE WITH ZIG-ZAGS EVERY INCH OR TWO.
11. SHORT BREAK LINE		A HEAVY WAVY LINE. IT CUTS OFF A NARROW LONG OBJECT.	A WIDE LINE.

LINE CONVENTIONS—WORKING DRAWING

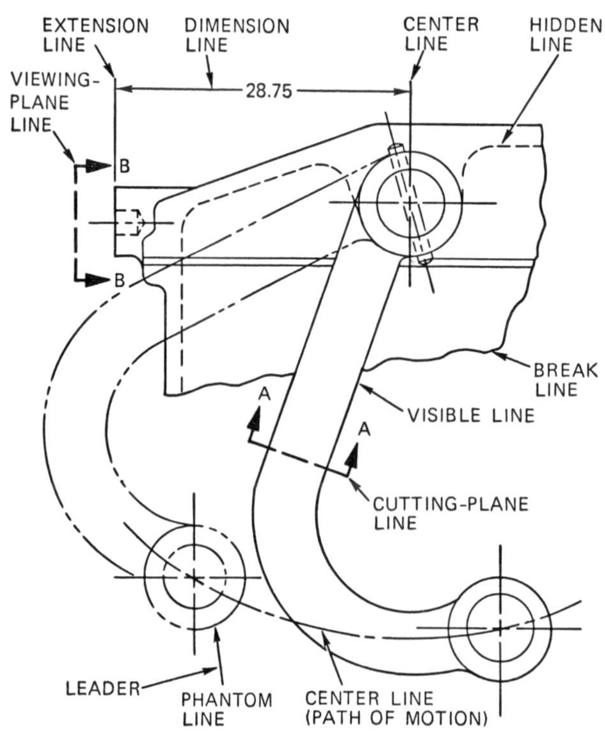

TYPES OF LINES AND THEIR USES IN DRAFTING

A clear understanding of the purpose of each of the various types of lines used in drafting will remove many of the doubts as to the thickness and quality of each type of line.

HEAVY LINES

The picture is the important thing. Since the visible edge (object line) and the hidden edge are the two types of lines used to create the picture, these lines should be heavy. They must be made with a firm stroke.

Visible edge (object line)

Hidden edge line

(Many drafters prefer to make the hidden edge line thinner than the visible edge line.)

THIN LINES (one fourth as thick as a visible edge)

Since all of the remaining lines are used to explain the picture further, they may be regarded as symbols. All of them must be made thin so as not to destroy the effect of the picture.

Center line

Extension line

Dimension line
(With extension lines)

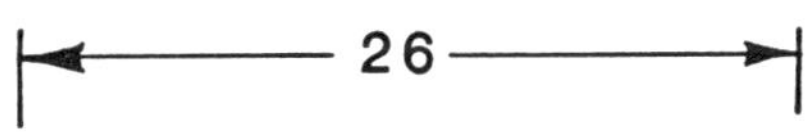

Adjacent parts

Section lining

Knurling

LINES THAT EXIST BUT SHOULD NOT BE SEEN

There are several other lines which are absolutely essential to the making of a mechanical drawing, and yet must never be seen on the finished drawing.

Projection lines

Guide lines

Construction lines

EXCEPTIONS

Cutting plane line

The cutting plane line is made very heavy in order not to confuse it with a center line.

Border line

The heaviest line is used as a border or frame on a drawing.

LINES USED TO DENOTE A BREAK IN A SURFACE

(These lines become part of the visible edge of an object. They must be drawn as thick as a visible edge.)

Denoting a short break  (Drawn freehand)

Denoting a long break  (Straight edge, plus freehand break)

Denoting a break in wood  (With the grain)

LINES USED ON FIVE COMMON SYMBOLS

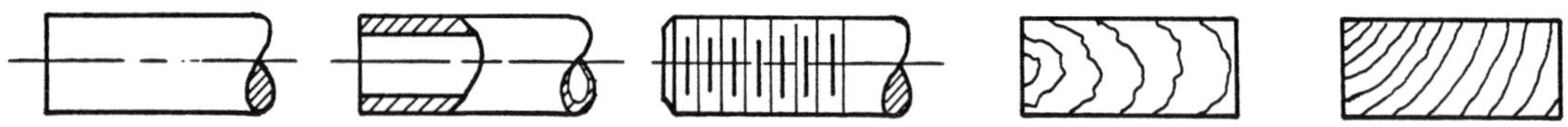

Break in a cylindrical rod	Break in a pipe or tube	Symbol for a thread	With the grain (wood)	Across the grain (wood)

EXAMPLES OF THE USE OF LINE CONVENTIONS

In engineering drawings, the line weight (or thickness) must be correct.

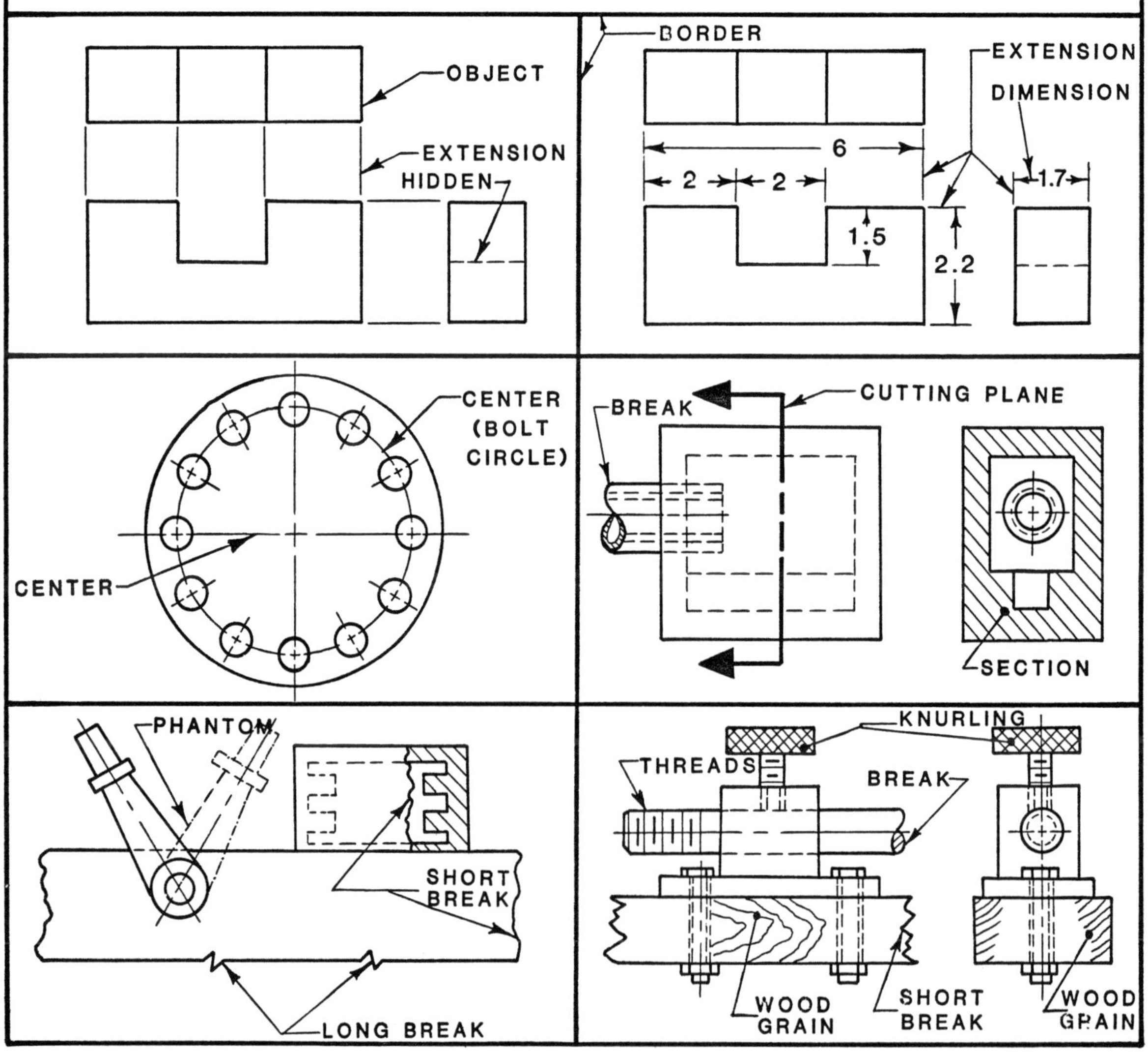

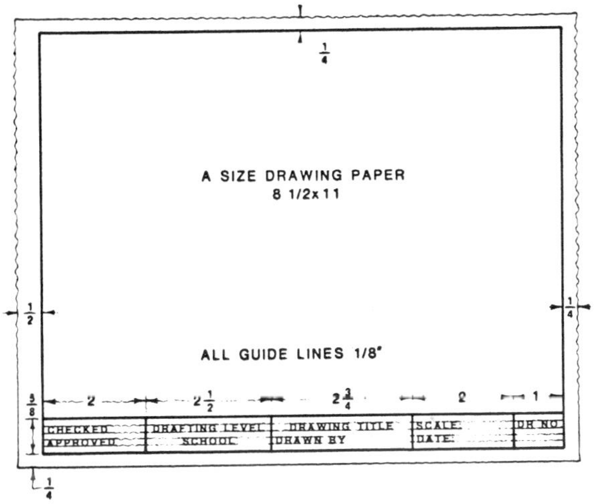

UNIT 5

SHEET LAYOUTS

STANDARD SHEET LAYOUTS FOR MECHANICAL DRAFTING

Before starting a drawing, the drafter must make the sheet layout for the drawing. This consists of the border, title blocks, and lettering. Although the sheet layout size and style may vary, this important information must always be given. All sheets layouts for one set or series of drawings must be the same format. This makes handling, checking, and storing more convenient.

The two sizes of drawing paper most often used in the classroom are the A size (8½″ × 11″), and B size (11″ × 17″). All lettering on these layouts should be ⅛″ in height.

Examples of different sheet layouts are shown in this unit.

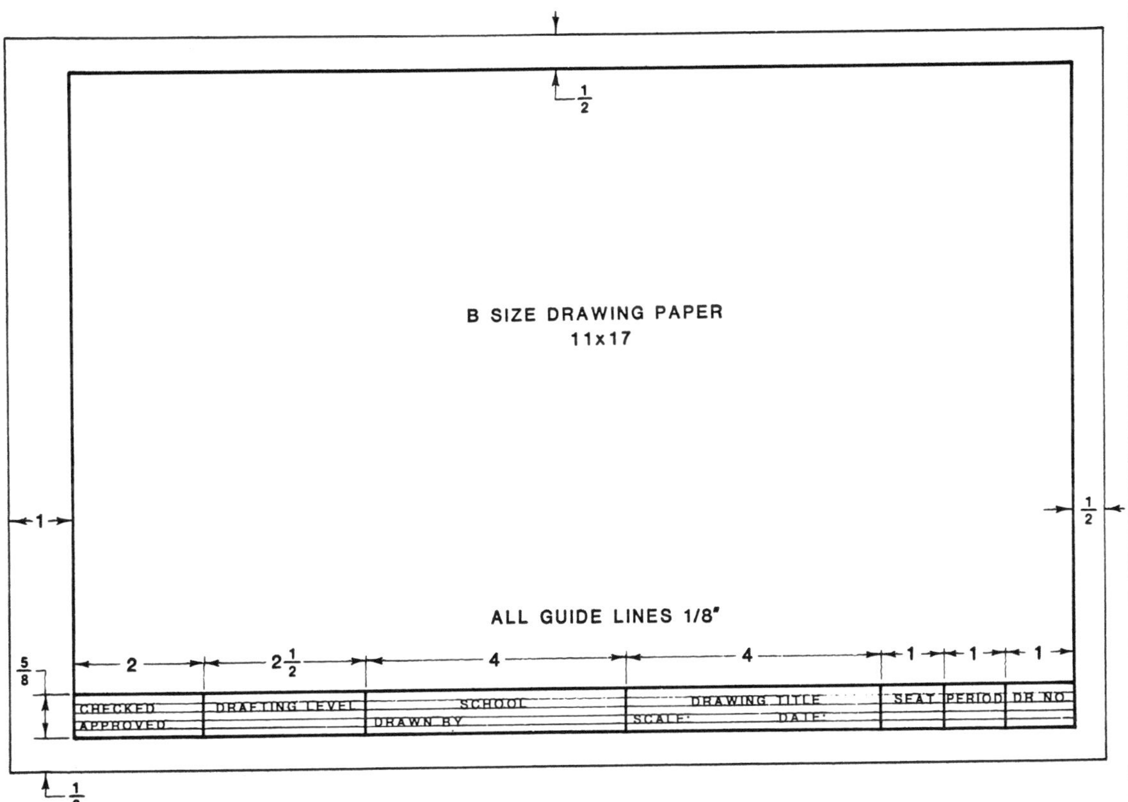

A SIZE DRAWING PAPERS

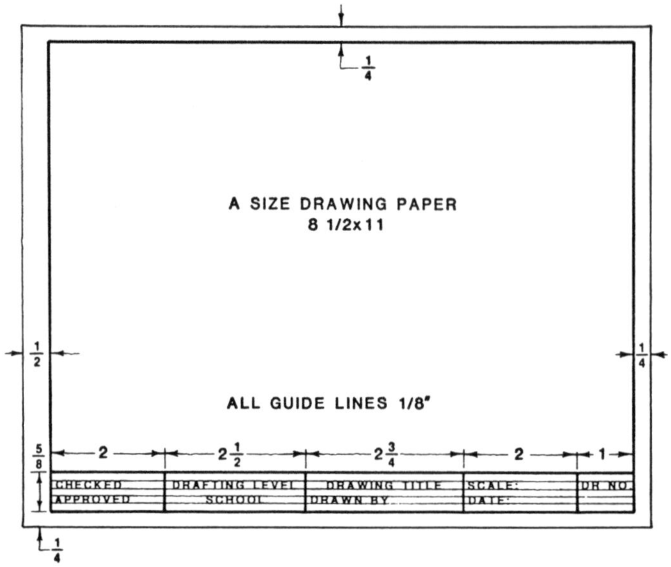

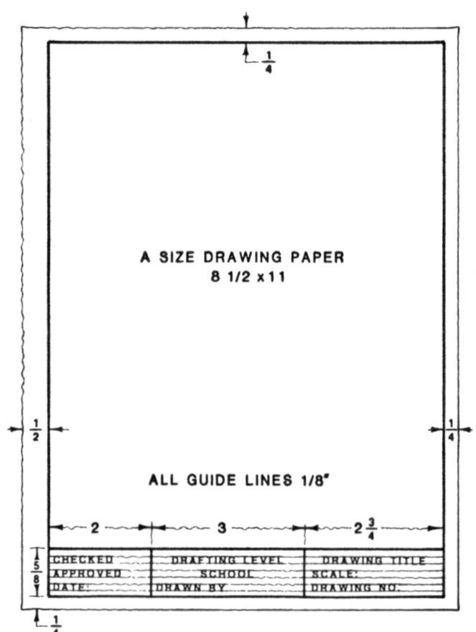

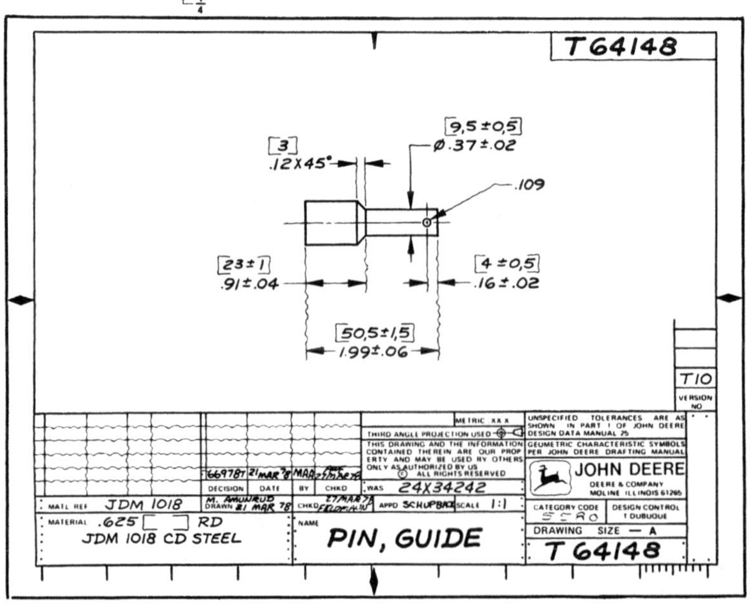

AN EXAMPLE OF AN A SIZE DRAWING AND FORMAT USED IN INDUSTRY

ROUTINE MATTERS

The plate layouts and the arrangement of the lettering box are routine matters.

PLATE LAYOUTS AND BORDERS FOR 8½" X 11" PAPER

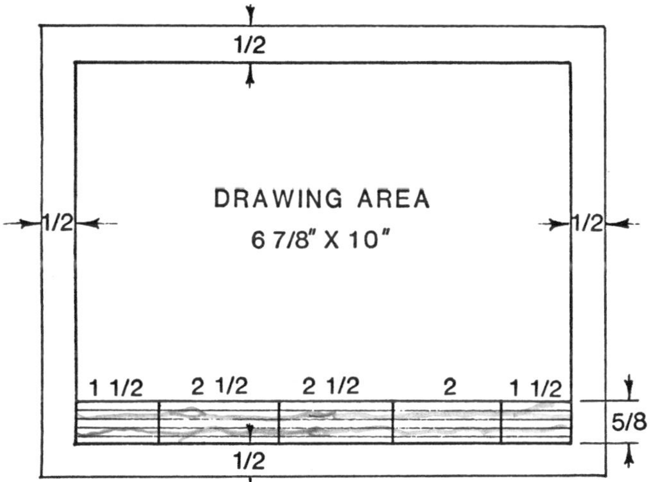

① A SIZE DRAWING PAPER
HORIZONTAL LAYOUT

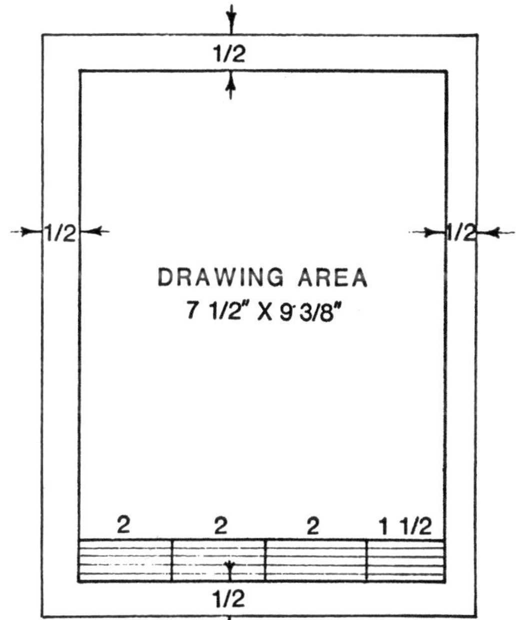

② A SIZE DRAWING PAPER
VERTICAL LAYOUT

TYPICAL VERTICAL LAYOUT LETTERING BOX

The lettering box consists of five 1/8" horizontal spaces drawn above the lower border line.

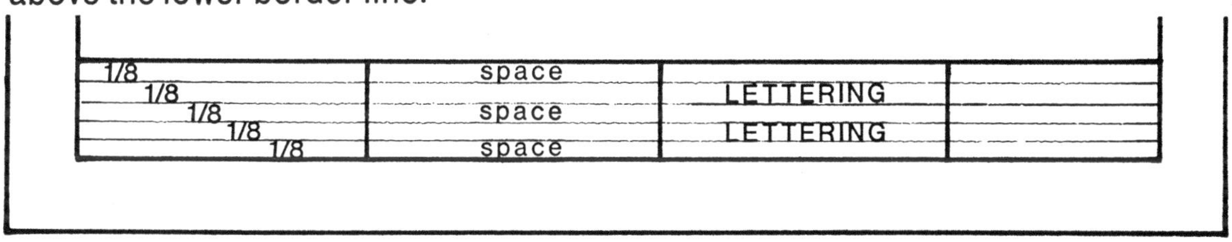

The completed lettering box (Make the lettering guidelines very light.)

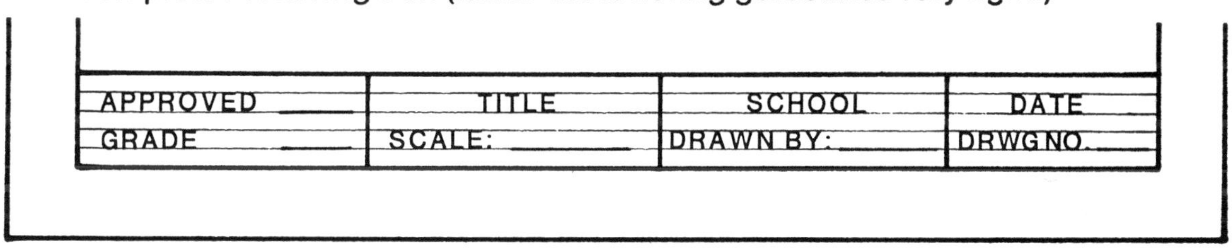

(See Unit 4 concerning the thickness of borders and guidelines.)

THE WORKING AREA

The working area is the space in which you will make your drawing. You can make several drawings on each page. You will be told to use a horizontal or vertical layout in each case.

It is a sign of good craftsmanship to see that each sheet layout is accurate and true. Center each drawing within its own frame.

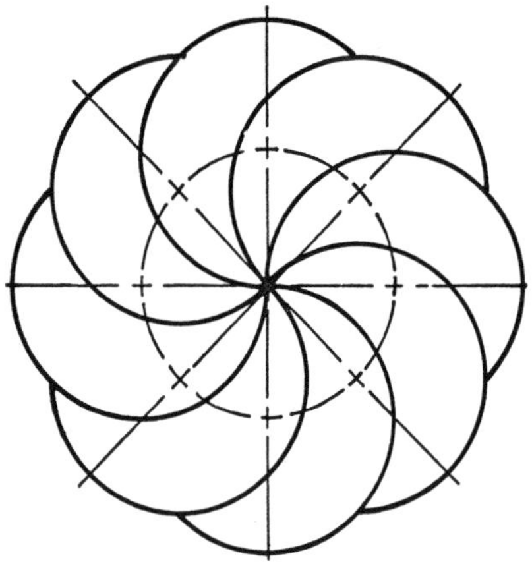

UNIT 6

BASIC DRAFTING SKILLS

POSITIONING THE DRAFTING INSTRUMENTS

(1)

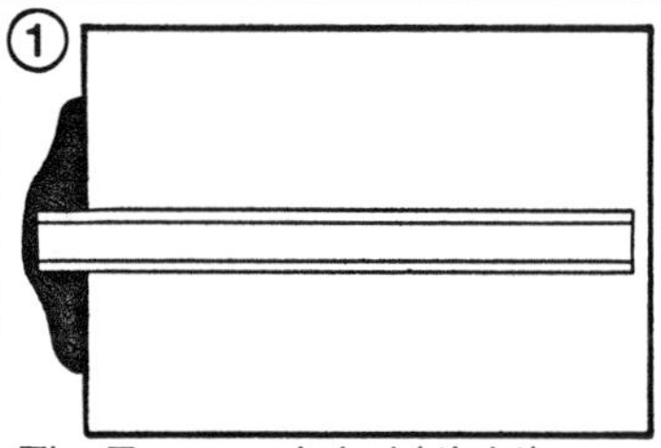

The T-square is held tightly against the edge of the drafting board.

(2)

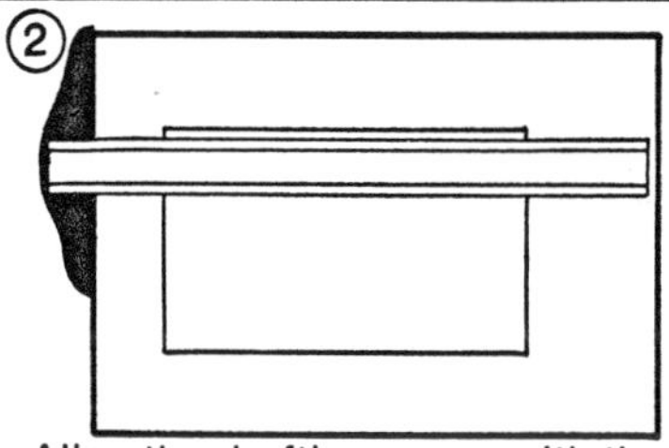

Align the drafting paper with the T-square and tape down each corner.

(3)

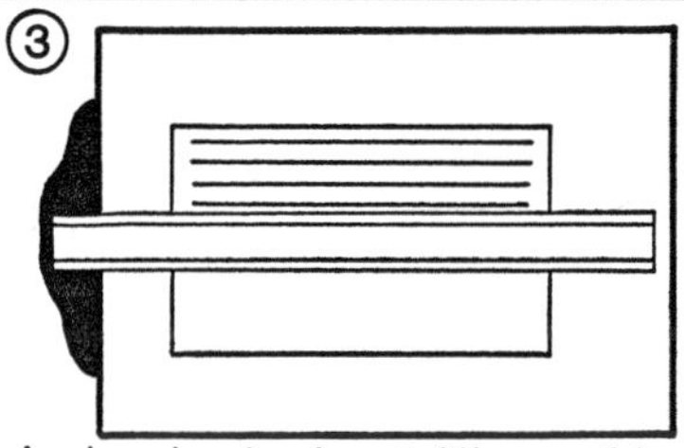

In drawing horizontal lines, slide the T-square to each position.

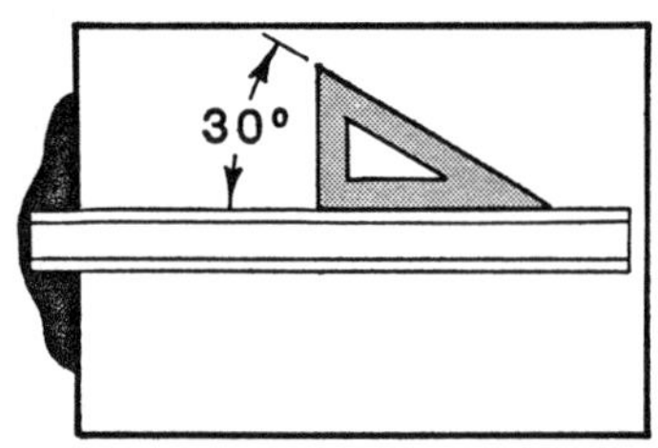

Drawing 30° lines

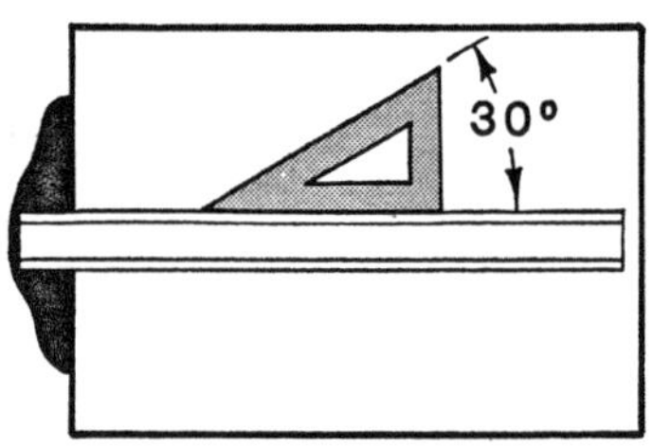

Drawing 30° lines

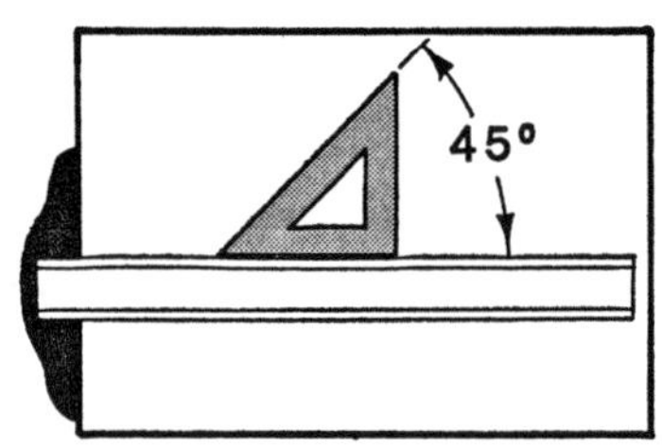

Drawing 45° lines

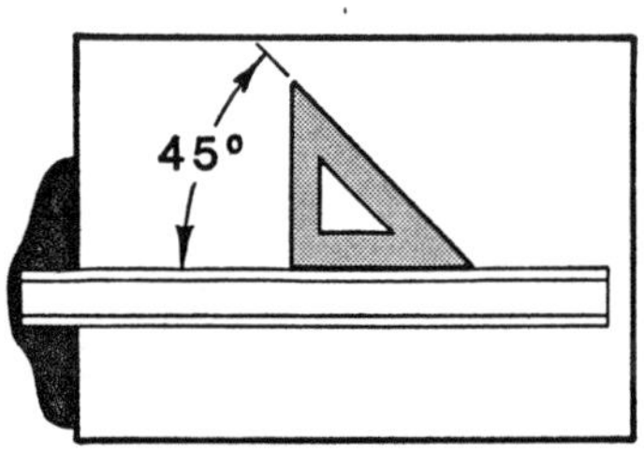

Drawing 45° lines

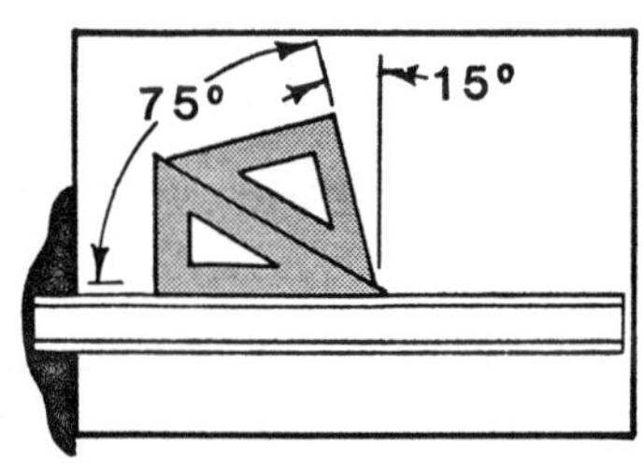

Drawing 15° and 75° lines

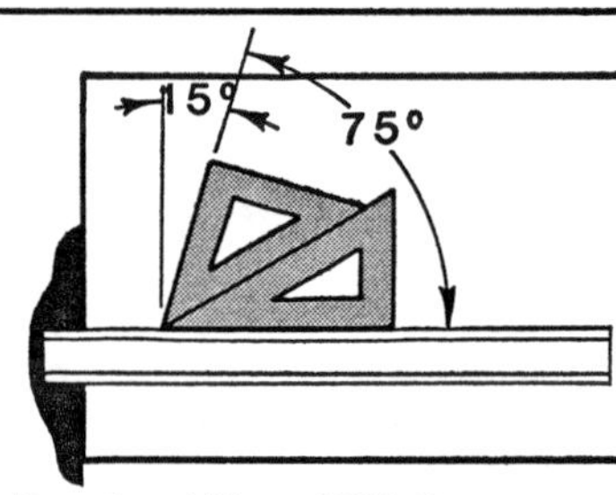

Drawing 15° and 75° lines

(1)

Irregular curve. Plot the curve and sketch it in lightly.

(2)

Place an irregular curve on the part of the line and darken the line as shown.

(3) Move the irregular curve to a new position, overlapping the darkened section. Repeat steps 2 and 3 until the drawing is completed.

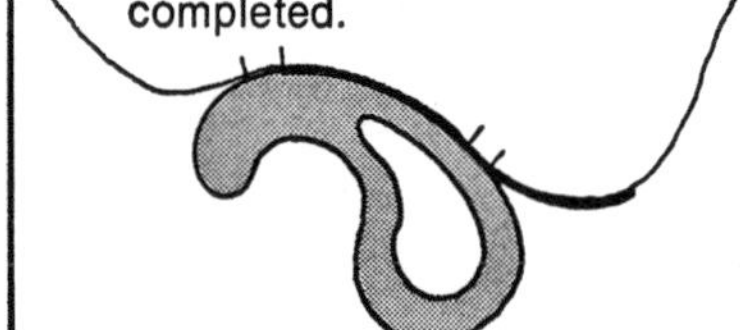

(Overlapping sections of lines must be smooth.)

THE COMPASS

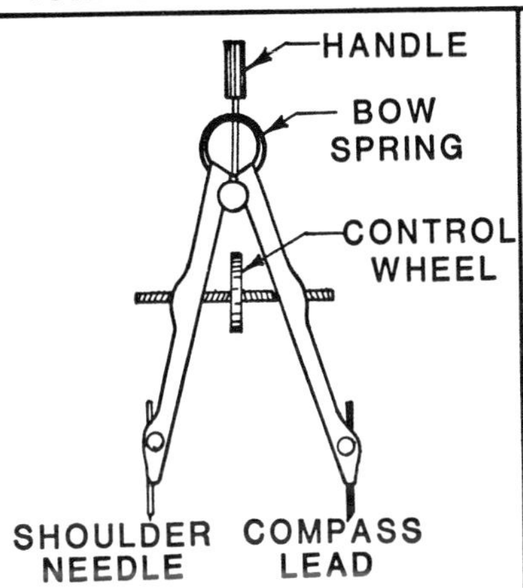

BOW COMPASS

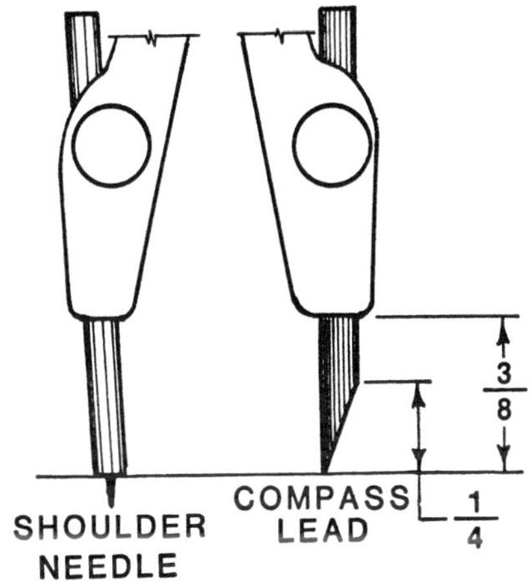

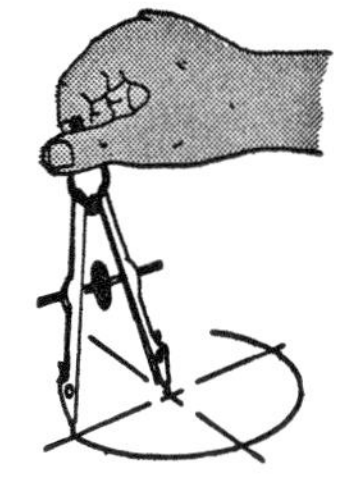

Spin the compass between the thumb and the index finger. Tilt the compass in the spin direction.

SETTING THE COMPASS

①

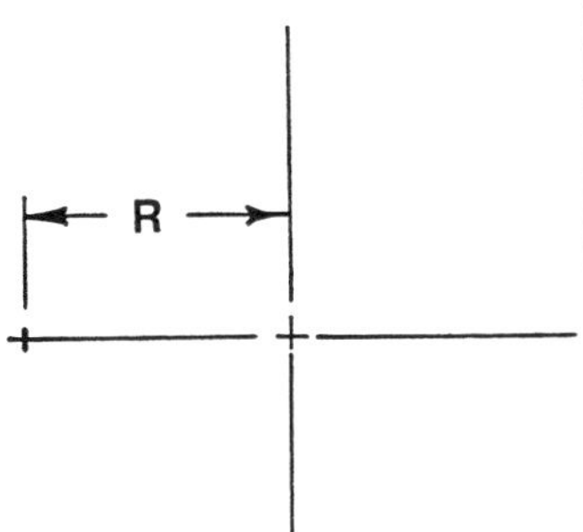

Layout the center line and measure the radius.

②

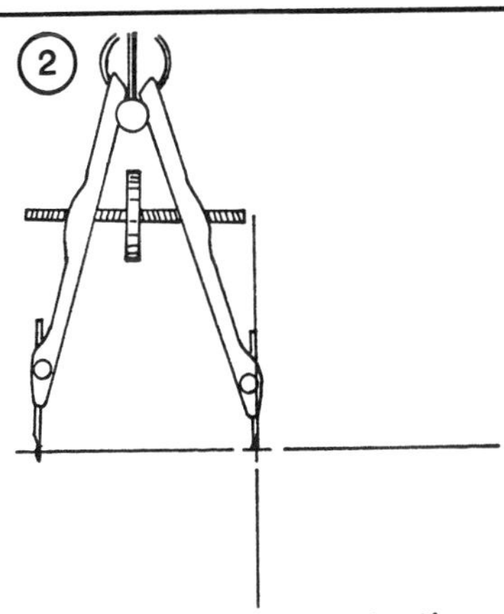

Set the compass to the radius.

③

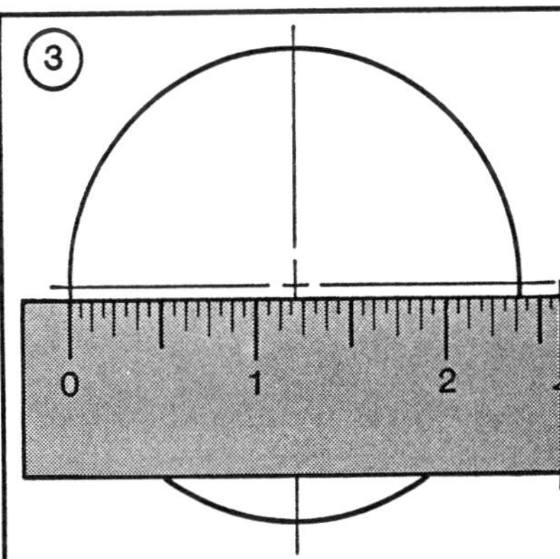

Draw the circle and measure the diameter for accuracy.

PARTS OF A CIRCLE

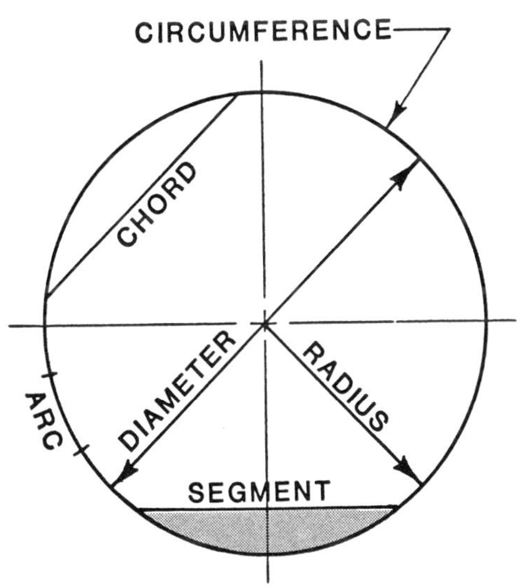

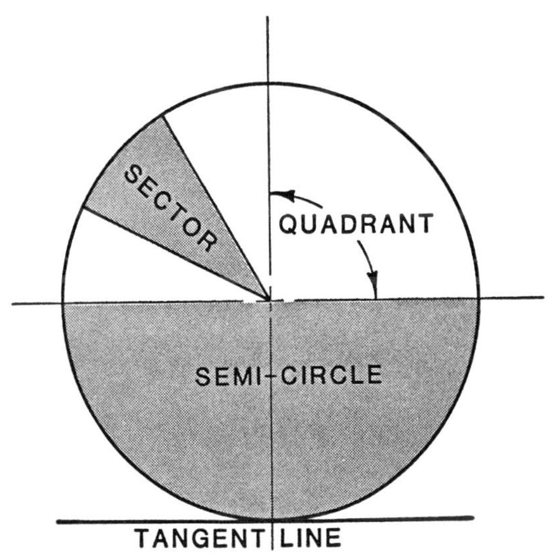

ACCURACY WITH A FREE TRIANGLE

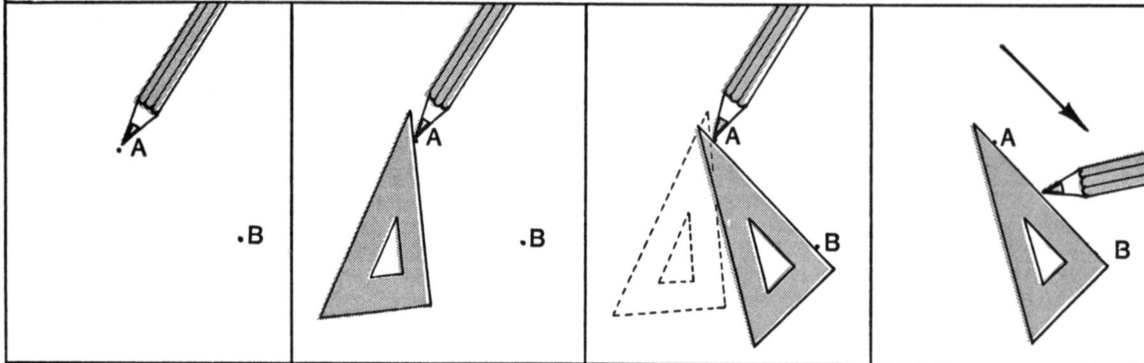

(1) Put the pencil on point "A."	(2) Move the triangle against the pencil.	(3) Swing the triangle until it touches "B" and the pencil.	(4) Draw "AB" with a downward stroke.

THE BASIC METHOD FOR ACCURATE DRAWING

1. Always measure correctly.
2. Use a small dot or line as a marker.
3. Place the pencil point on the mark from which you intend to draw.
4. Move the drawing tools into position so that the tools touch the pencil.
5. Draw the line.

ASSIGNMENT: LINE PRACTICE

Make a copy of the twelve panels shown below.
(Do not letter any of the directions given.)

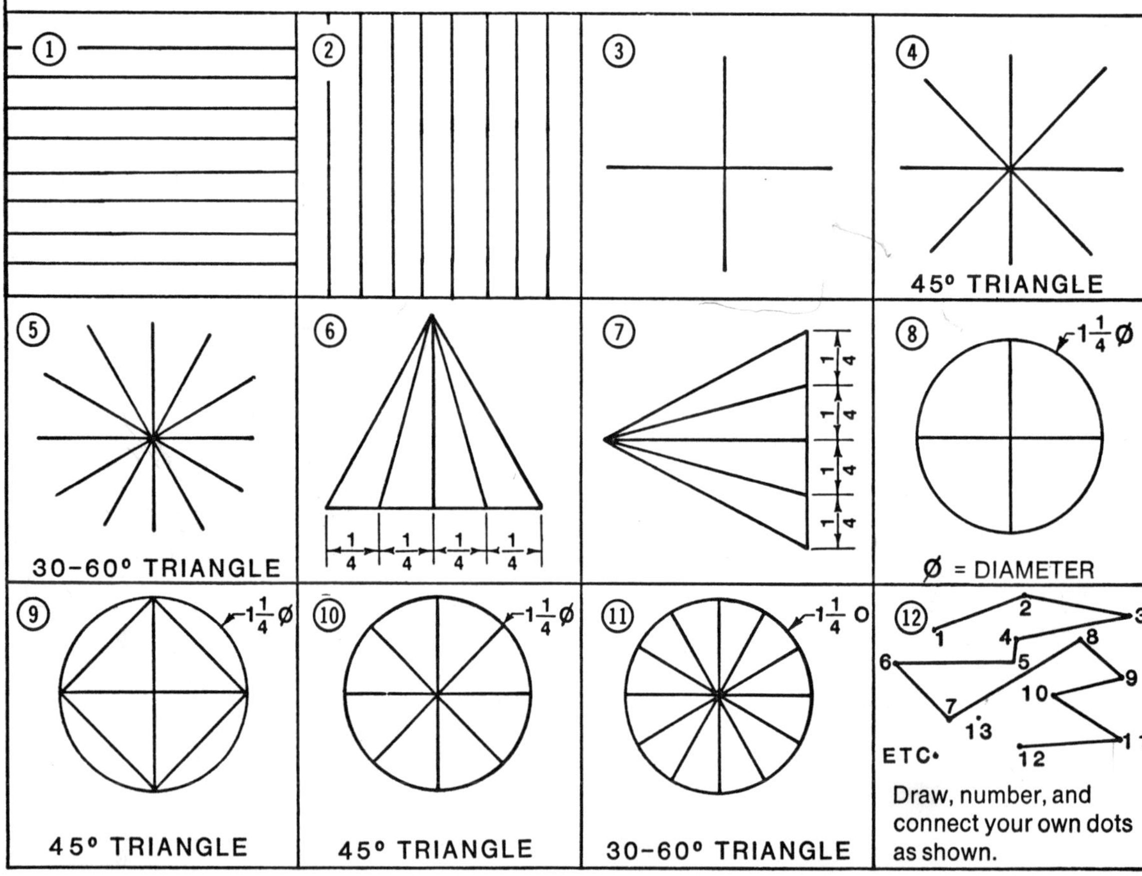

INSTRUMENT DRAWING EXERCISES

UNIT 7

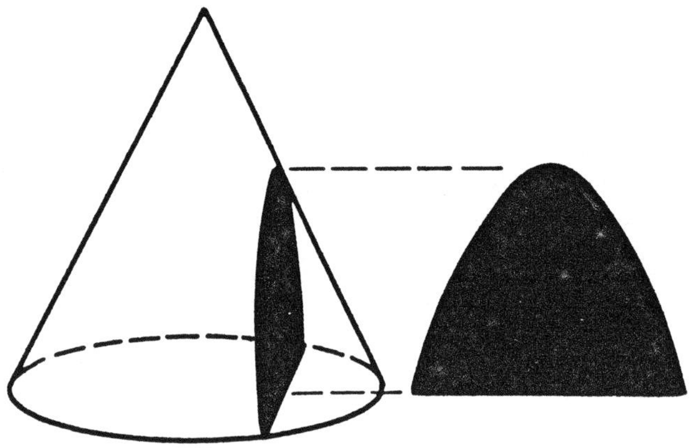

GEOMETRIC CONSTRUCTIONS

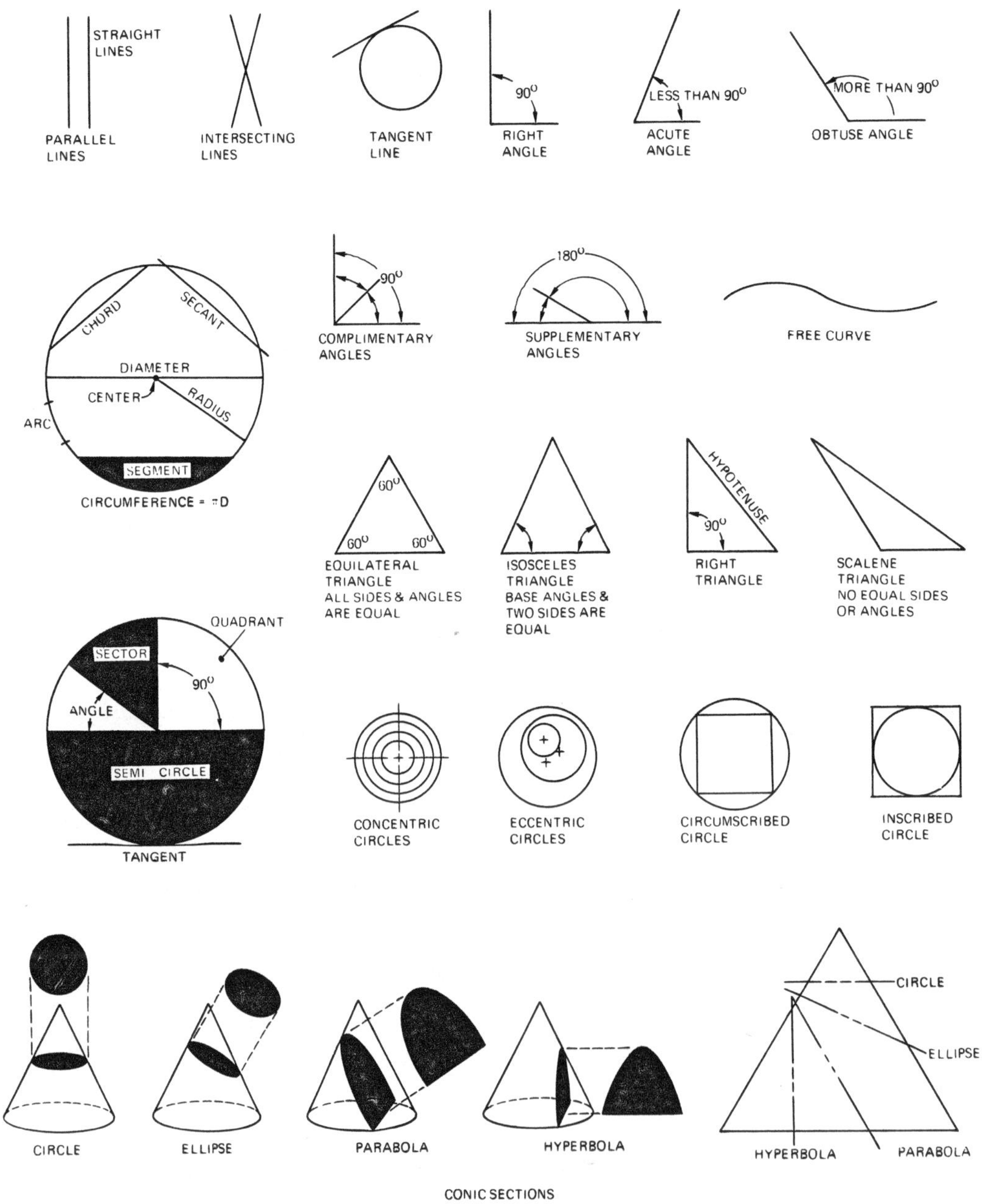

TWO-DIMENSIONAL GEOMETRIC FORMS

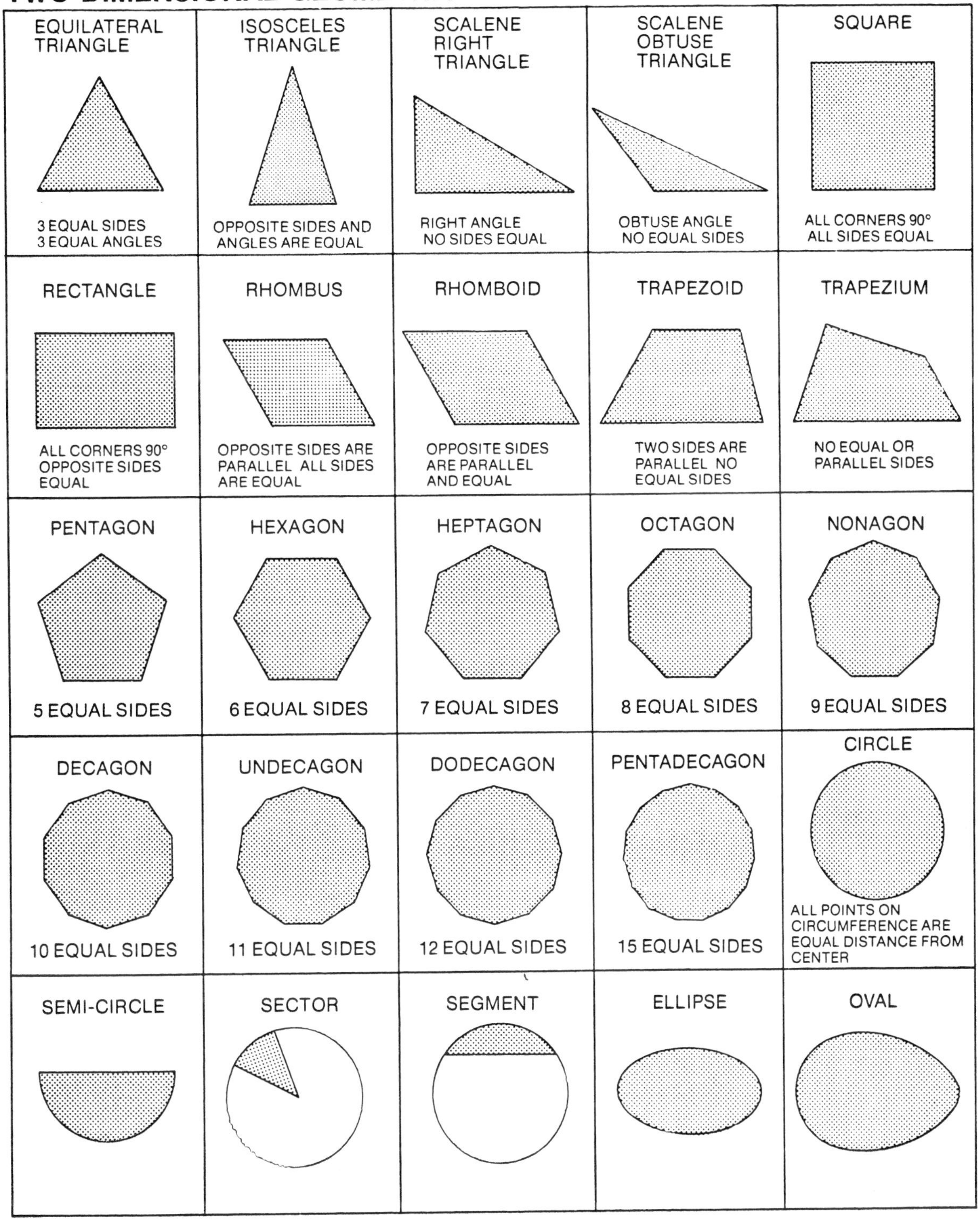

THREE-DIMENSIONAL GEOMETRIC FORMS

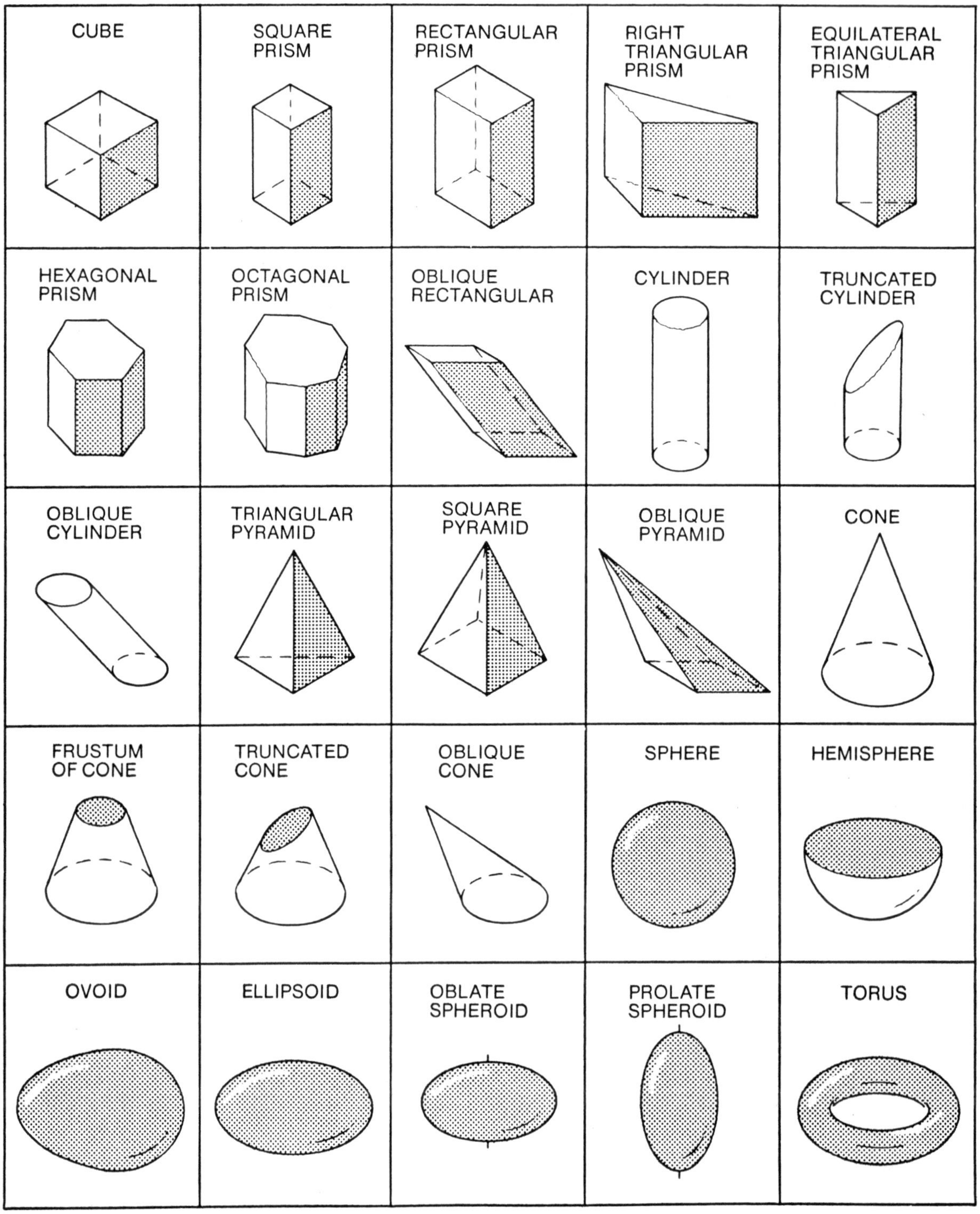

CENTERING YOUR DRAWING

A drafter considers neatness second only to accuracy. He or she considers a well-centered drawing as part of one's drafting skill.

CENTERING YOUR DRAWING		
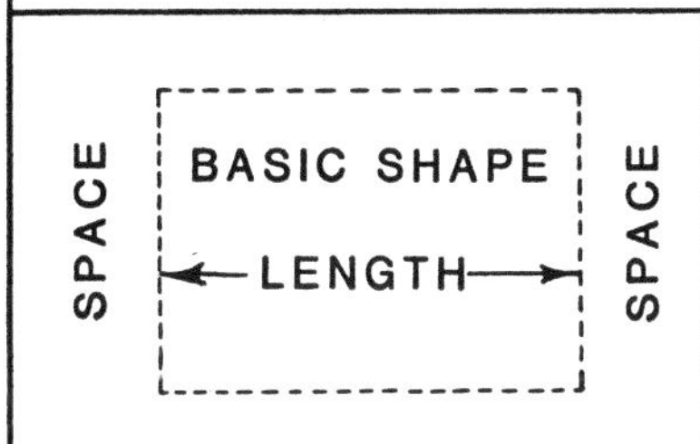	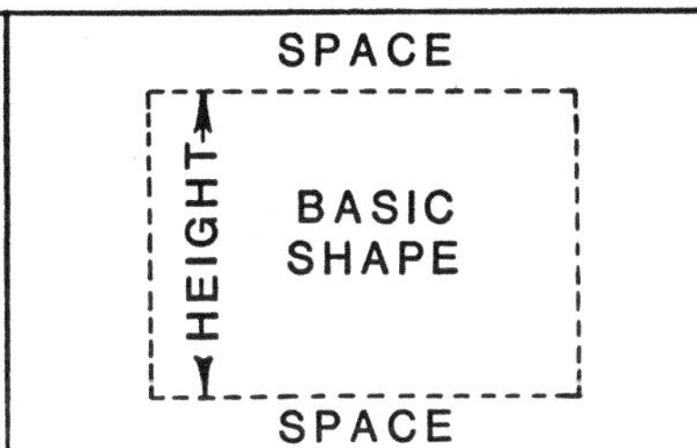	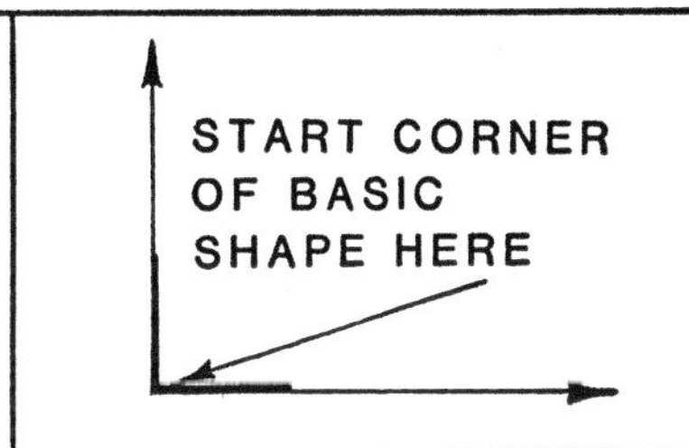
(1) Subtract the length of the basic shape from the length of the drawing area. Leave an equal space on each side of the basic shape.	(2) Subtract the height of the basic shape from the height of the drawing area. Leave equal spaces at the top and the bottom.	(3) After your calculations have been completed, locate your starting point by careful measurements.

ASSIGNMENT

Make a centered drawing of one of the following drawings.

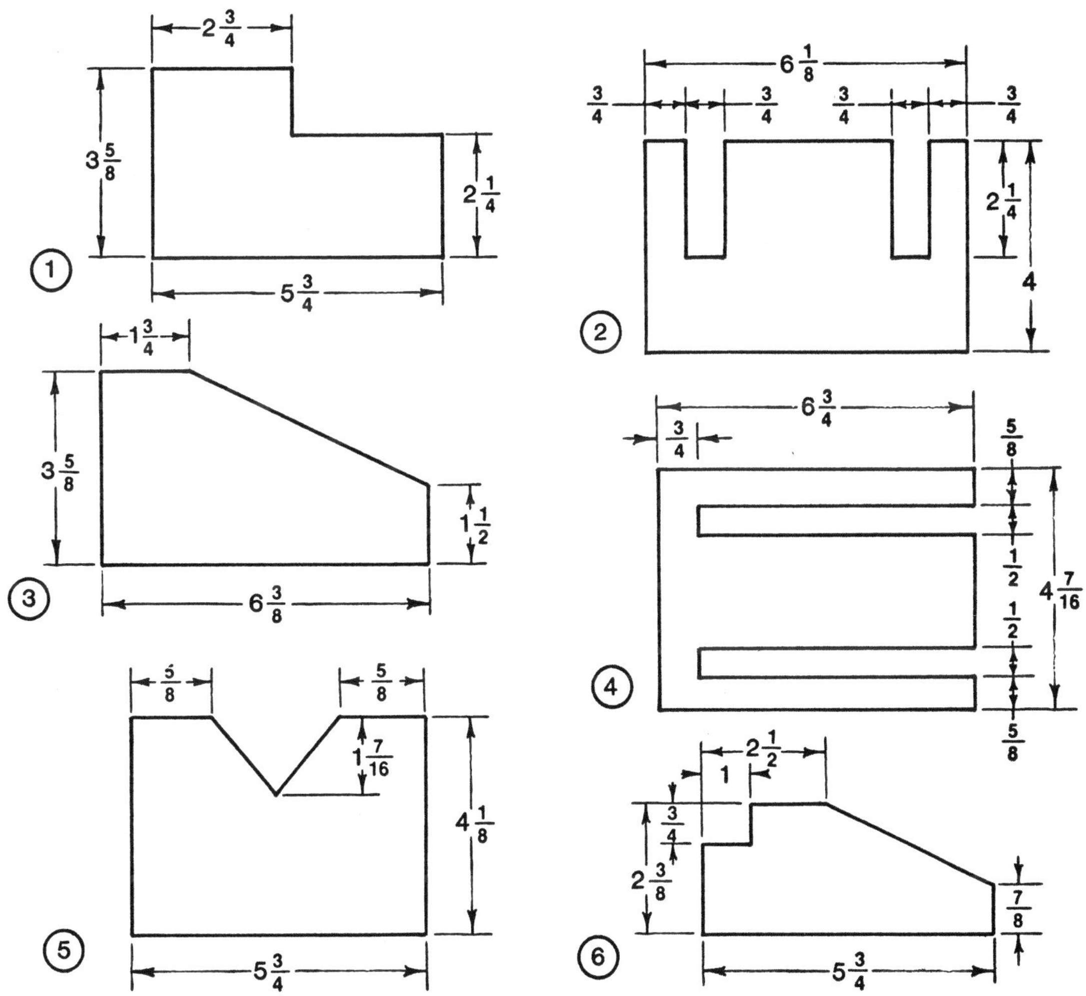

STUDY THESE DRAWINGS

Study each object shown below in order to discover the plan used in the arrangement of the holes and other circular parts.

In each case below, the plan of each object has been revealed.

Compare each object in column one with the related plan in column two.

Very often the entire hole or circle is not used; however, you must learn to visualize the object in terms of related holes or circles.

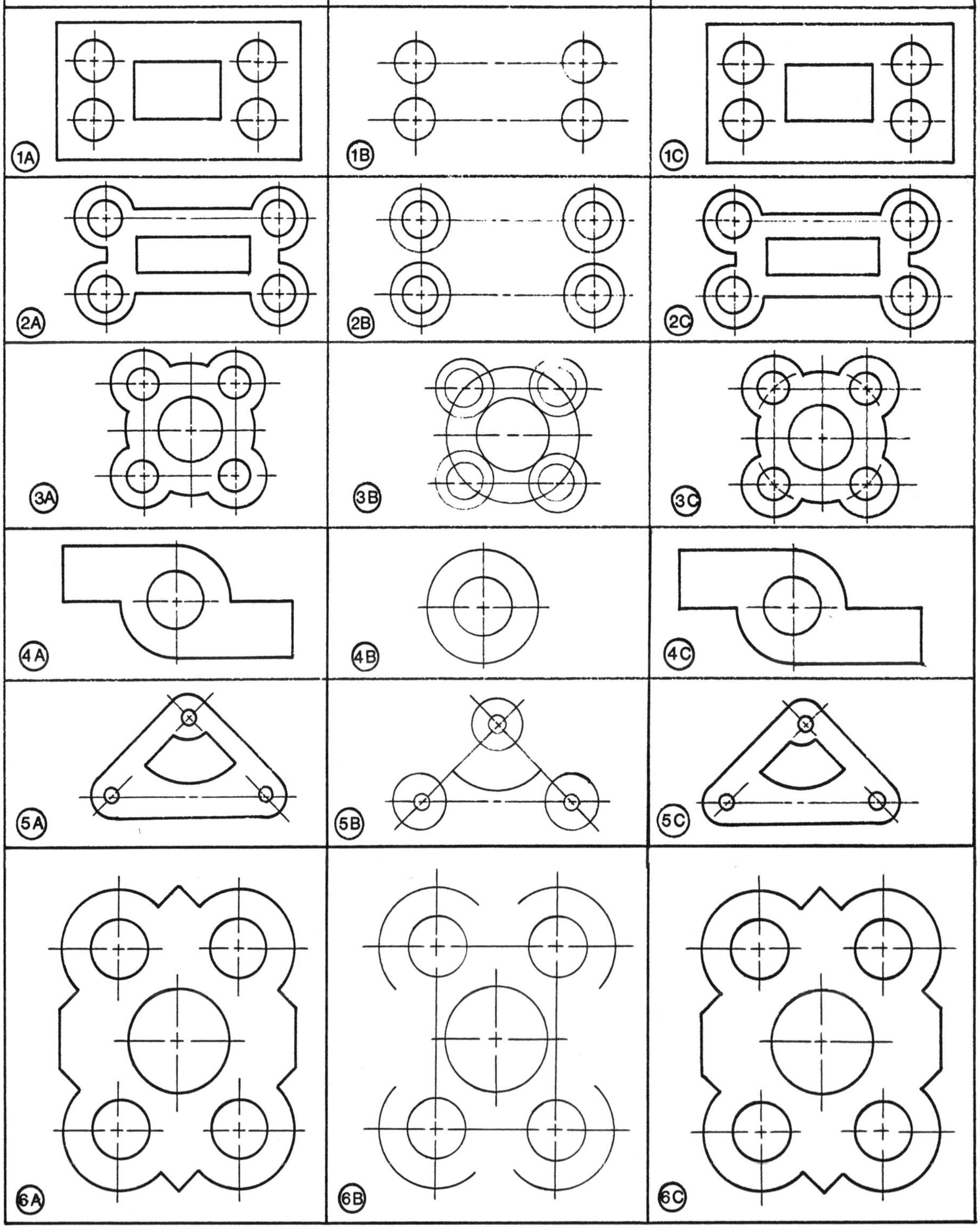

DRAWING A LINE TANGENT TO A CIRCLE AT A GIVEN POINT

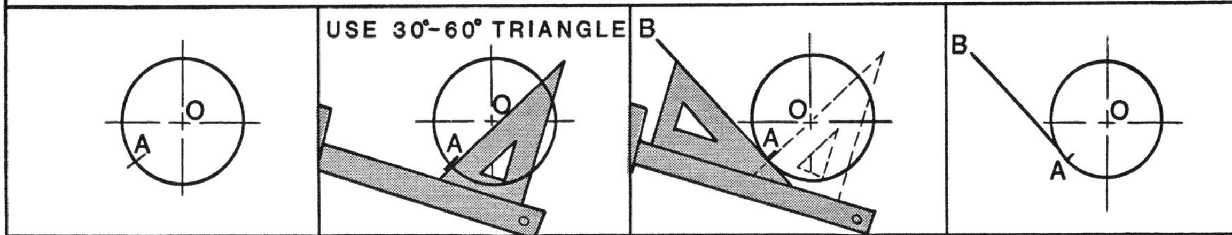

① Problem: Draw a tangent to circle "O" at point "A."	② Arrange the triangle and T-square exactly as shown.	③ Hold the T-square firmly. Move the triangle to the position as shown. Draw the tangent.	④ "AB" is tangent to circle "O" at point "A." (A 45° triangle may also be used.)

DRAWING TWO TANGENTS TO CIRCLE FROM POINT OUTSIDE CIRCLE

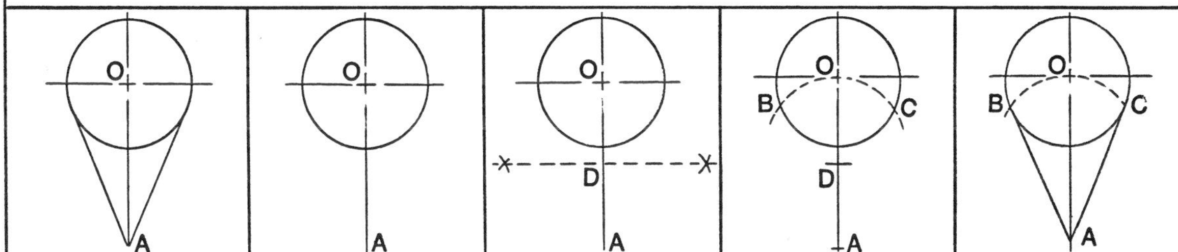

① To draw two tangents to the circle from point "A":	② Draw a line from center "O" to point "A."	③ Bisect "OA" in order to locate point "D."	④ With "OD" as a radius and "D" as the center, swing arc "BC."	⑤ Draw lines "AC" and "AB."

DRAWING AN ARC TANGENT TO TWO LINES (ACUTE ANGLE)

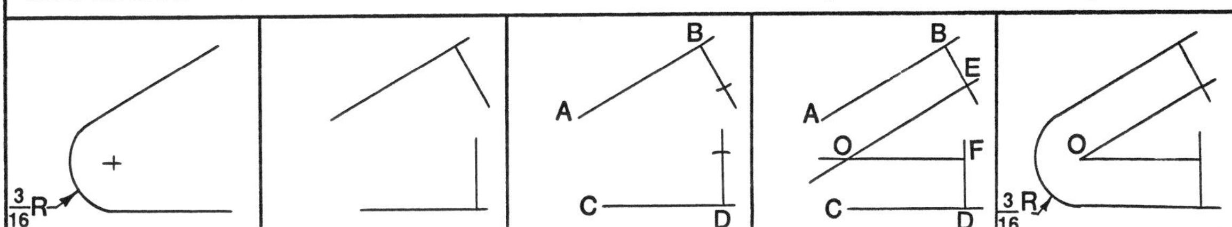

① To make the construction shown above:	② Draw perpendiculars as shown.	③ Swing the radius of the 3/16" arc from "B" and "D."	④ Draw "OE" parallel to "AB" and "FO" parallel to "CD."	⑤ Swing the 3/16 R arc from "O" as required.

DRAWING AN ARC TANGENT TO TWO LINES (OBTUSE ANGLE)

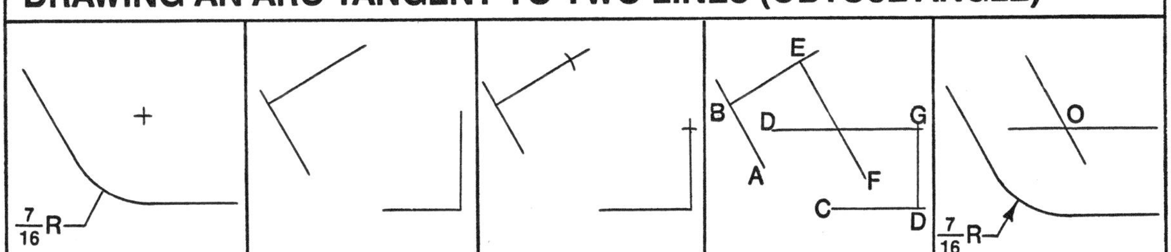

① To make the construction shown above:	② Draw perpendiculars as in ② above.	③ Swing the 7/16 R from "B" and "D" as in above.	④ Draw "EF" parallel to "AB" and "DG" parallel to "CD."	⑤ Swing the 7/16 R arc from "O" as required.

DRAWING ONE OR TWO TANGENTS TO TWO CIRCLES

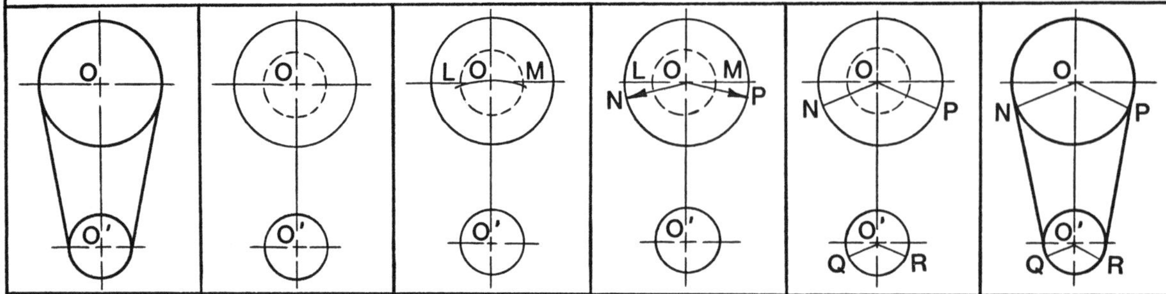

(1) To make the construction above:	(2) Repeat circle "O'" at "O."	(3) Use radius "O'O". Swing arc "LOM" from "O."	(4) Extend "OL" to "N" and "OM" to "P."	(5) Draw "O'Q" parallel to "ON." Draw "O'R" parallel to "OP."	(6) Draw tangent lines "NQ" and "PR."

A WORKING KNOWLEDGE OF ANGLES

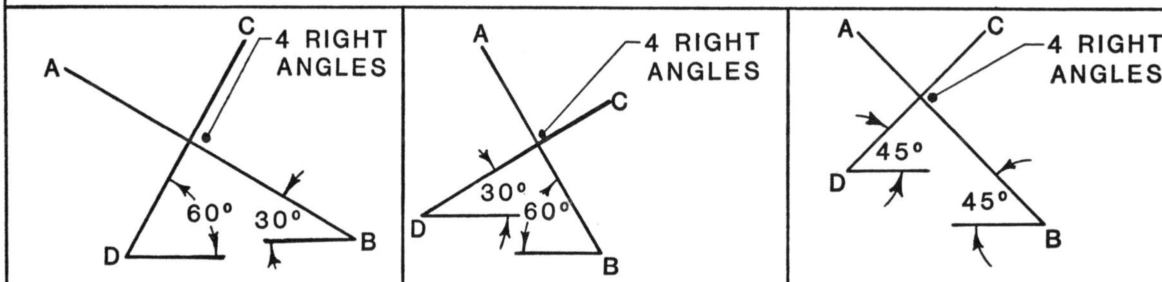

(1) The 30° line (AB) will intersect a 60° line (CD) at right angles.	(2) The 60° line (AB) will intersect a 30° line (CD) at right angles.	(3) These 45° lines intersect each other at right angles.

MEASURING FOR EVENLY SPACED OBLIQUE LINES

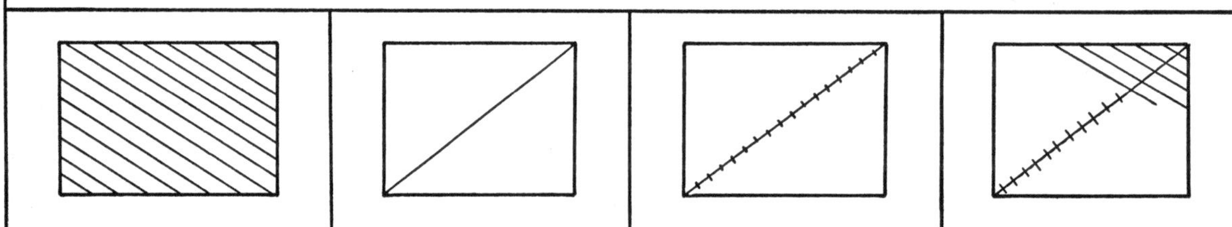

(1) To draw the above:	(2) Draw the diagonal as shown.	(3) Measure the spaces on the diagonal.	(4) Draw through the marks.

HOW TO USE A PROTRACTOR

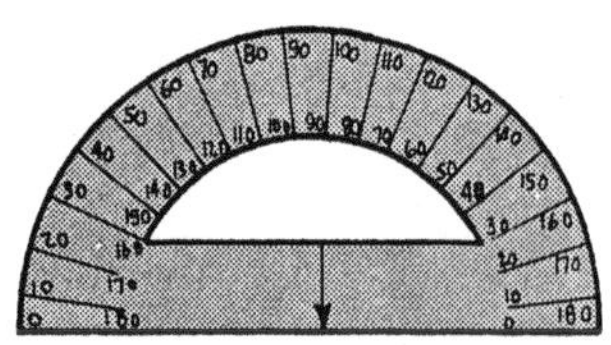

PROTRACTOR

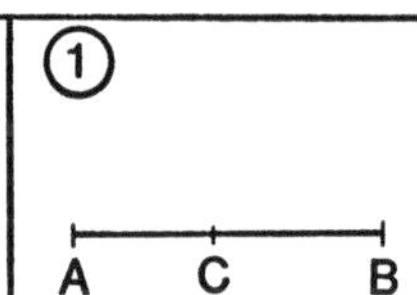

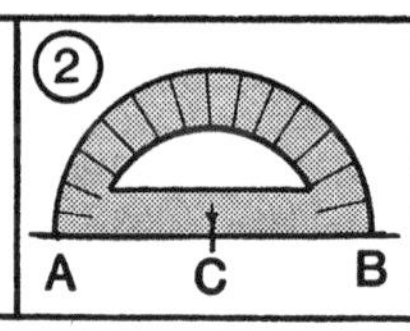

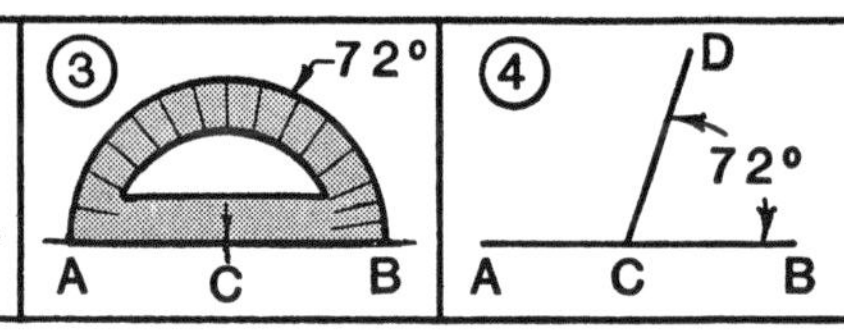

(1) Draw line "AB," mark point "C."
(2) Place protractor along the line "AB," with arrow at point "C."
(3) Mark graduated point 72 for an angle of 72°.
(4) Angle "DCB" is an angle of 72°.

DIVIDING A LINE INTO A GIVEN NUMBER OF PARTS (METHOD 1)

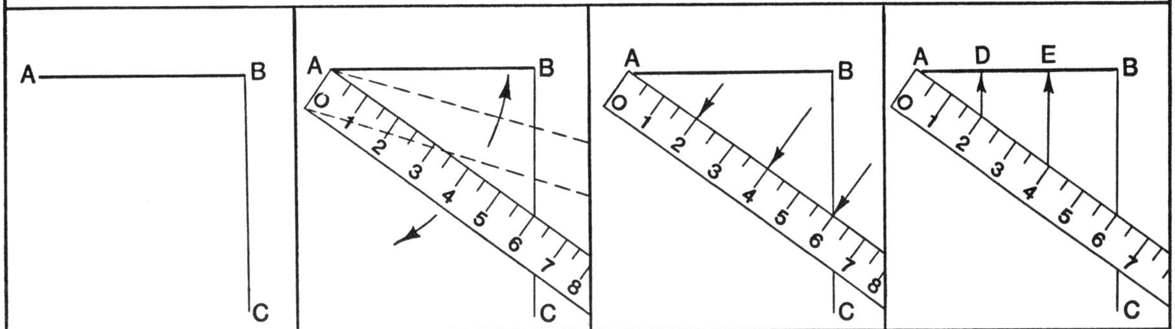

(1) To divide line "AB" into three equal parts, draw vertical line "BC":

(2) Place the end of the ruler on "A." Swing the ruler up or down until:

(3) Some number divisible by 3 (6) touches line "BC." Mark 2 and 4.

(4) Project upward vertically from 2 and 4. "AD" = "DE" = "EB."

DIVIDING A LINE INTO A GIVEN NUMBER OF PARTS (METHOD 2)

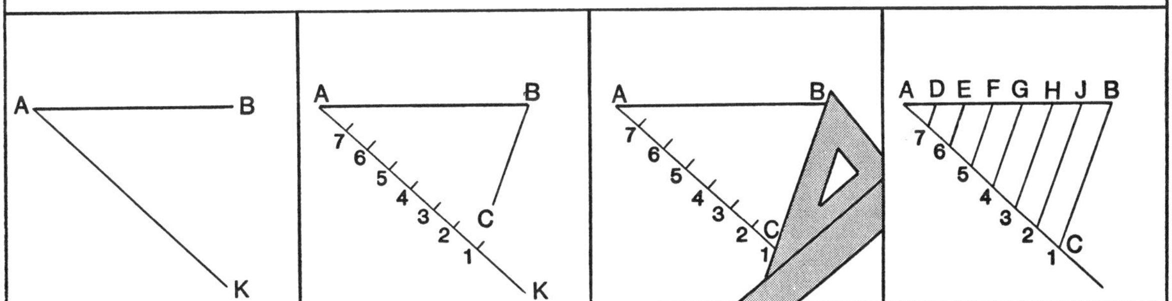

(1) To divide "AB" into seven equal parts: Draw "AK" at any angle to "AB."

(2) Starting at "A," measure seven equal spaces along "AK." Draw a line from the seventh mark (C) through "B."

(3) Draw lines parallel to "CB" from 1, 2, 3, 4, 5, and 6.

(4) "AD" = "DE" = "EF" = "FG," etc.

FINDING FRACTIONAL PARTS OF ANY DISTANCE

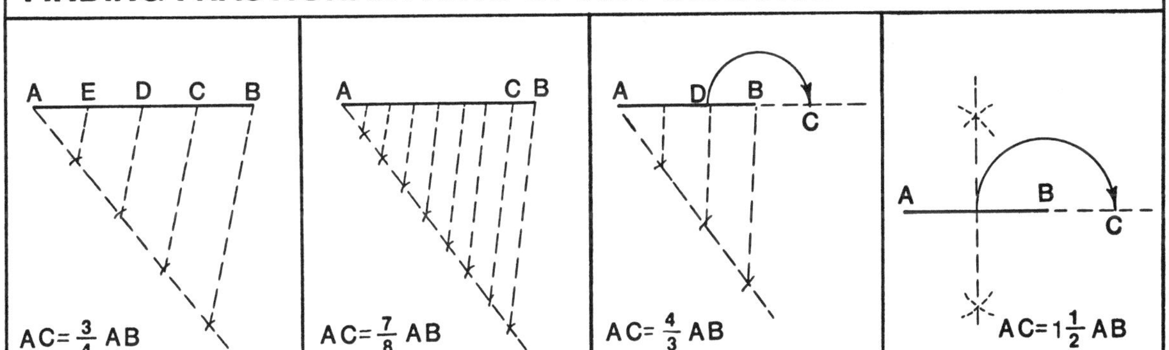

(1) To find 3/4 of "AB": Divide "AB" into 4 parts. Three of the parts equals three fourths.

(2) To find 7/8 of "AB": Divide "AB" into 8 parts. Seven of the parts equals seven-eighths.

(3) To find 4/3 of "AB": Divide "AB" into 3 parts. Swing "BD" to "C." "AC" = 4/3 of "AB."

(4) To find 1½ "AB": Divide "AB" into 2 parts. Swing "BD" to "C." "AC" will then be 1½ "AB."

DRAWING A LINE PARALLEL TO A GIVEN LINE

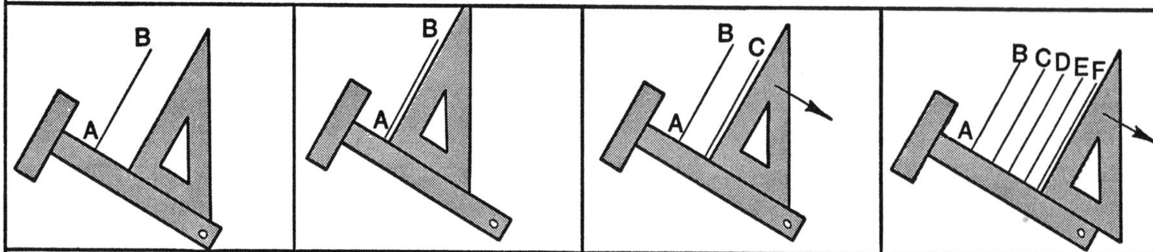

(1) Turn the T-square upside down. Arrange the triangle as shown.

(2) Move both tools as one until the triangle is parallel to "AB."

(3) Hold the T-square steady. Slide the triangle as shown. Draw a line from "C."

(4) Repeat (3) in order to draw other lines parallel to "AB."

DRAWING A LINE AT RIGHT ANGLES TO A GIVEN LINE

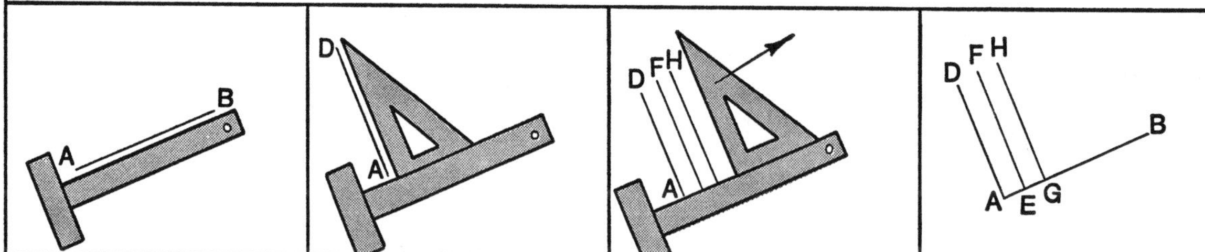

(1) Turn the T-square upside down. Align the T-square blade with line "AB."

(2) Place the right angle of the triangle against blade at "A." Draw "AD."

(3) Slide the triangle along the blade as shown. Draw lines at "D," "F," "H."

(4) Lines "DA," "FE," "HG," are all perpendicular to line "AB."

PERPENDICULARS AND PARALLELS

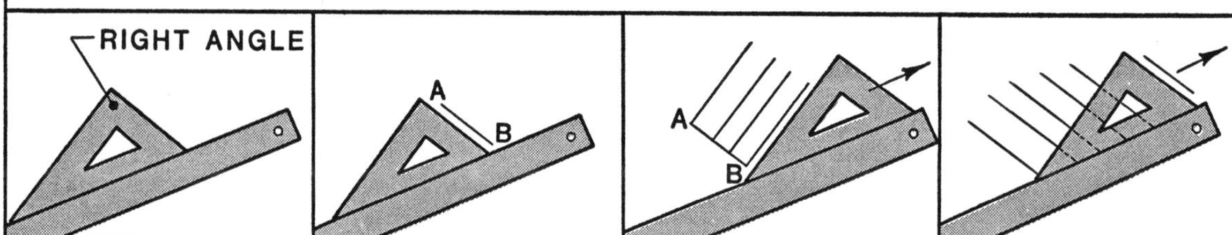

(1) Place the right angle of the triangle at the top as shown.

(2) Align the triangle with "AB" as shown.

(3) Draw perpendiculars to "AB" in the manner indicated.

(4) Draw parallels to "AB" in the manner indicated.

CONSTRUCTING A ROUNDED CORNER

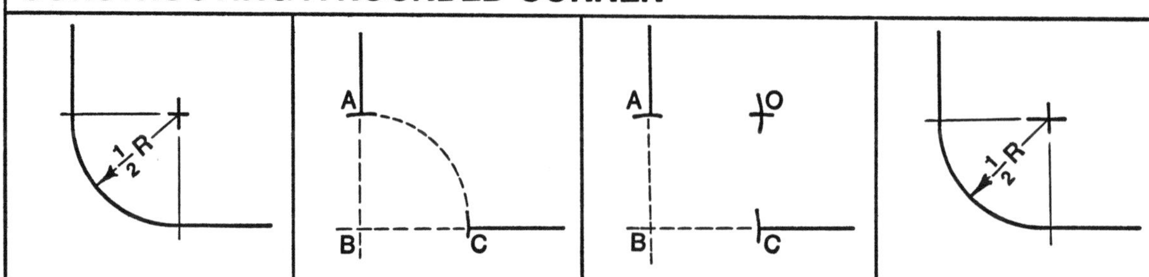

(1) This is a rounded corner having a radius of 1/2".

(2) With a radius of 1/2", swing "BA" and "BC" from center "B."

(3) With a radius of 1/2", swing "AO" from "A," swing "CO" from "C."

(4) Construct all rounded corners in this manner.

CONSTRUCTING AN ANGLE EQUAL TO A GIVEN ANGLE

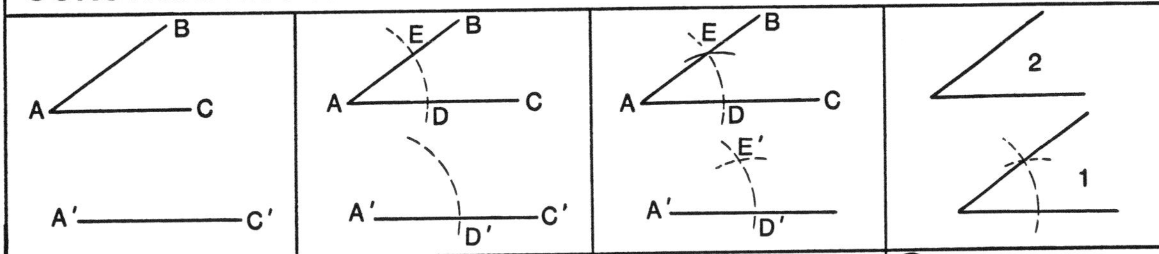

(1) To construct an angle equal to angle "BAC" at "A′":	(2) Swing arc "ED" from "A." Repeat the same arc from "A′."	(3) Take radius "DE." Swing this radius from "D′" to cross at "E′."	(4) Angle 2 equals angle 1.

CONSTRUCTING A SQUARE AROUND A CIRCLE

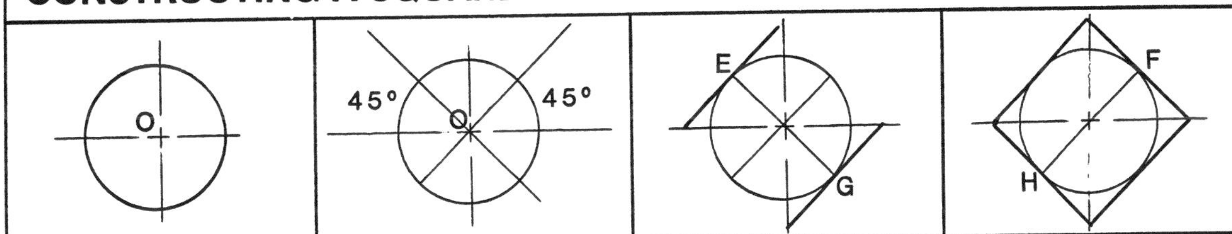

(1) Draw the horizontal and the vertical center lines.	(2) Draw 45° lines through center "O."	(3) Draw 45° lines which touch the circle at "E" and at "G."	(4) Draw 45° lines which touch the circle at "F" and at "H."

CONSTRUCTING A HEXAGON AROUND A CIRCLE

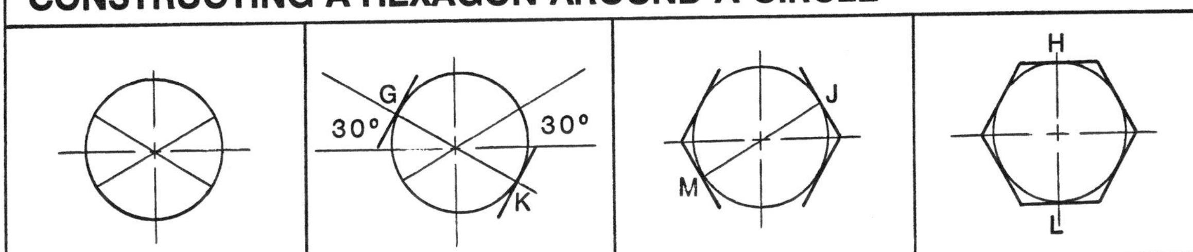

(1) Draw two 30° lines as shown.	(2) Draw two 60° lines which touch the circle at "G" and at "K."	(3) Draw two 60° lines which touch the circle at "J" and at "M."	(4) Draw two horizontal lines which touch the circle at "H" and at "L."

CONSTRUCTING A HEXAGON WHEN ONE SIDE IS GIVEN

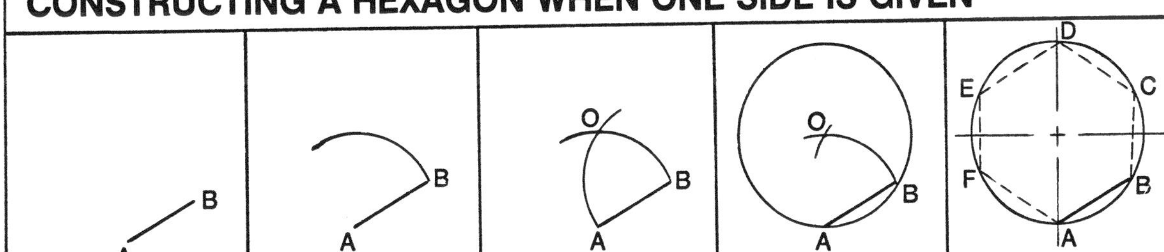

(1) Given side "AB." Take "AB" as a radius.	(2) Swing "AB" to the left from point "A."	(3) Swing "AB" from point "B." Mark center "O."	(4) Using radius "AB," swing a circle from point "O."	(5) Lay off "AB" around the circle. Complete as shown.

CONSTRUCTIONS YOU MUST KNOW

BISECTING A LINE (A PERPENDICULAR BISECTOR)

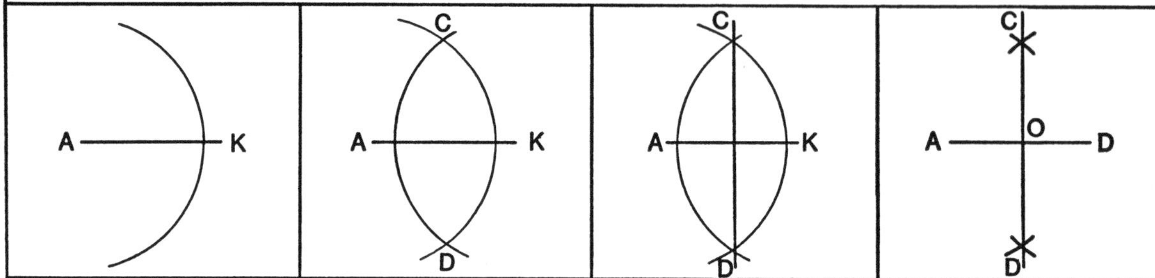

① Using "A" as a center, swing an arc.	② Using "K" as a center, swing an equal arc.	③ Draw a line through "D" and "C."	④ "AO" = "OK", and "CD" is perpendicular to "AK."

BISECTING AN ANGLE

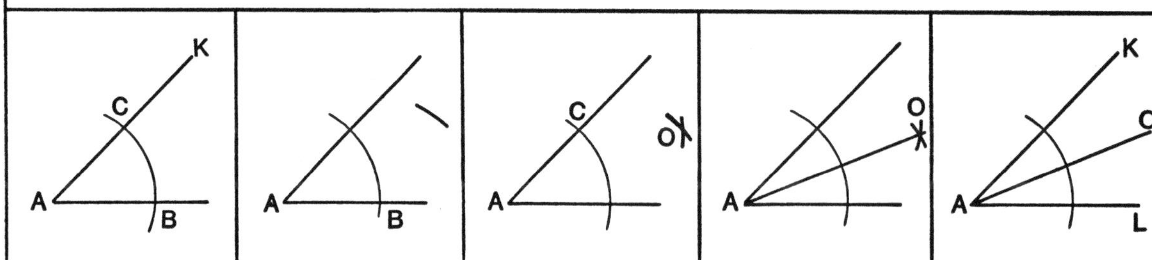

① Swing arc "CB" from "A."	② Swing an equal arc from "B."	③ Swing an equal arc from "C."	④ Draw a line through "O" and "A."	⑤ Angle KAO equals angle LAO.

CONSTRUCTING A HORIZONTAL SQUARE

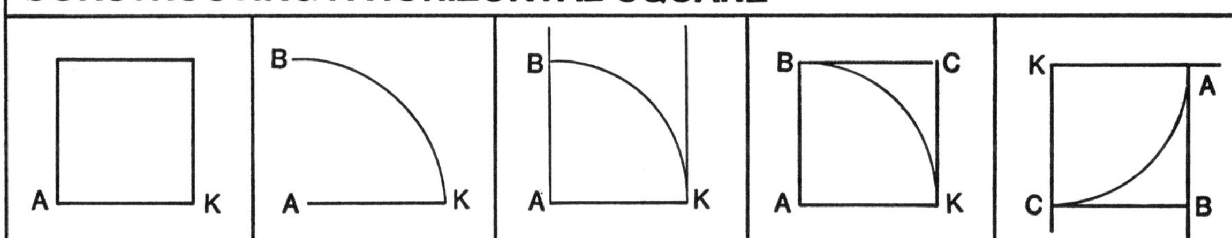

① To construct a 5/8" square on line "AK."	② With "A" as a center, swing a 5/8" radius.	③ Draw vertical lines from "A" and from "K."	④ Draw a horizontal line through "B."	⑤ The same construction drawn below the line "AK."

REVOLVING A SQUARE (TWO METHODS)

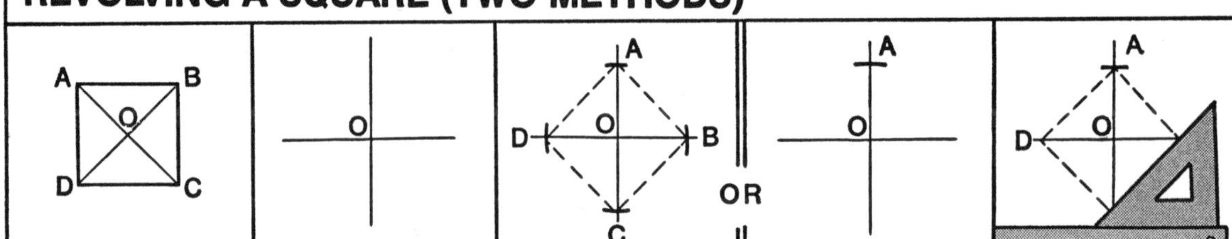

① Draw the diagonals. Take "OA" as a radius.	② Draw a vertical and a horizontal line through "O."	③ From "O," swing "OA," "OB," "OC," "OD." Join A, B, C, D.	④ Second method: With "O" as center and "OA" as radius, locate "A." Then, using a 45° triangle, draw the sides of the square.

CONSTRUCTING A HEXAGON USING A COMPASS

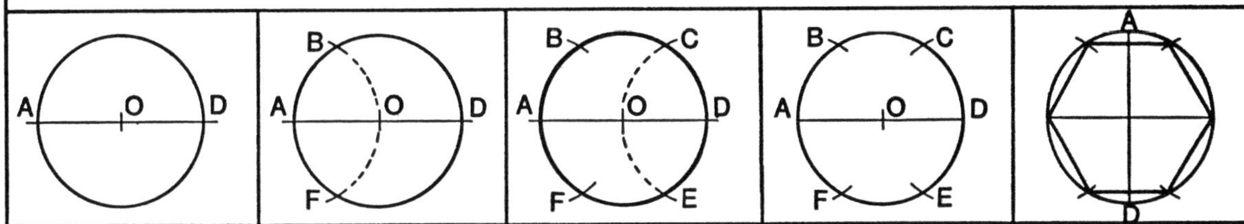

① Draw a circle. Draw the horizontal diameter.

② Swing the radius from "A" to "B" and "F."

③ Swing the radius from "D" to "C" and "E."

④ Join "A" and "B," "B" and "C," etc., with a "free" triangle.

⑤ Try it with a vertical diameter.

CONSTRUCTING A HEXAGON USING A 30°-60° TRIANGLE

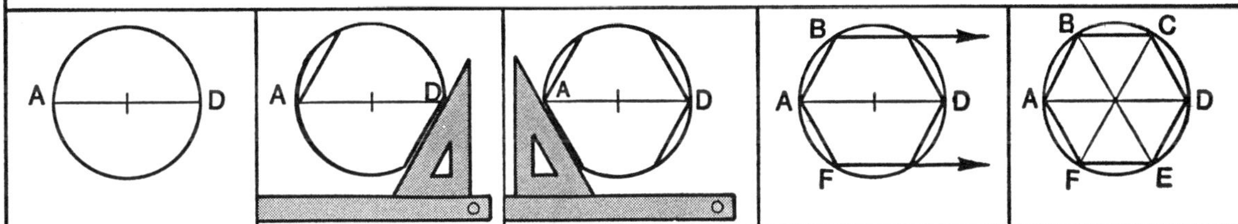

① Draw a circle. Draw diameter "AD."

② Draw a 60° line from "A" and from "D."

③ Draw a 60° line from "A" and from "D."

④ Draw horizontal lines from "B" and from "F."

⑤ . . . or . . . draw 60° lines through the center of the circle.

REPEATING A SINGLE MEASUREMENT USING A COMPASS

When a measurement, such as "AB," is to be repeated several times, repeat it by taking "AB" as a radius and swinging from "A" to "B," "B" to "C," etc.

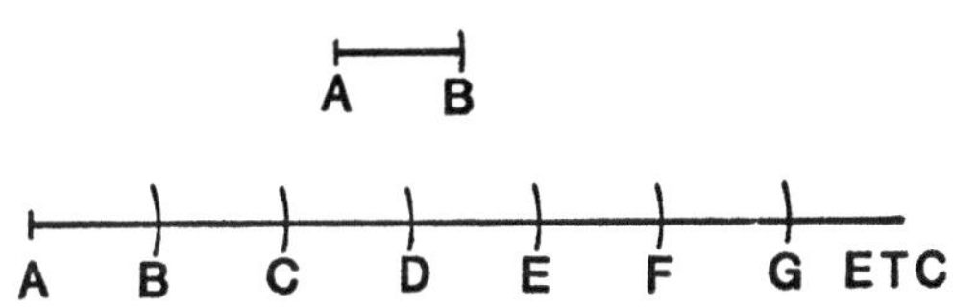

ASSIGNMENT

Do the constructions as indicated in each panel.

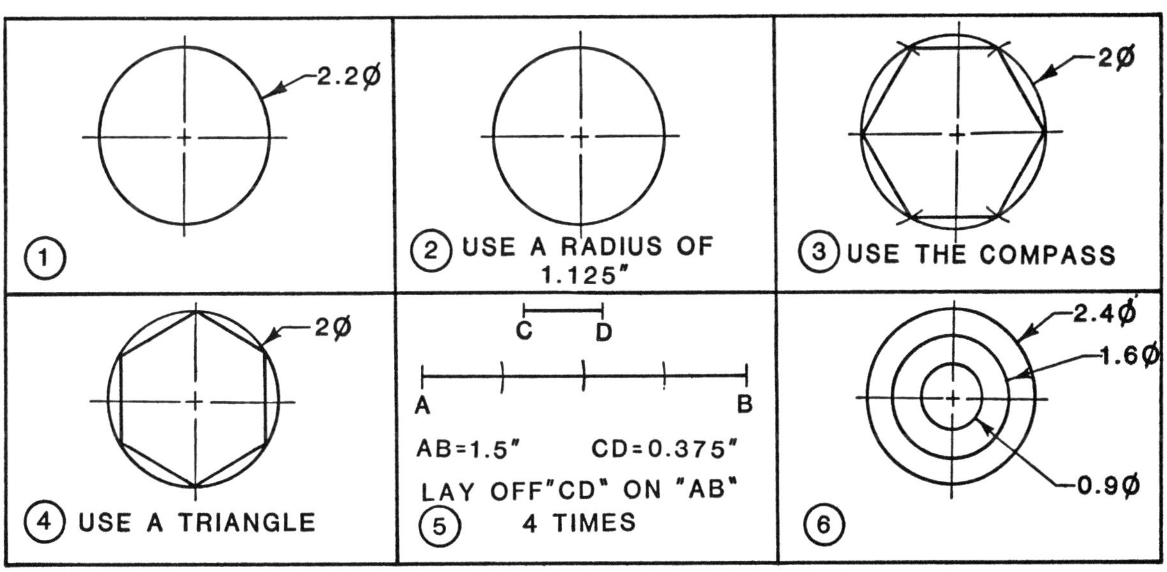

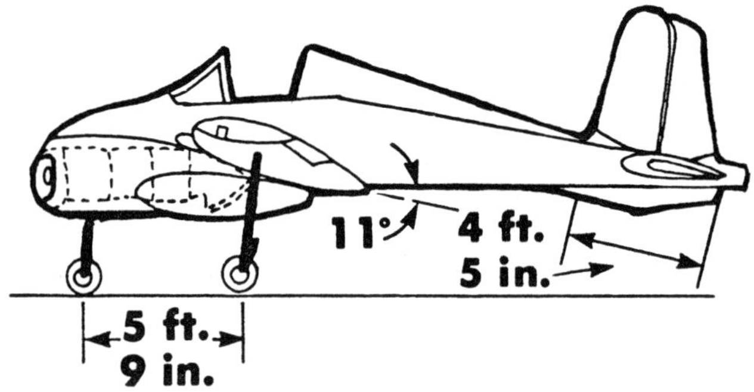

UNIT 8

MEASURING SCALES

There are several types of measuring scales used in mechanical drafting. To become proficient in all areas of mechanical drawing, a thorough knowledge of the scales and their uses is necessary.

THE MECHANICAL ENGINEER'S SCALE

The units of measurement are the inch and fractional part of the inch.

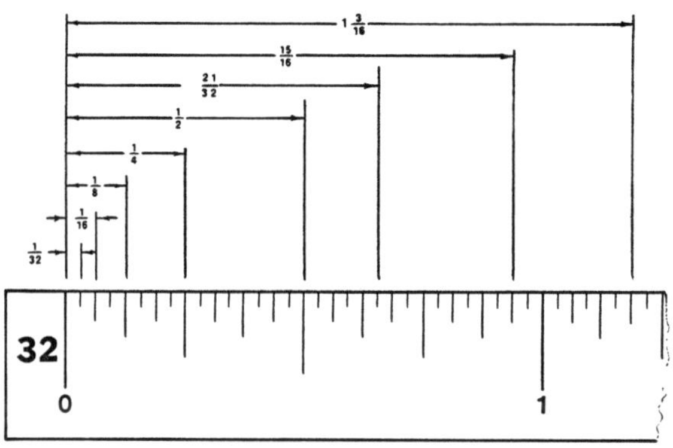

THE CIVIL ENGINEER'S SCALE

The units of measurement are the inch and the decimal part of an inch.

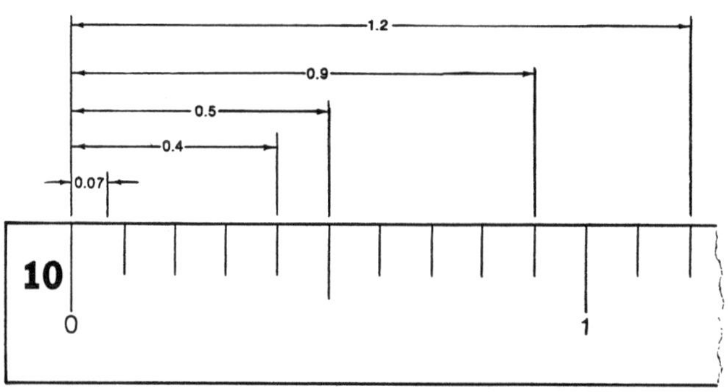

THE ARCHITECT'S SCALE

The scale: 1/4″ = 1′-0″ is the scale used by architects to draw floor plans. The units of measurement are the foot and inch.

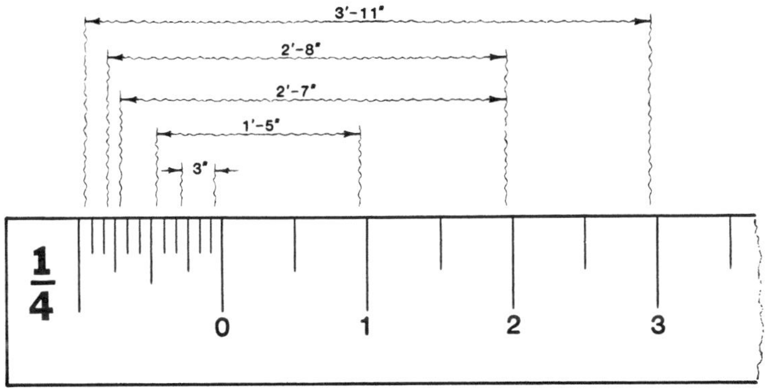

THE METRIC SCALE

The only unit of measurement for metric scales in engineering drawings is the millimeter.

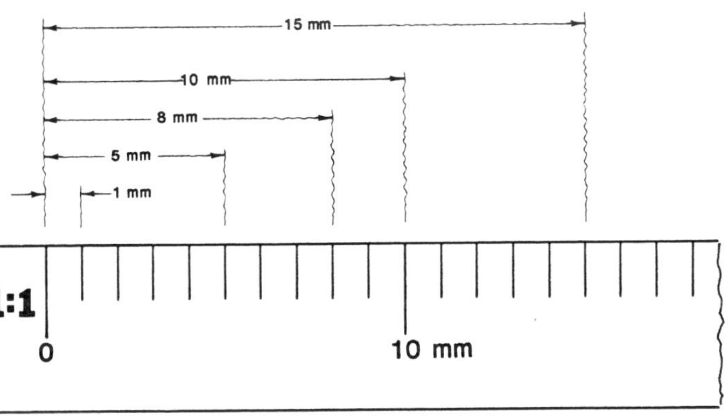

DIFFERENT SHAPES OF DRAFTING SCALES

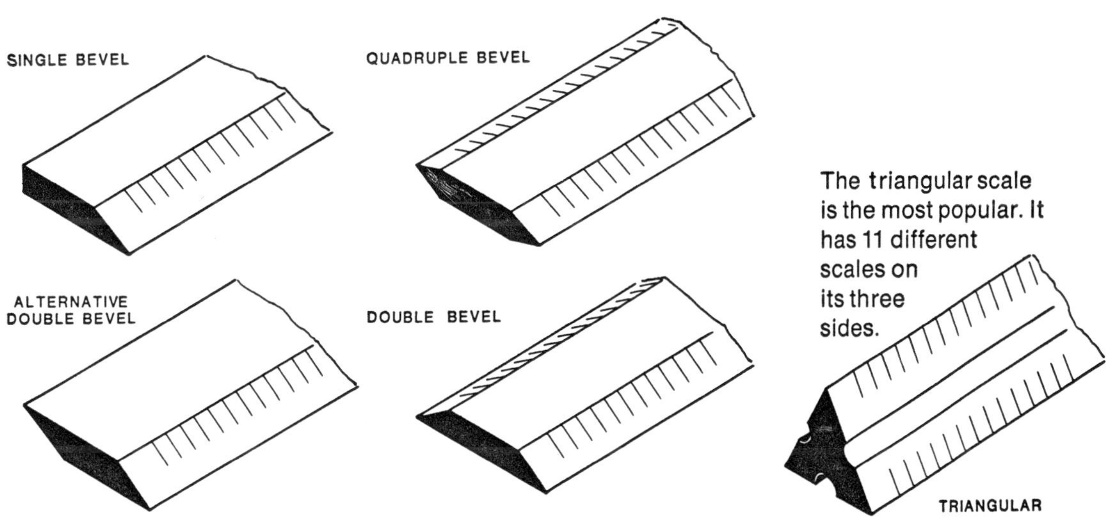

The triangular scale is the most popular. It has 11 different scales on its three sides.

SCALING UP

When a drawing is made the exact size as the actual object, it is a full-size drawing. Its scale is given as scale 1:1. If the full-size drawing is too small for the drawing format, it should be scaled up in size to fit the drawing format. An example is scale 5:1.

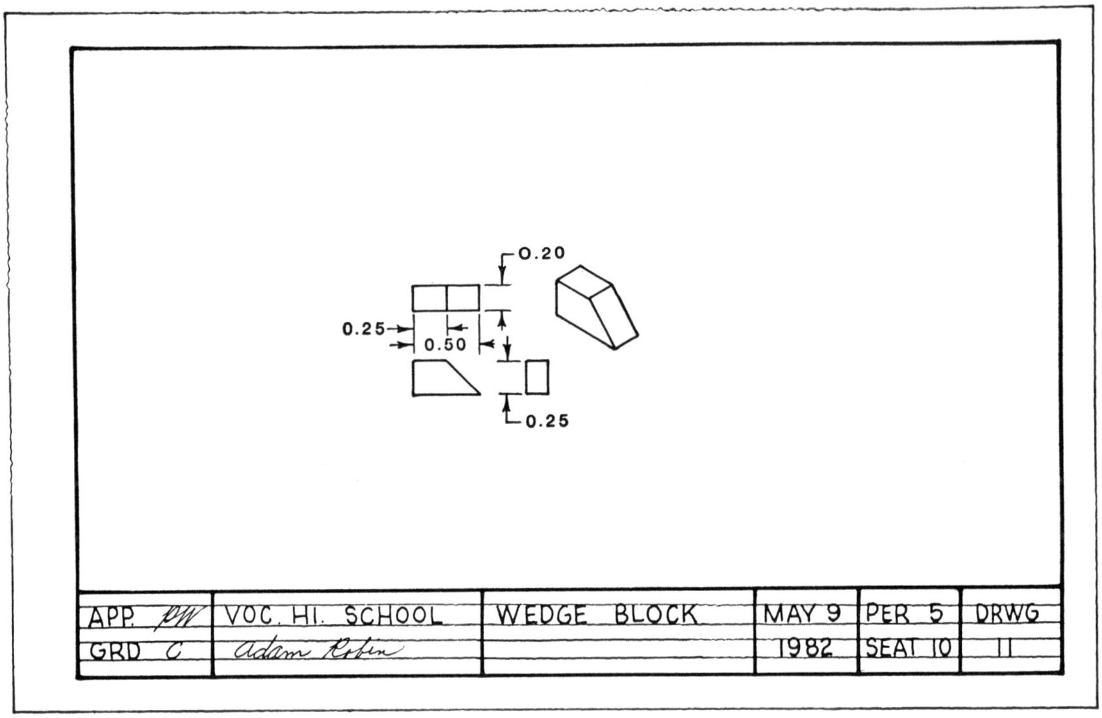

scale: full-size

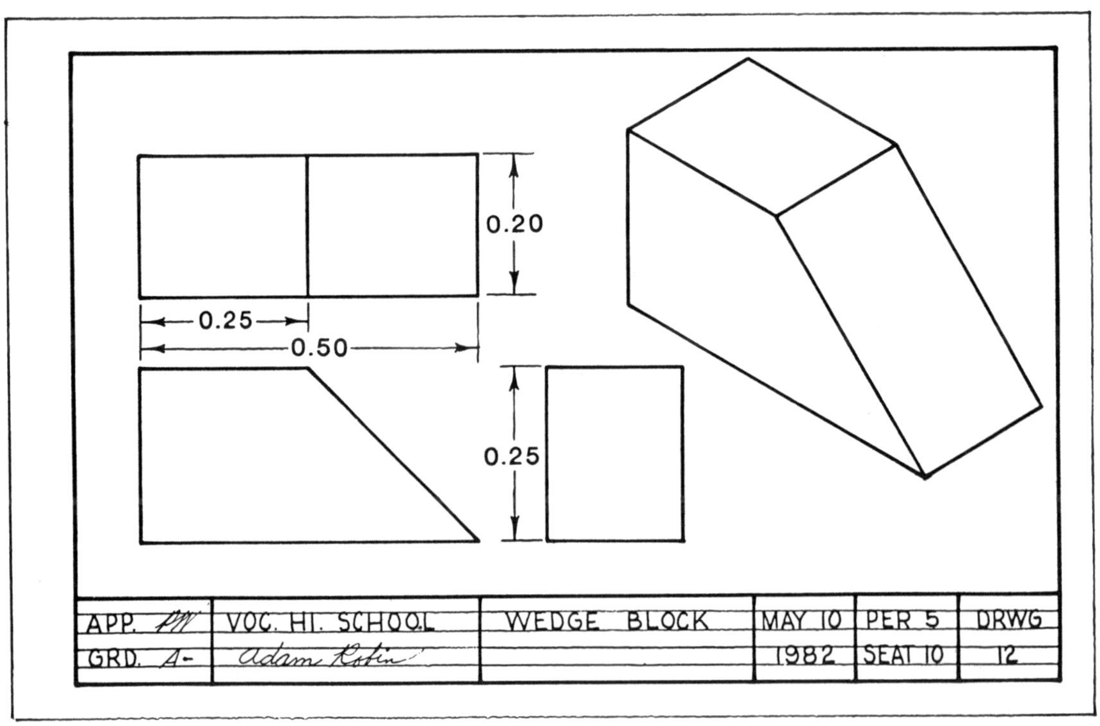

scale: 5:1

SCALING DOWN

If the size of the actual object is too large to draw full size on the drawing format, the drawing must be scaled down. An example is 1/8″ = 1′-0″.

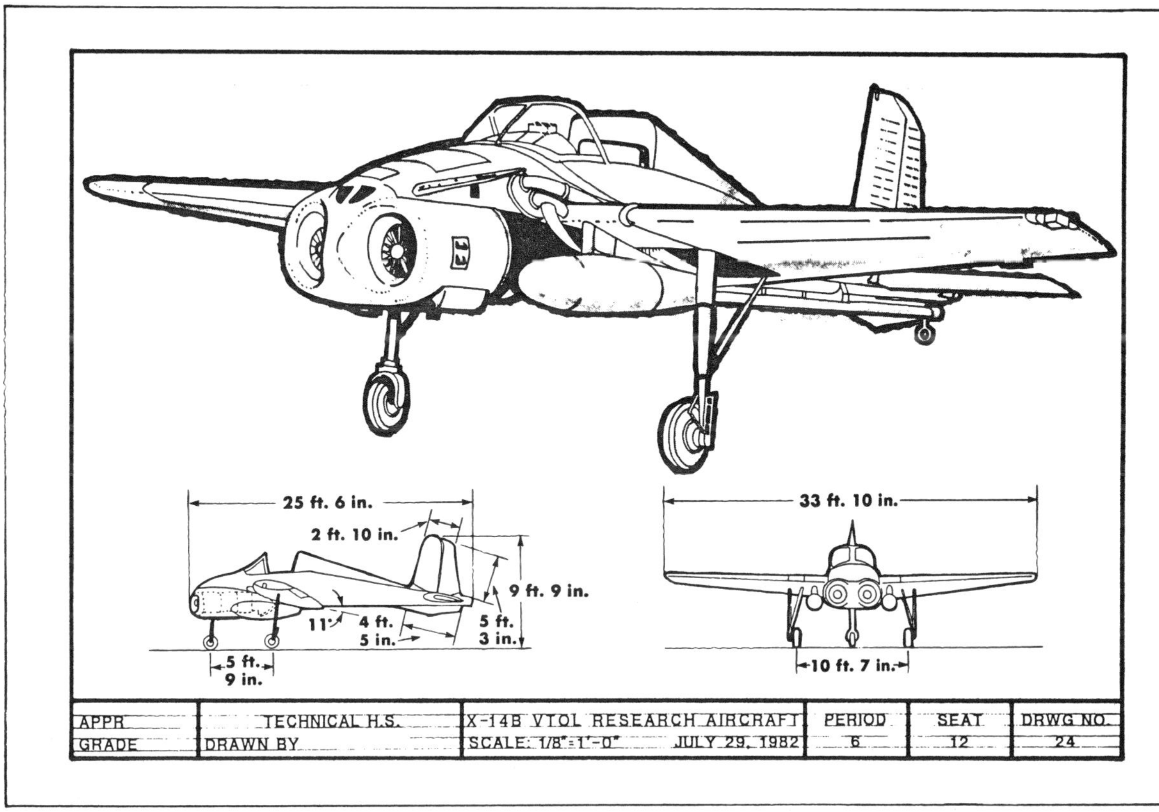

SCALE SELECTIONS

The table is a general list for selecting a scale for an engineering drawing.

Scale Selection for Engineering Drawings			
Type of Drawing	Mechanical Engineer's Scale	Civil Engineer's Scale	Metric Scale
The object to be drawn is slightly smaller than the drawing format.	Full size	1:1	1:1
The object to be drawn is slightly larger than the drawing format.	½″ = 1″	1:2	1:2
The object to be drawn is a good deal larger than the drawing format.	⅛″ = 1″	1″ = 10″ 1:10	1:10
The object to be drawn is small.	1″ = ½″ (2:1)	2:1	2:1
The object to be drawn is very small.	1″ = ⅛″ (8:1)	10:1	10:1

TYPICAL SCALES USED WITH THE MECHANICAL ENGINEER'S DRAFTING SCALE

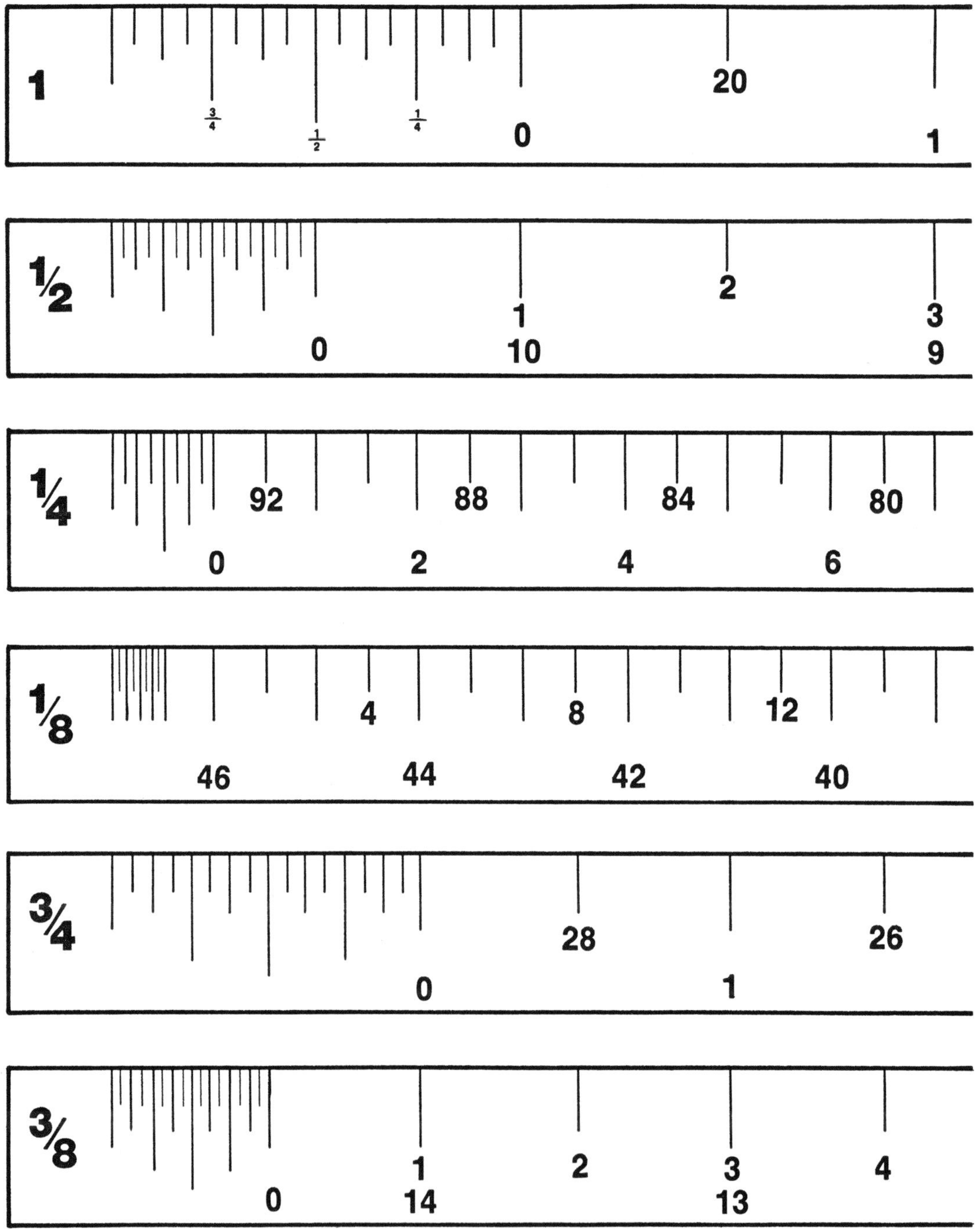

TYPICAL SCALES USED WITH THE CIVIL ENGINEER'S SCALE

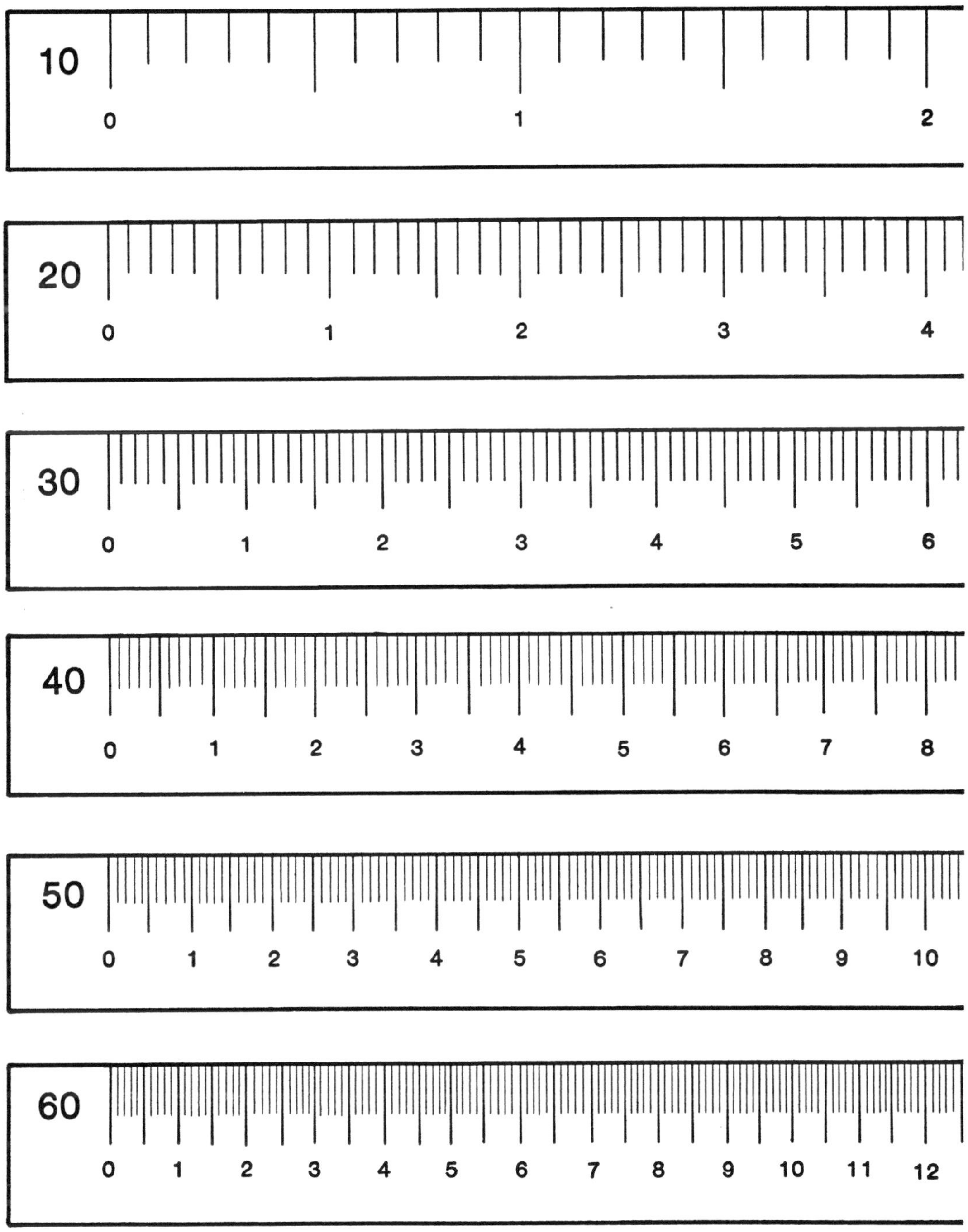

TYPICAL SCALES USED WITH THE ARCHITECT'S SCALE

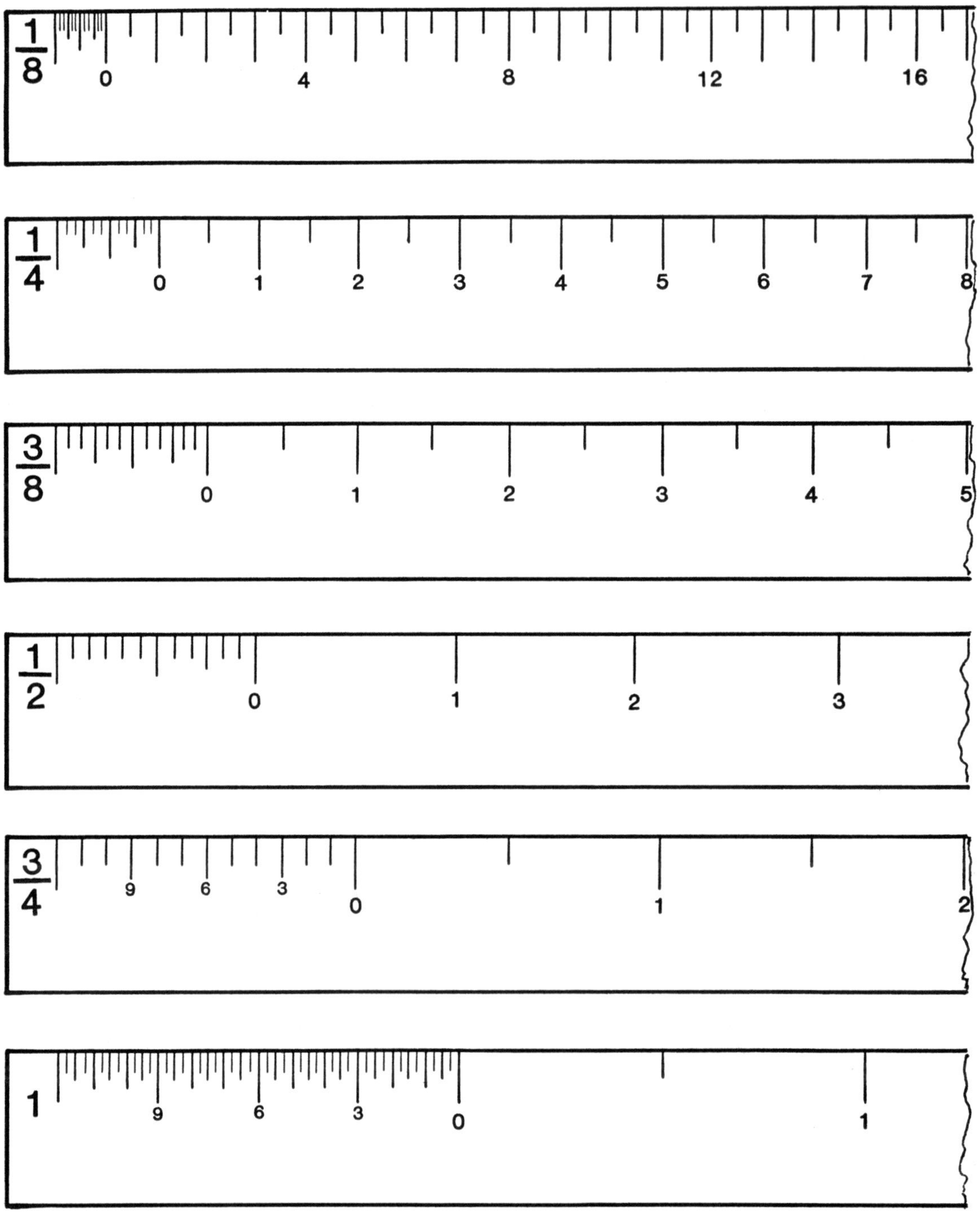

TYPICAL SCALES USED WITH THE METRIC SCALE

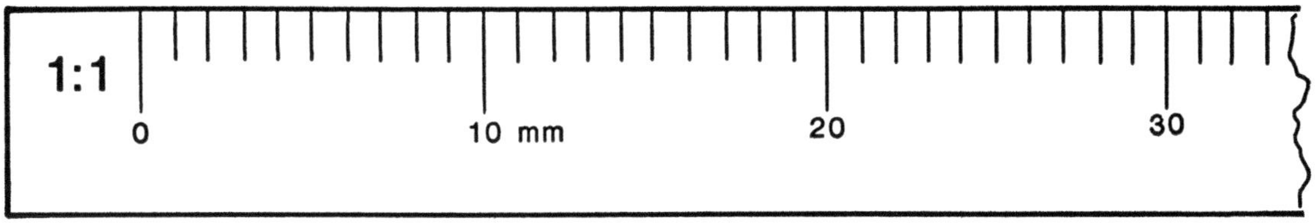

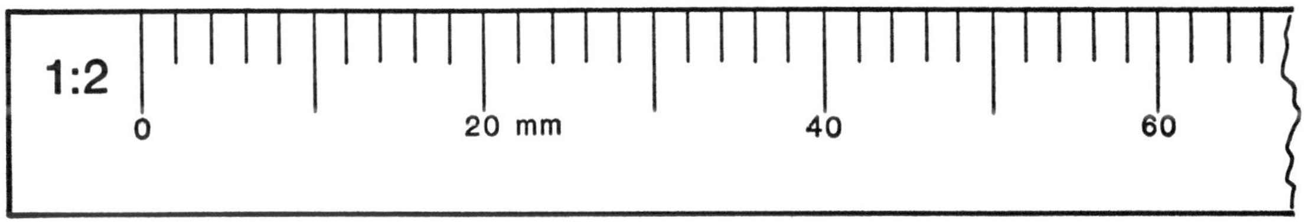

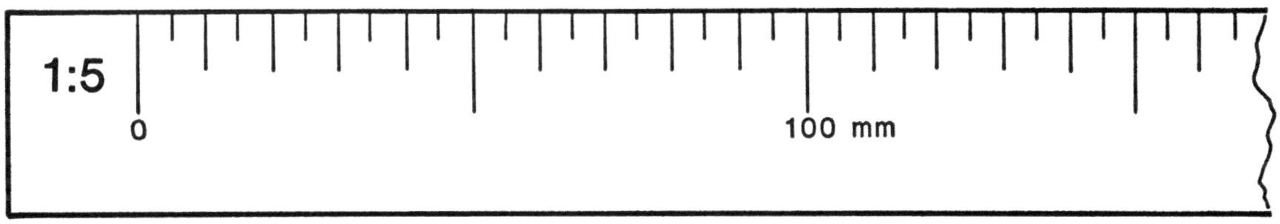

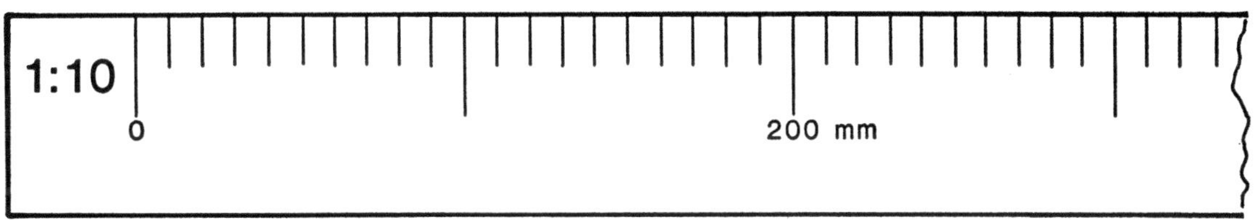

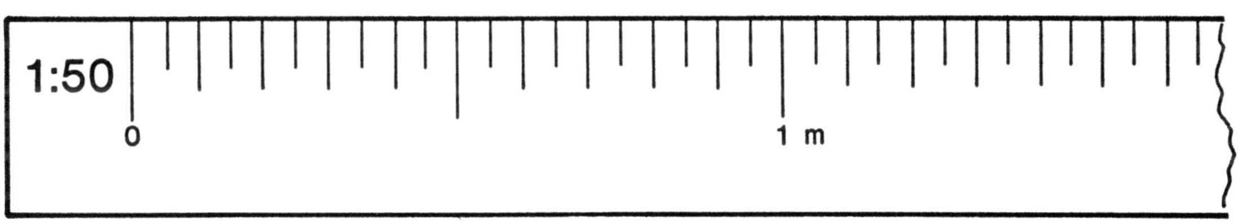

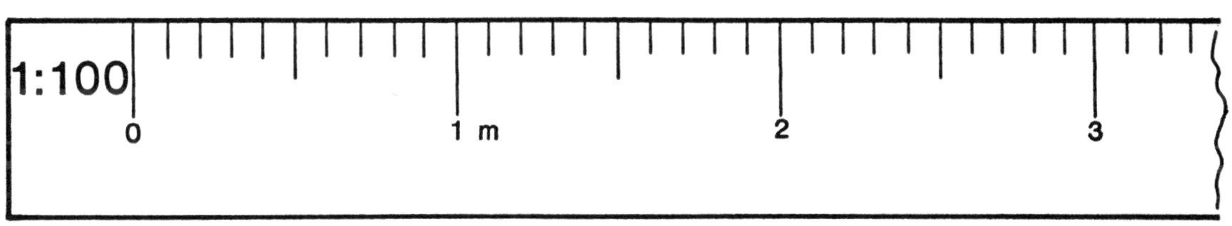

Measure the length of each line with the correct scale. Write your answers on a separate piece of paper. Correct your answers and remeasure your wrong answers.

MECHANICAL ENGINEER'S SCALE TEST

Measure to the closest 1/16".

SCALE:	
FULL	1
FULL	2
1/2"=1"	3
1/2"=1"	4
1/4"=1"	5
1/4"=1"	6
1/4"=1"	7
1/8"=1"	8
1/8"=1"	9
1/8"=1"	10
3/8"=1"	11
3/8"=1"	12
3/8"=1"	13
3/4"=1"	14
3/4"=1"	15
3/4"=1"	16

CIVIL ENGINEER'S SCALE TEST

Measure to the closest hundreds for inches and the closest tenths for feet.

SCALE:	
FULL	1
FULL	2
FULL	3
1"=10'	4
1"=10"	5
1"=20'	6
1"=20"	7
1"=30'	8
1"=30'	9
1"=40'	10
1"=40"	11
1"=50'	12
1"=50"	13
1"=50'	14
1"=60'	15
1"=60"	16

ARCHITECT'S SCALE TEST

Measure to the closest inch.

SCALE:	
1/8"=1'-0"	1
1/8"=1'-0"	2
1/4"=1'-0"	3
1/4"=1'-0"	4
1/4"=1'-0"	5
1/4"=1'-0"	6
3/8"=1-0"	7
3/8"=1-0"	8
1/2"=1'-0"	9
1/2"=1'-0"	10
3/4"=1'-0"	11
3/4"=1'-0"	12
3/4"=1'-0"	13
1"=1'-0"	14
1"=1'-0"	15
3"=1'-0"	16

METRIC SCALE TEST

Measure to the closest millimetre.

SCALE:	
1:1	1
1:1	2
1:1	3
1:2	4
1:2	5
1:5	6
1:5	7
1:10	8
1:10	9
1:50	10
1:50	11
1:50	12
1:50	13
1:100	14
1:100	15
1:100	16

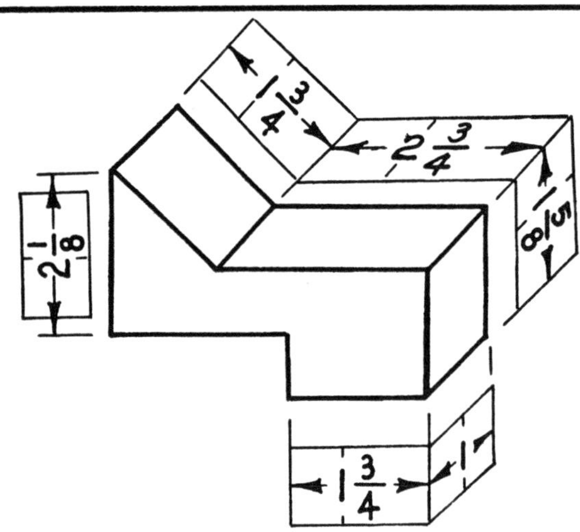

UNIT 9

DIMENSIONING

Before a product can be manufactured, there must be a drawing describing its shape. To complete the drawing so the fabricator can correctly build the product, dimensions and notes must be placed on the drawing specifying the sizes, locations, and other general information needed for all the parts.

In this unit the rules, principles, and methods of dimensioning mechanical drawings are defined. Most of the dimensioning standards were developed by ANSI (American National Standards Institute). As you study this unit, you should learn the dimensioning symbols, become skillful in dimensioning techniques, and understand how to dimension and note machine operations.

Dimensioning is a system of lines, symbols, numerical values, and notations. Because the number of shapes and forms of industrial products are unlimited, basic rules for dimensioning are needed. Each dimension should be logically placed, using the principles of dimensioning as guidelines.

A finished drawing is a complete set of instructions for manufacturing an item. If the fabricator has to estimate, guess, or scale off the drawing, the drawing is unfinished. This will probably cause costly errors in manufacturing.

BASIC PARTS OF A DIMENSION AND ITS PLACEMENT

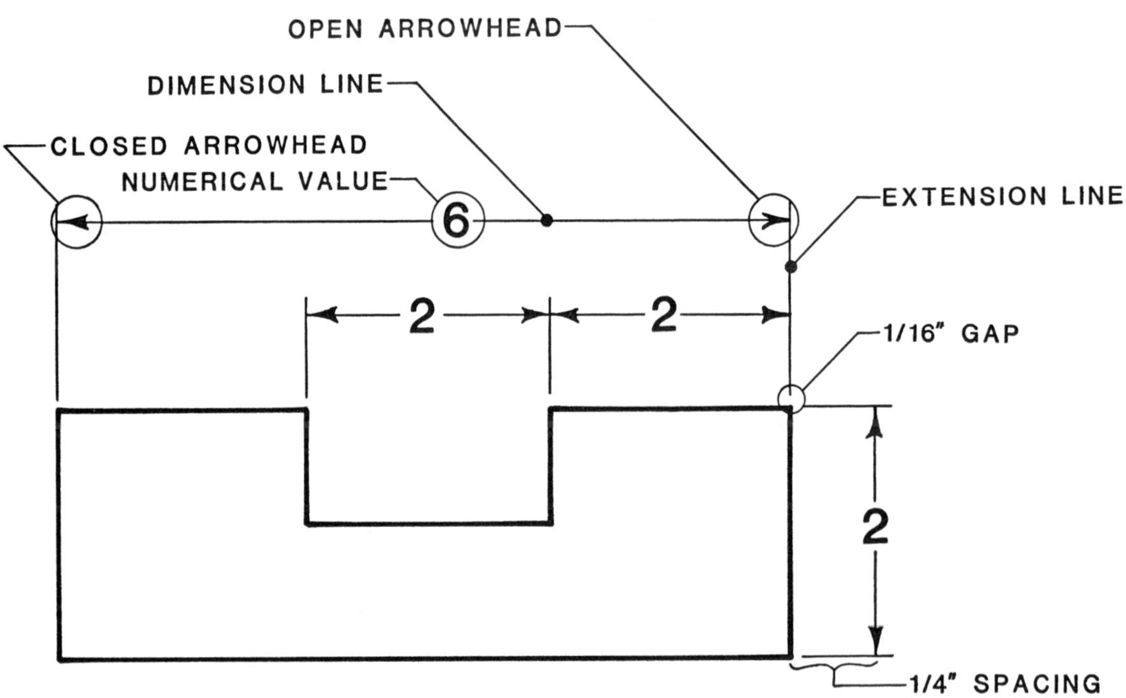

THE MECHANICS OF DIMENSIONING

Much can be gained by a knowledge of the mechanics of dimensioning. A knowledge of how extension lines must be drawn, how arrowheads must be made, and an understanding of the physical side of dimensioning will make dimensioning easier for the beginner.

EXTENSION LINES ARE USED IN DIMENSIONING

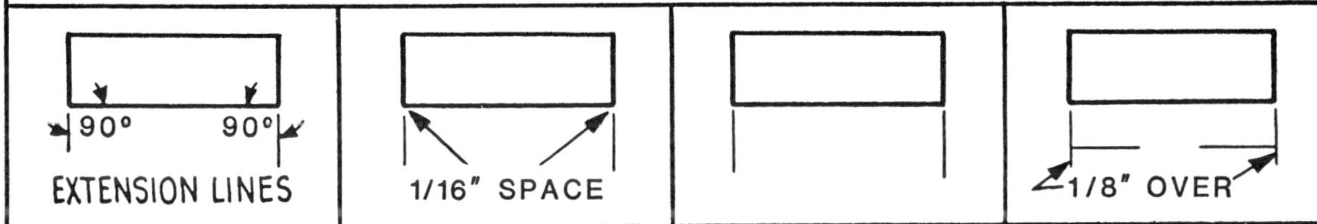

① Extension lines must be drawn at 90° to the surface dimensioned.	② Leave a 1/16" space between the object and the extension line.	③ An extension line is only one-fourth as thick as an object line.	④ Extension lines should project only 1/8" beyond the arrowhead.

STUDY THE PURPOSE OF THESE EXTENSION LINES

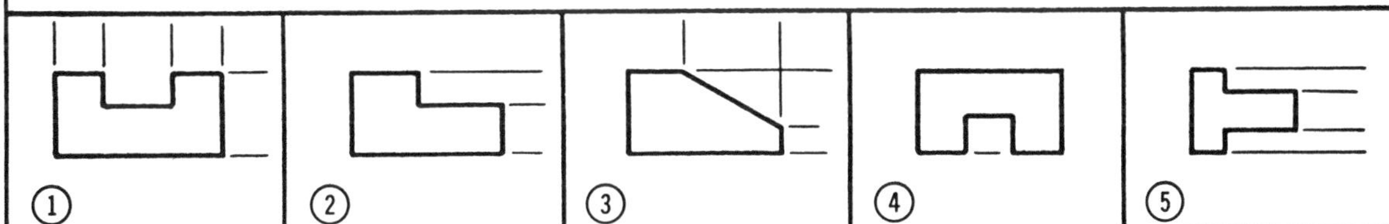

Extension lines enable the drafter to extend a surface to a point outside of the drawing for the purpose of dimensioning.

DIMENSION LINES AND EXTENSION LINES

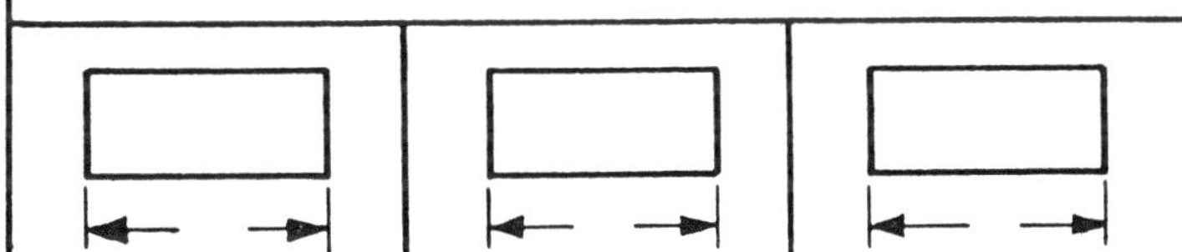

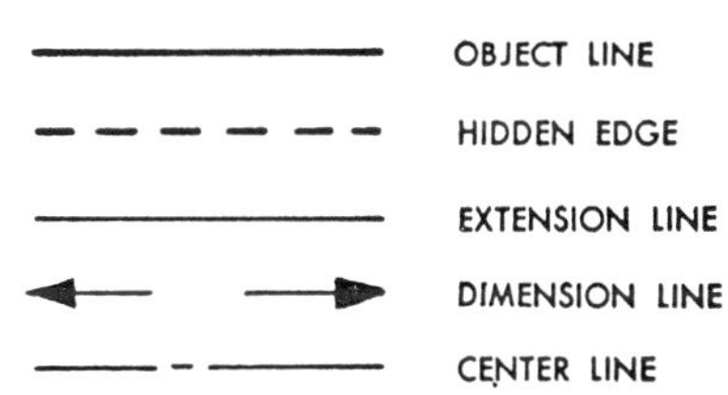

① Dimension lines are the arrow shafts used in dimensioning.	② Dimension lines and extension lines are the same thickness.	③ Dimension lines must always touch the extension lines.	④ Study the comparative thickness of these lines.

IMPORTANT!

1. Extension lines and dimension lines should be one-fourth as thick as object lines so that they will not be mistaken for part of the outline of an object.
2. Extension lines should never touch a drawing. Leave a 1/16" space between the extension line and the outline of the drawing.
3. Never draw extension lines more than 1/8" beyond the last arrowhead.

ARROWHEADS

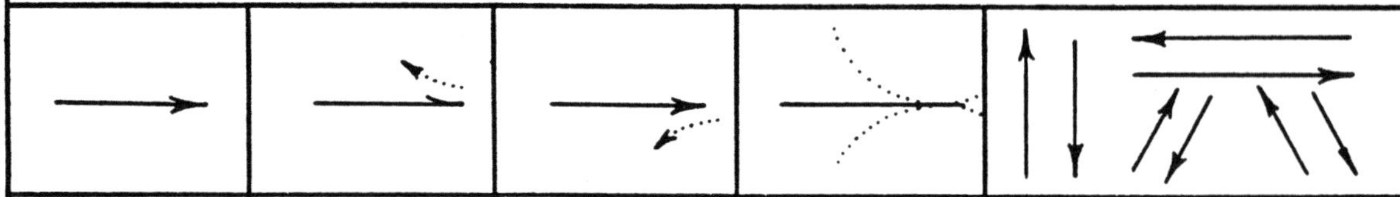

(1) Actual size of an arrowhead.	(2) Start as a curved line.	(3) Finish as a curved line.	(4) Consider each side as part of a circle.	(5) Actual size of arrowheads. (Practice making many of these.)

PLACING THE DIMENSIONS CORRECTLY

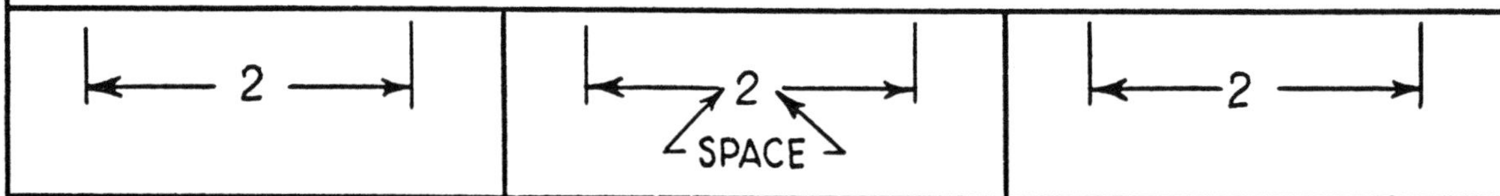

(1) The numeral must be 1/8" high.	(2) Leave a space between the numeral and the dimension line.	(3) Center the number (when possible).

FRACTIONS

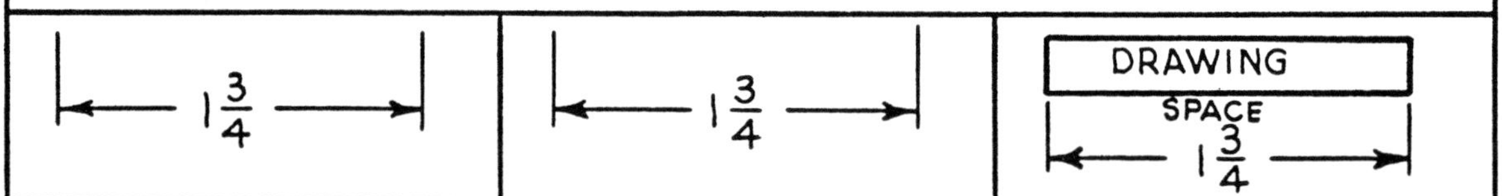

(1) The fraction line must be in line with the dimension line. (It should be made with the T-square.)	(2) Each numeral of the fraction must be nearly as large as the whole number (1/8").	(3) Leave a space of no less than 1/4" between the drawing and the nearest numeral.

DIMENSIONING SITUATIONS

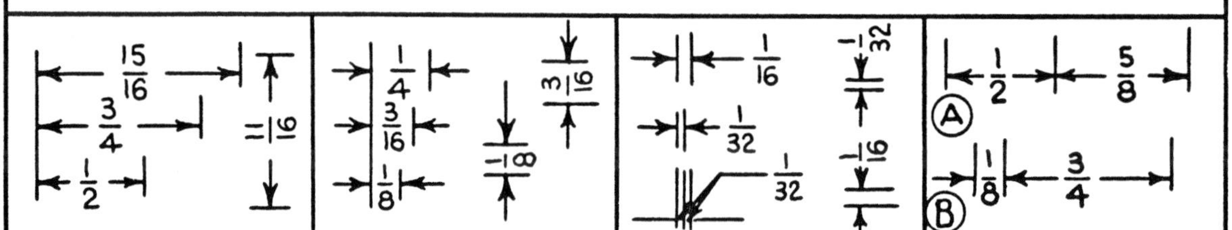

(1) When there is room enough between the extension lines (more than 1/4"), place the numbers and arrowheads as shown.	(2) For dimensions from 1/2" to 1/8", use this method of arranging the numerals and the arrowheads.	(3) In small spaces (less than 1/8"), use this method.	(4) (A) Dimensions can be placed in line (series). (B) Dimensions in areas with little space can be placed between arrowheads.

Dimensions must read from the bottom or from the right-hand side of the sheet.

BASIC PRACTICES IN DIMENSIONING

Here are several basic practices in dimensioning that you should learn.

BASIC PRACTICES IN DIMENSIONING

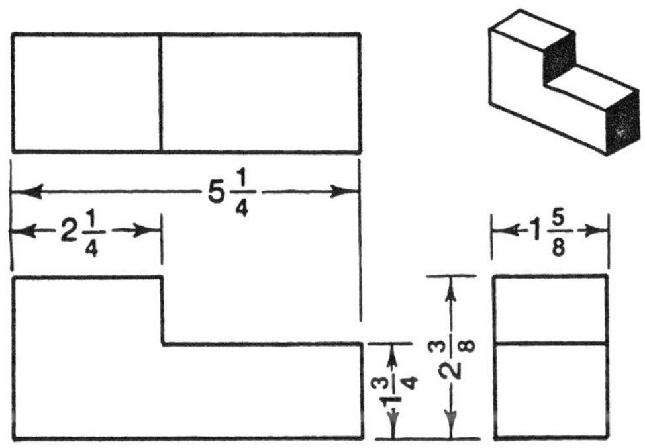

ALIGNED SYSTEM

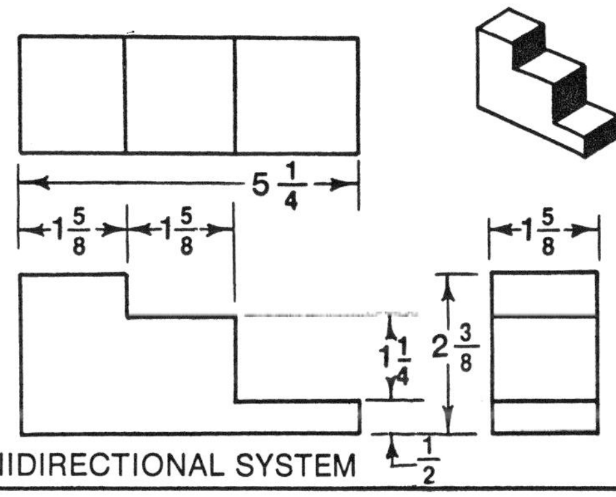

UNIDIRECTIONAL SYSTEM

(1) Dimensions must be read from the bottom or from the right-hand side of a drawing in the aligned dimensioning system. In the unidirectional system, all dimensions are read from the bottom.

(2) When possible, place dimensions between two views. The dimension will be of service to both views. Space between views should be 1 or more inches.

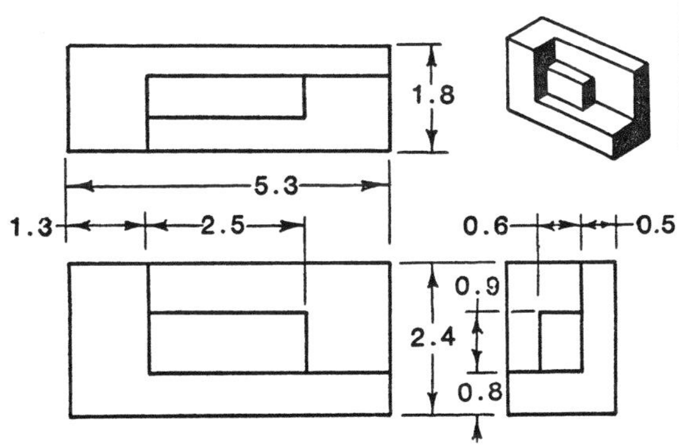

(3) Use the profile view (outline view) for indicating the size of most of the parts of an object.

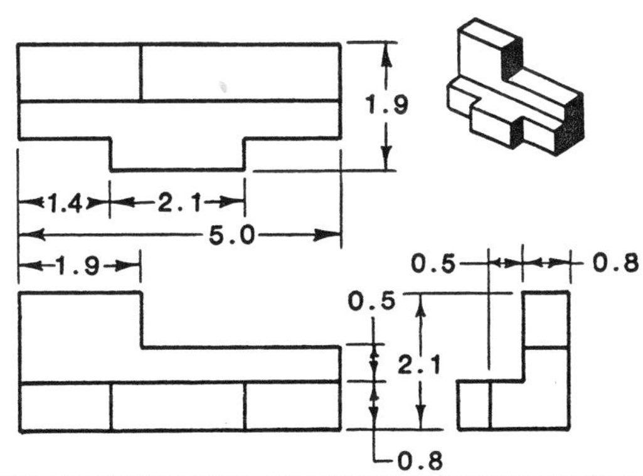

(4) Place the smaller dimensions nearer to the view (but never closer than 1/4"). Place larger dimensions on the outside.

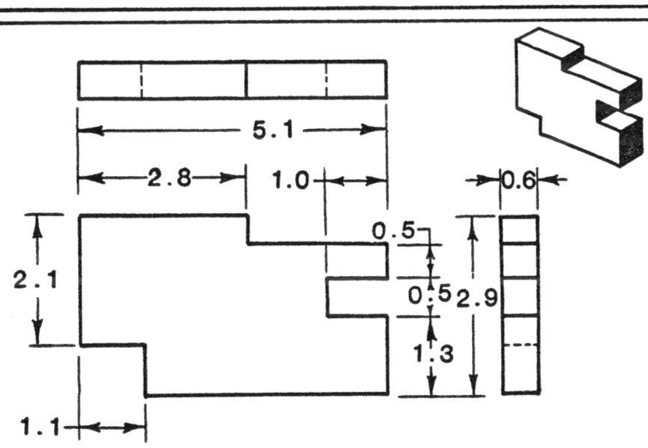

(5) You may dimension on the left side and at the bottom of a drawing only if it is necessary.

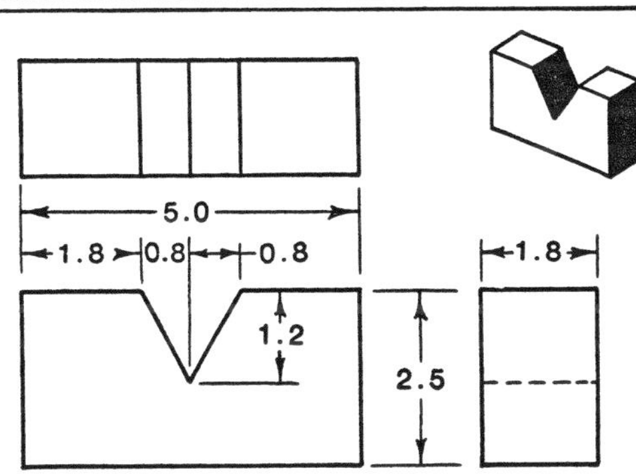

(6) You may dimension on the drawing only if it adds to the clarity of the dimensioning, or if there is no space for dimensions on the sides of the drawing.

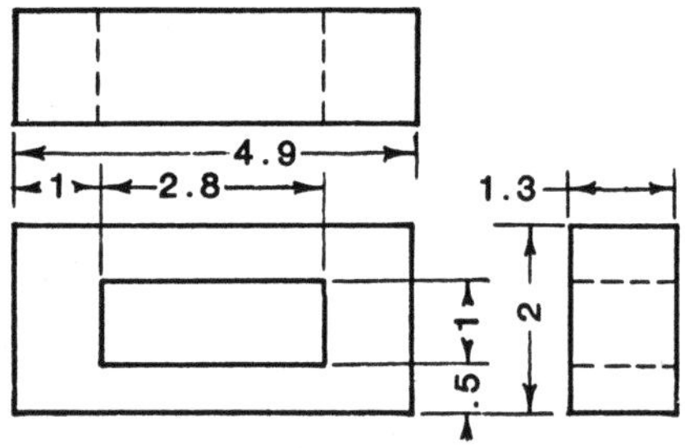

(7) Dimension to a hidden edge only when dimensioning cylinders.

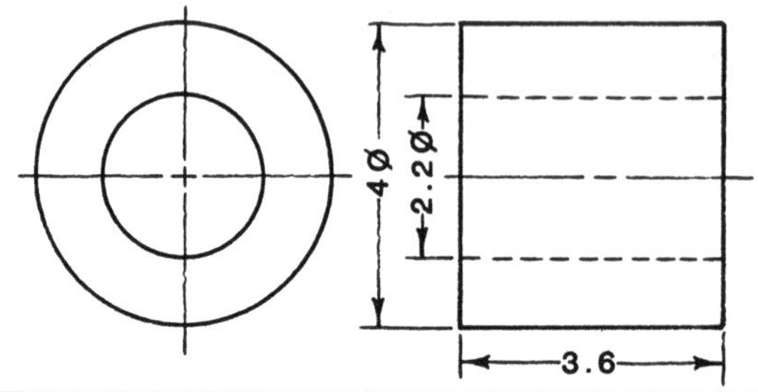

(8) Dimension the profile view of a cylinder giving the necessary diameters and the length.

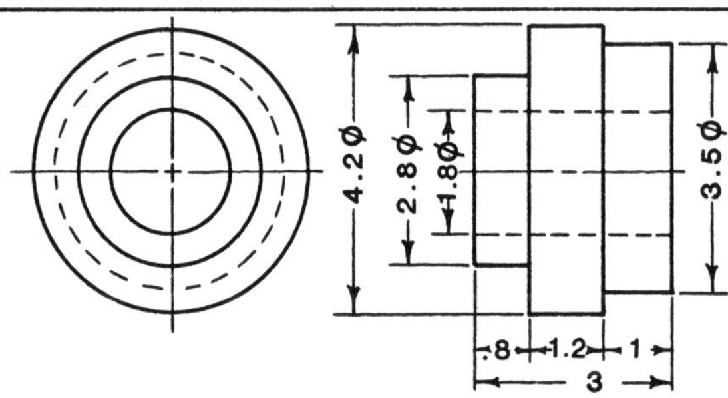

(9) Dimension the profile view. Never letter a numeral directly on a center line. Stagger the numbers when necessary.

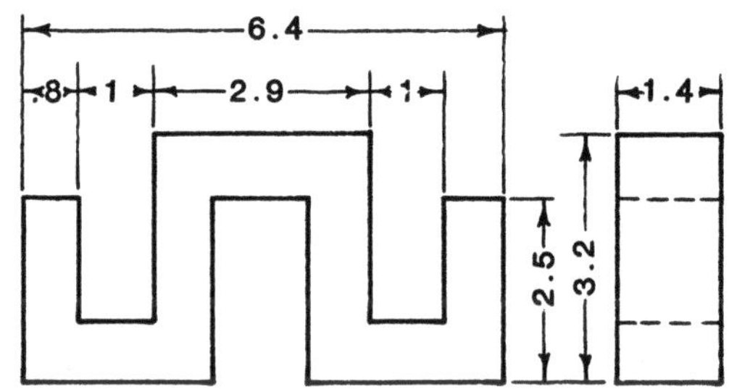

(10) Indicate the overall length, height, and thickness of a rectangular object.

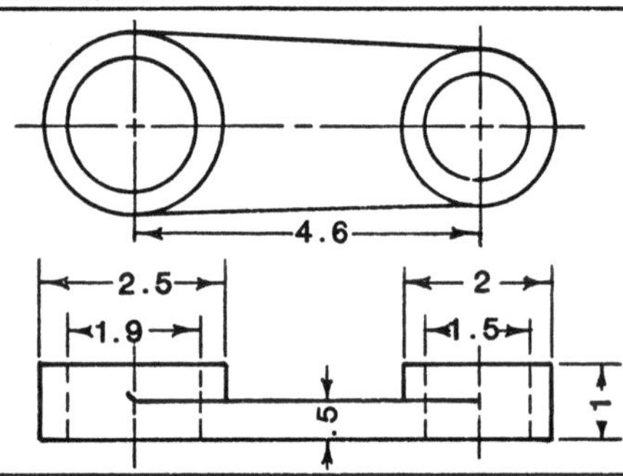

(11) When an object has rounded ends, indicate the distance between the centers. Do not give the overall length.

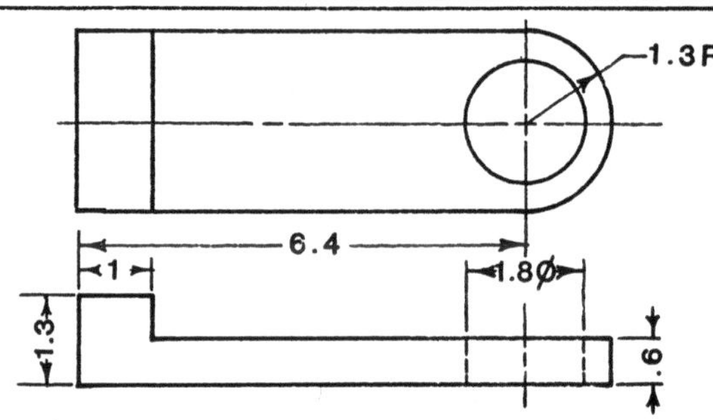

(12) When an object has one rounded end, locate the center line of the arc as shown. Do not give the overall length.

DO NOT OVERDIMENSION

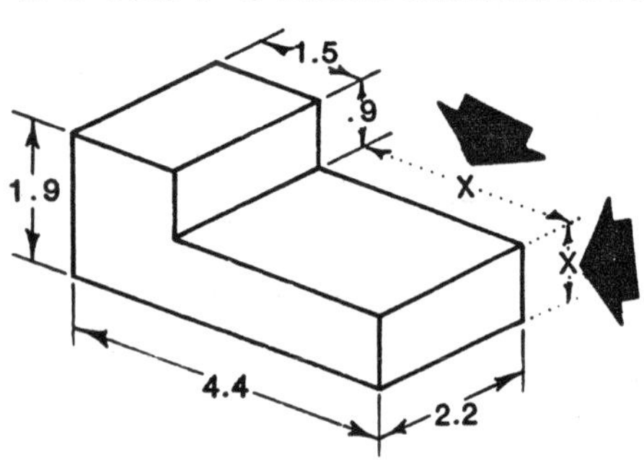

Many drafters follow the practice of not overdimensioning. For example: In the notched block shown to the left, both adjoining dimensions of the notched part need not be given. To give either one is sufficient. Other drafters believe that all dimensions should be given. Your teacher will tell you which system to use.

THE BASIC IDEA BEHIND DIMENSIONING

There are two things that dimensioning must do. First, dimensioning must give the **size** of every part of the object. Second, dimensioning must give the exact **location** of every part of the object.

SIZE AND LOCATION DIMENSIONS

SIZE DIMENSIONS

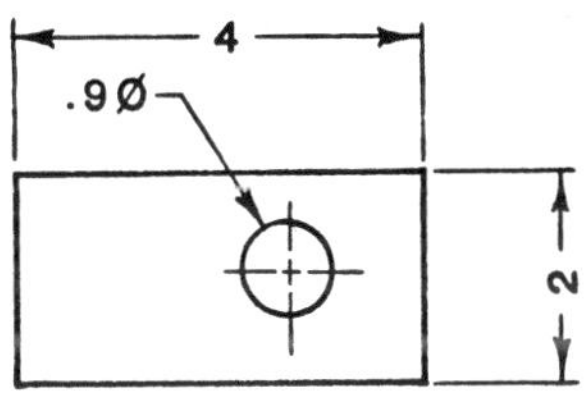

① On this figure, the length and height of the surface must be given. The size of the hole must be given as well.

LOCATION DIMENSIONS

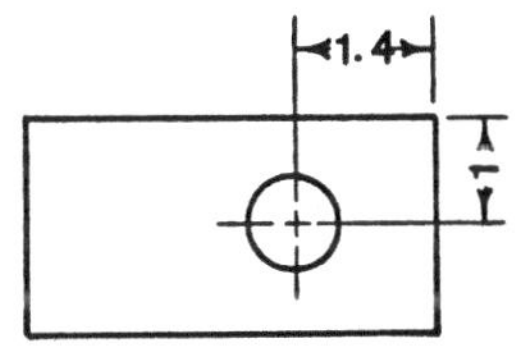

② The location of the center of the hole must be given by the use of two dimensions.

COMPLETE DIMENSIONING

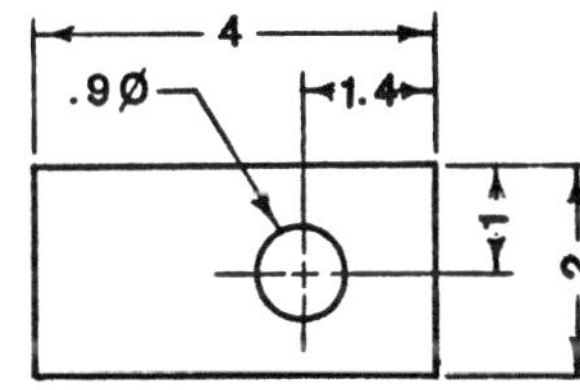

③ Five dimensions are needed to describe this surface.

EXAMPLES OF SIZE AND LOCATION DIMENSIONS

SIZE DIMENSIONS

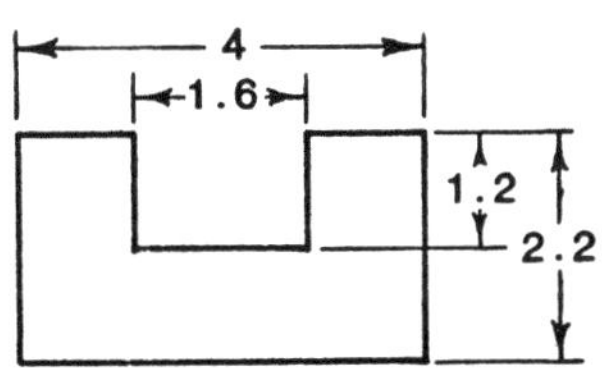

1A

LOCATION DIMENSIONS

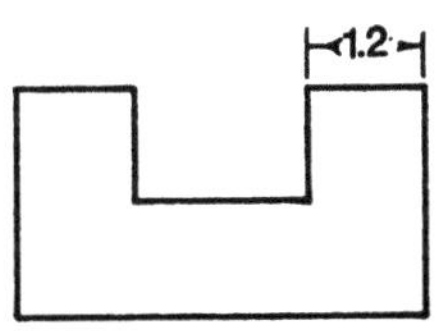

1B

COMPLETE DIMENSIONING

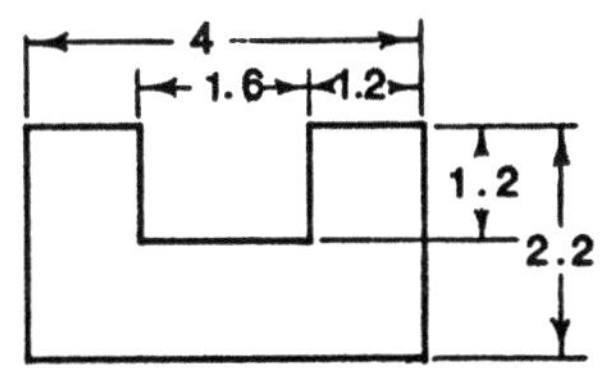

1C

SIZE DIMENSIONS

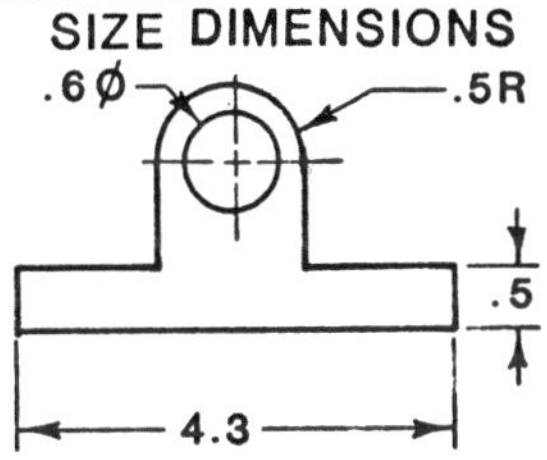

2A

LOCATION DIMENSIONS

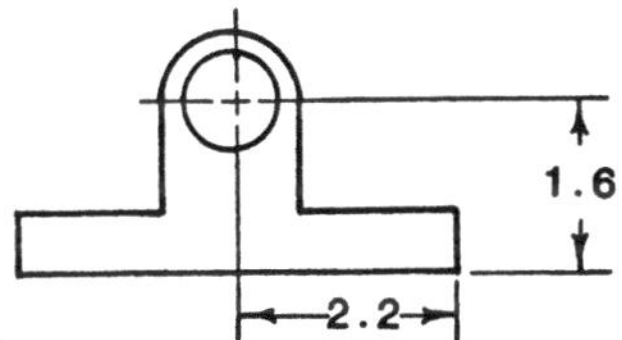

2B

COMPLETE DIMENSIONING

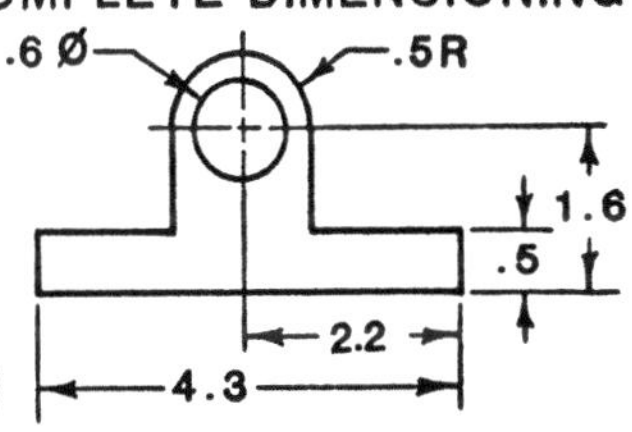

2C

SIZE DIMENSIONS

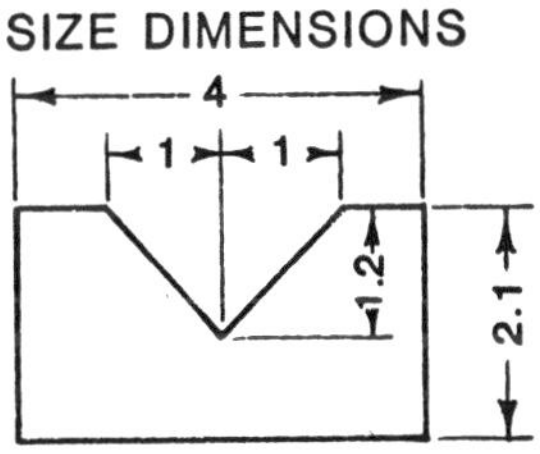

3A

LOCATION DIMENSIONS

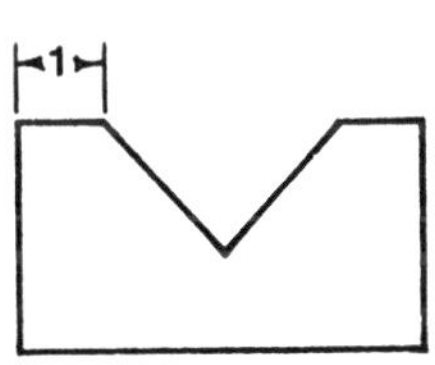

3B

COMPLETE DIMENSIONING

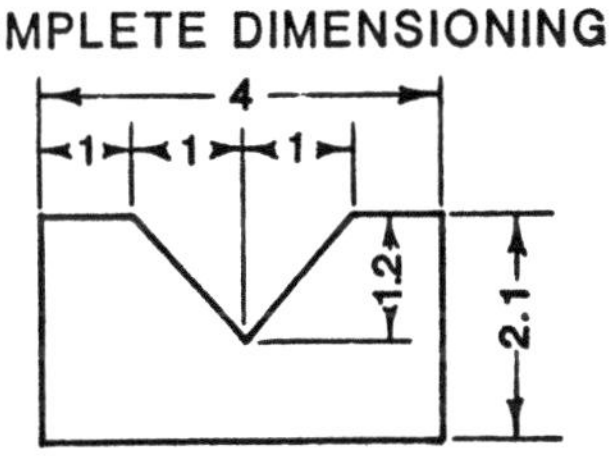

3C

REMEMBER

1. Place smaller dimensions near the drawing, but never closer than 1/4″.
2. Place larger dimensions on the outside, but have them 1/4″ away from the smaller dimensions.
3. If possible, dimensions should be aligned.
4. Side dimensions can be read from the bottom or from the right-hand side.

SUMMARY

1. Size Dimensions

A. The length, width, and height of the entire object must be given.

B. The length, width, and height of every detail must be given.

C. Do not repeat a dimension.

2. Location Dimensions

Every detail of an object must be located either in relation to the entire object or in relation to a previously located detail.

THE TWO METHODS OF DESCRIPTION (PICTURE AND DIMENSIONS)

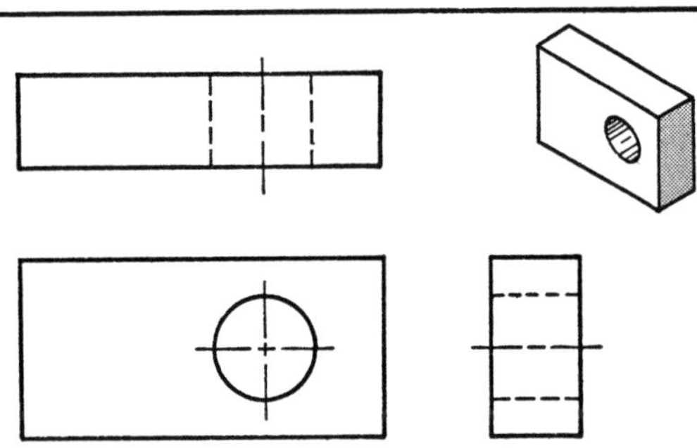

(1) A description by means of a three-view drawing and a sketch (pictorial description). This is not a complete description since the dimensions are missing.

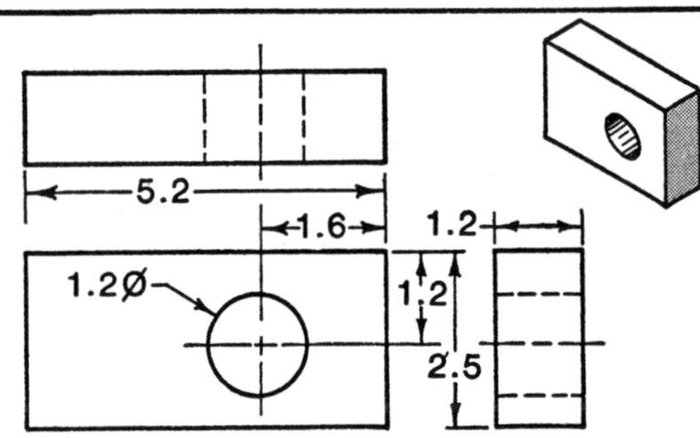

(2) The complete description:
a) The three-view drawing and sketch
b) The complete description consists of both the pictorial and mathematical descriptions.

ASSIGNMENT

Make a completely dimensioned three-view drawing of one of these objects. (See Unit 10.)

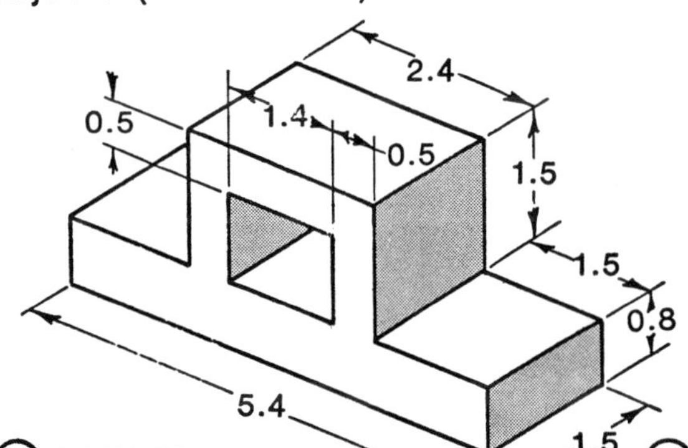

(1) HANGER

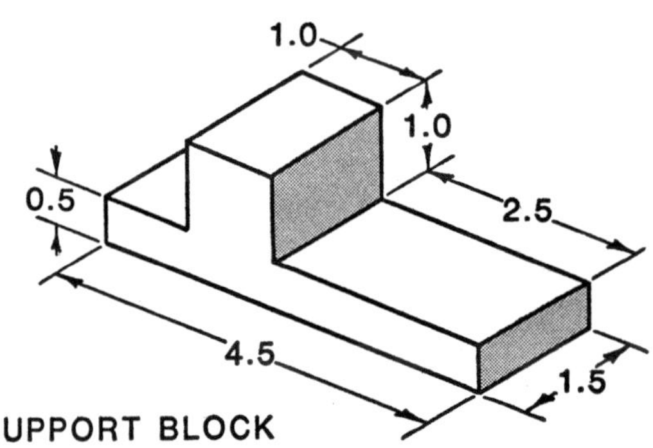

(2) SUPPORT BLOCK

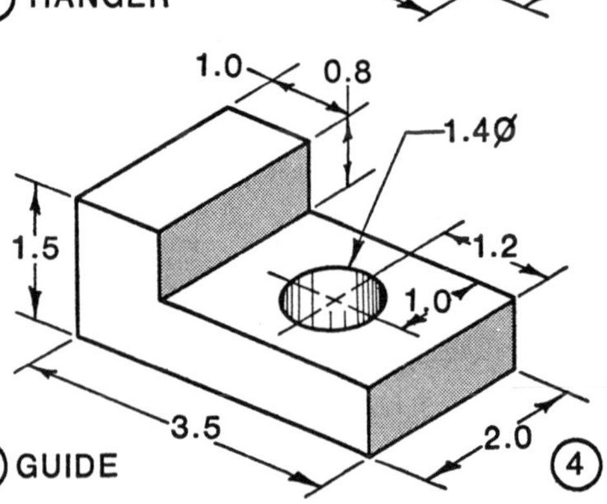

(3) GUIDE

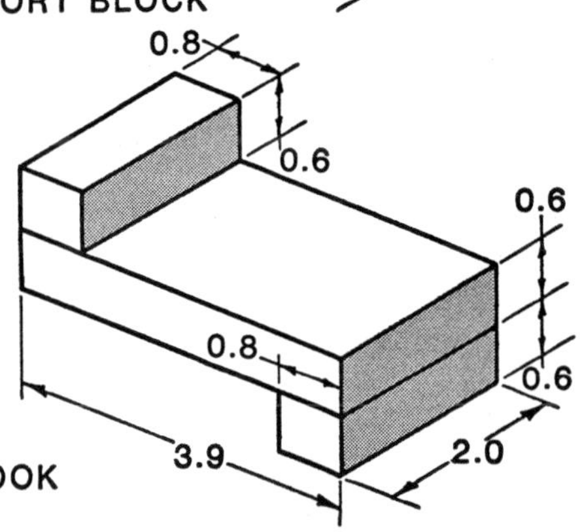

(4) BENCH HOOK

STUDY THESE EXAMPLES IN DIMENSIONING

When size dimensions are considered, the length, height, and thickness of the entire object must be given. These are called the overall dimensions. The length, height, and thickness of each minor detail must be given. Do not repeat a dimension.

SIZE DIMENSIONS SOMETIMES SERVE TWO PURPOSES

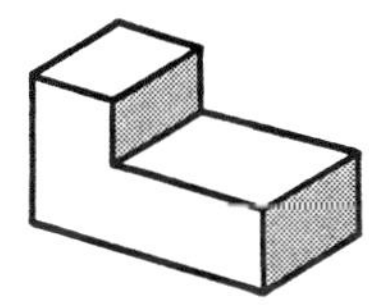	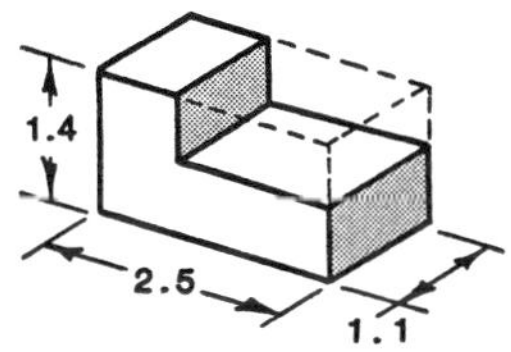	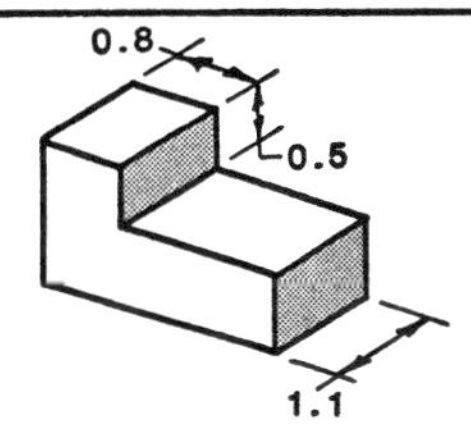	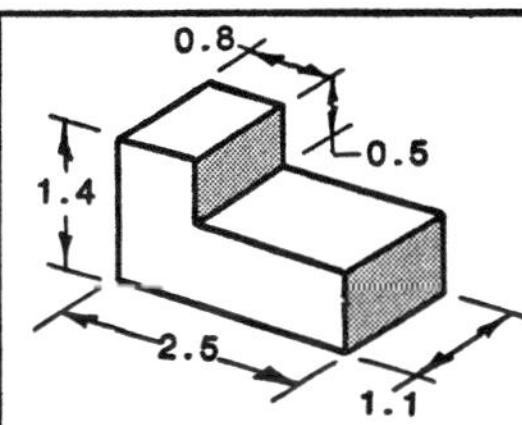
(1) This object may be considered as consisting of two parts.	(2) Give the overall dimensions of the entire basic shape.	(3) Give the height and thickness of the ridge. Do not repeat the length.	(4) The entire object and each of the two parts have been dimensioned.

HOW TO DIMENSION THIS TYPE OF OBJECT

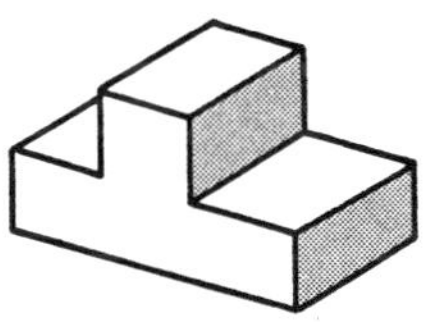	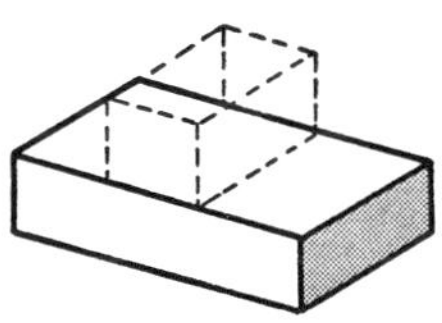	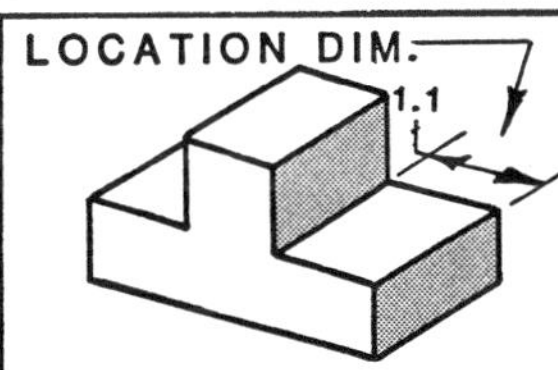	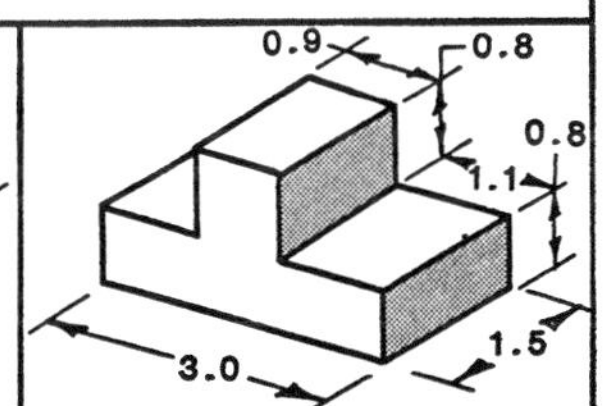
(1) Study the object.	(2) It consists of a flat block and a ridge.	(3) Locate the ridge (location dimension).	(4) Dimension the two shapes (size dimensions).

DIMENSIONING EMPTY OR CUTOUT SPACES

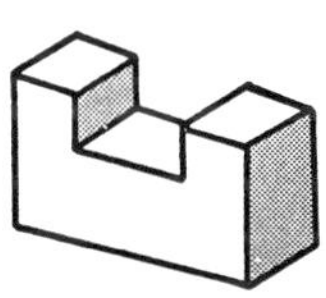	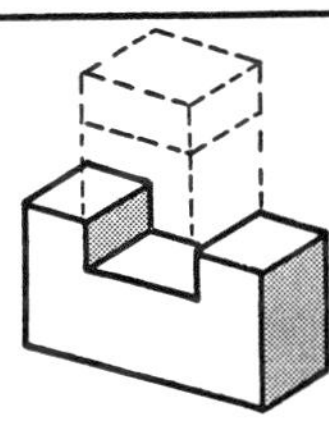	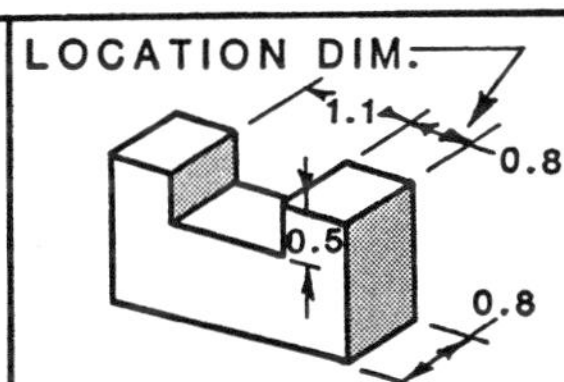	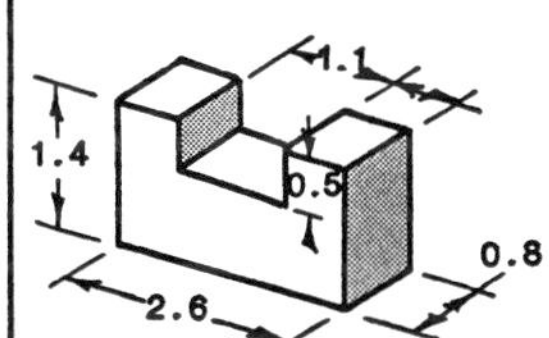
(1) The notch is an empty space. It should be dimensioned.	(2) An empty space may be regarded as having length, height, and thickness.	(3) The empty space is dimensioned for size and location.	(4) The completely dimensioned drawing.

ALIGN THE DIMENSIONS

The dimensions in one view should be aligned with the dimensions in another view. Study the example to the right and the other examples.

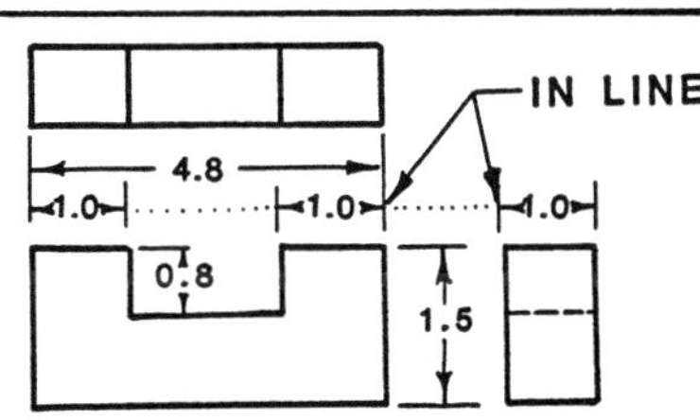

STUDY THESE EXAMPLES IN DIMENSIONING

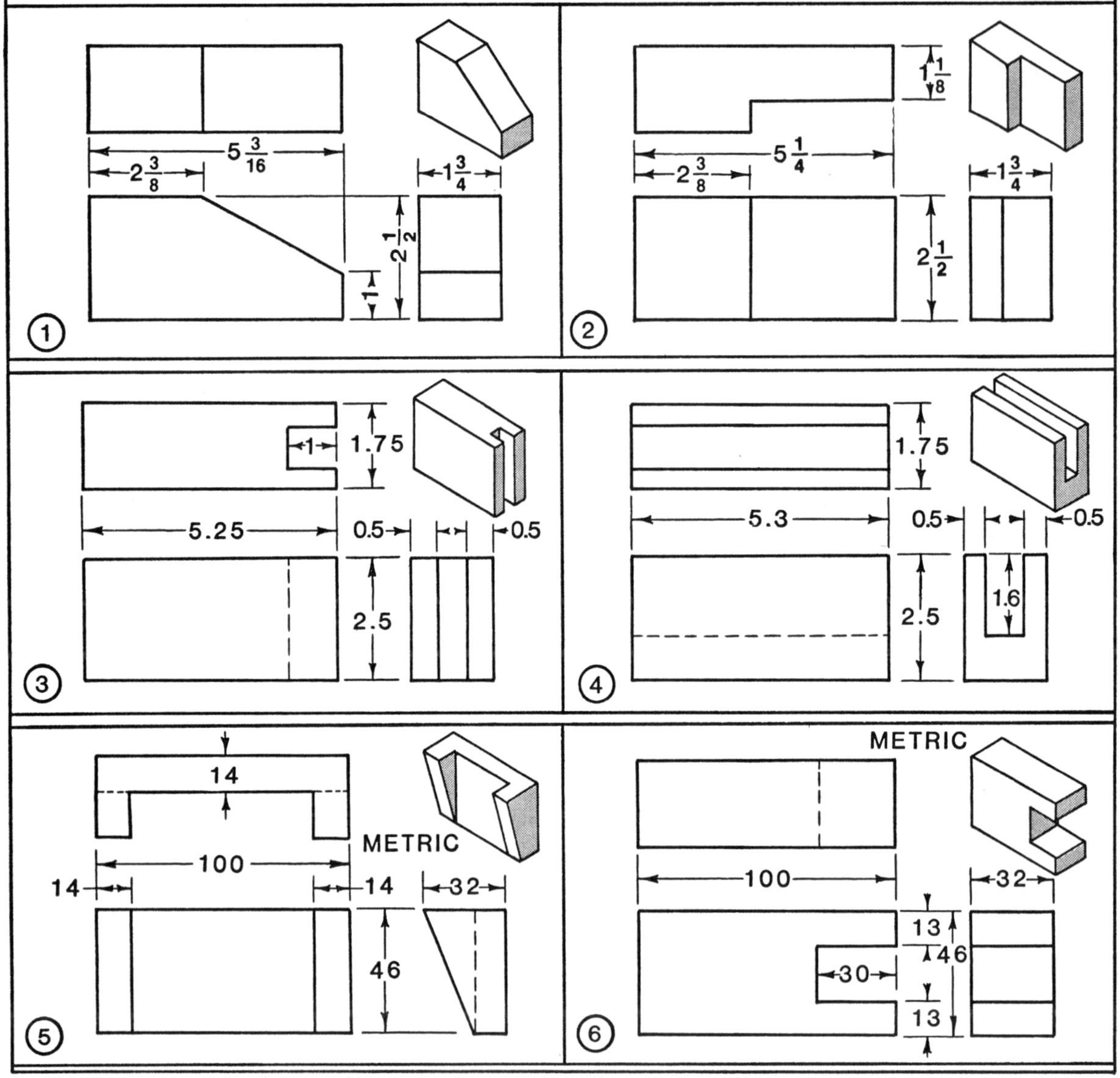

WORKING DRAWINGS

A working drawing is any drawing that has been completely dimensioned so that the object described could be manufactured. The term "working drawing" is usually applied to fully dimensioned orthographic projections. But any fully dimensioned drawing, if used for the purpose of making the object shown, can be considered to be a working drawing.

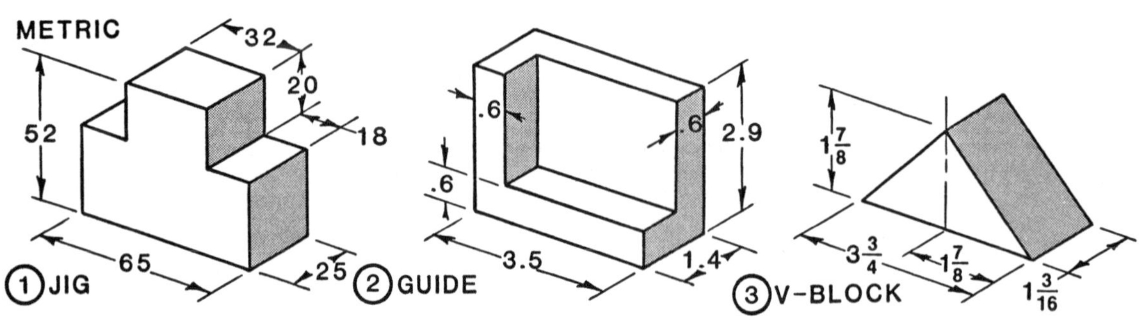

ASSIGNMENT Make a working drawing of one of these objects.

DIMENSIONING CIRCLES

The size of a circle is always indicated by giving the size of its diameter. The size of an arc is always indicated by giving the radius of the arc.

THE "LEADER"

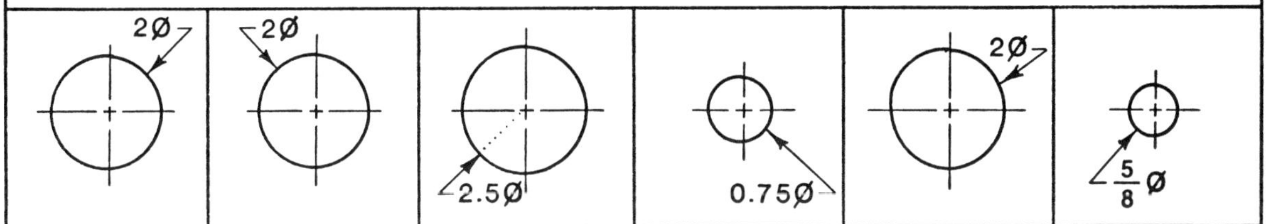

1. A leader is a bent arrow used to point out something.
2. The arrowhead must touch the circumference.
3. The arrow must aim for the center of the circle.
4. The bend in the leader should occur close to the numeral.
5. REMINDER: ϕ (for *diameter*) must always be used. It can be placed in front or behind the dimension.
6. The fraction bar and the leader must be in line. The leader must be centered with decimals.

SIZE DIMENSIONS OF CIRCLES

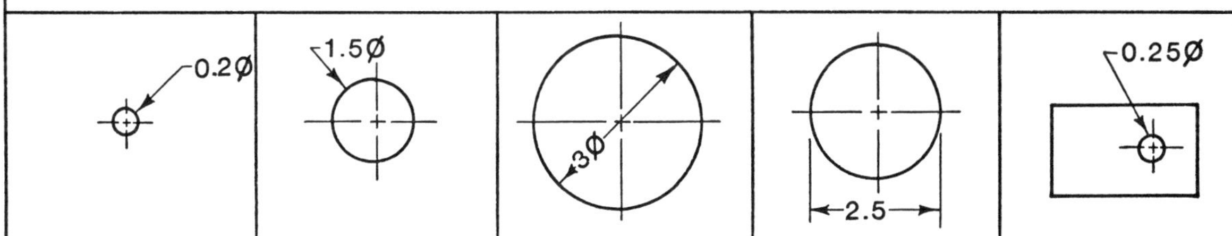

1. Use this method on very small circles.
2. Use this method on medium-size circles.
3. The diameter may be directly indicated as shown.
4. In this case, use of ϕ is optional.
5. The leader can be used to get the dimension clear of the drawing.

STUDY THE DIMENSIONING OF THIS DRAWING

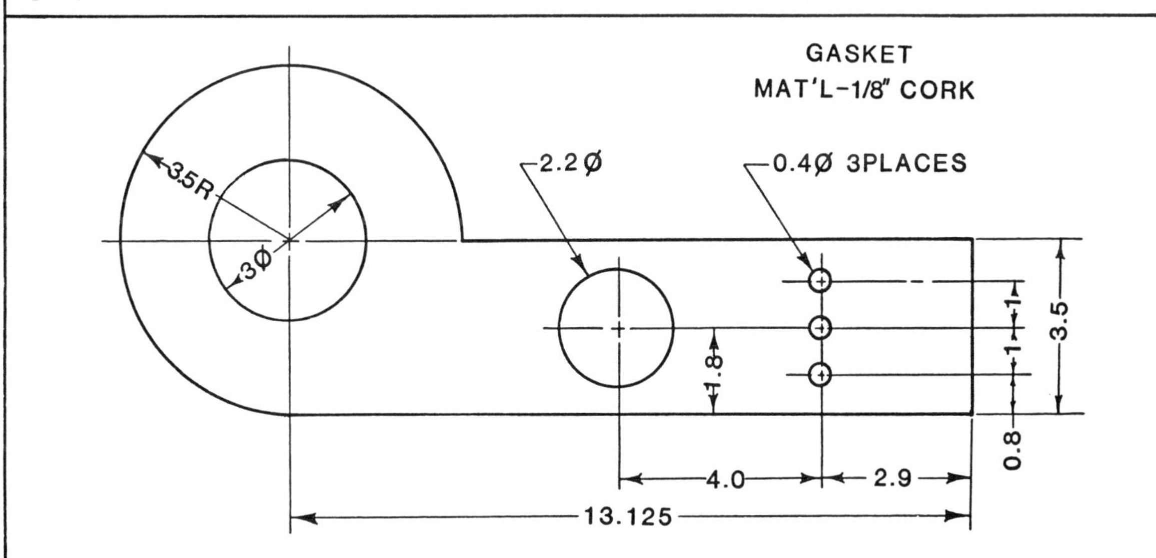

LOCATION DIMENSIONS FOR CIRCLES

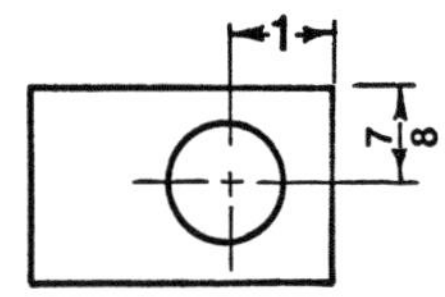

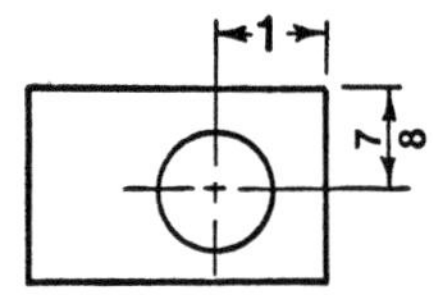

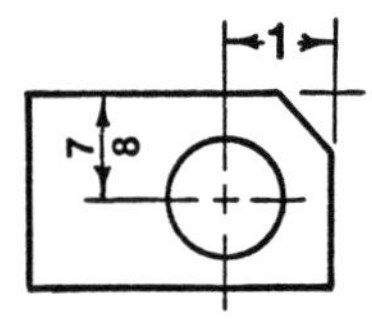

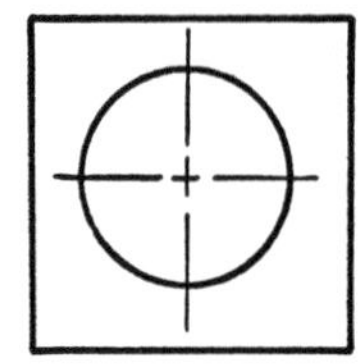

(1) Always locate a hole by locating its center.

(2) When possible, locate a hole in relation to one corner.

(3) If necessary, locate a circle from two points.

(4) If the location is obvious, you may sometimes assume that it is centered.

LOCATION BY MEANS OF DIMENSIONING FROM CENTER TO CENTER

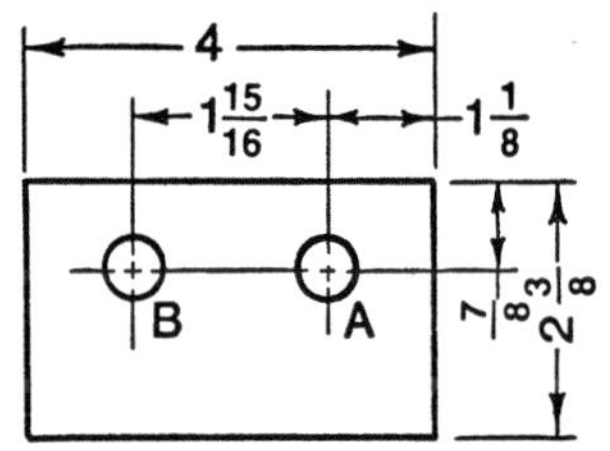

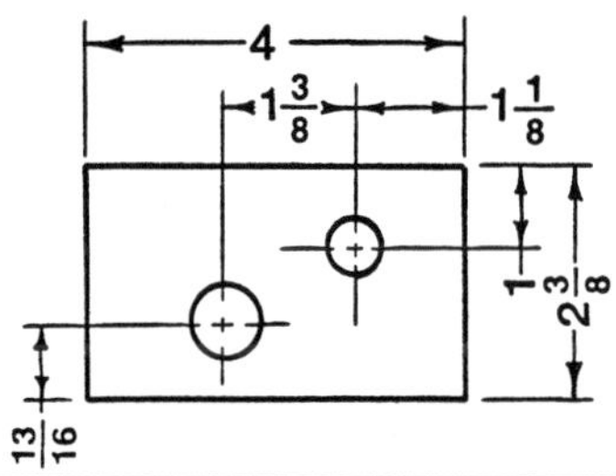

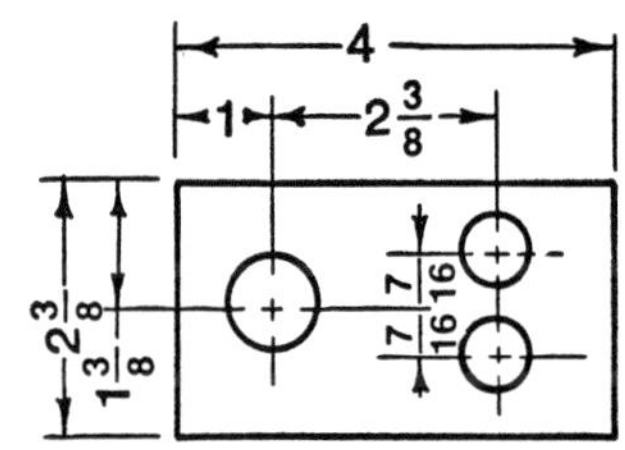

(1) "A" has been located in relation to the corner of the block. "B" has been located in relation to "A."

(2) The vertical center lines have been located in relation to the corner and to each other. The horizontal center lines have each been located in relation to an edge of the block.

(3) The distance between the center lines has been indicated in two directions.

NOTICE HOW THESE CENTER LINES HAVE BEEN LOCATED

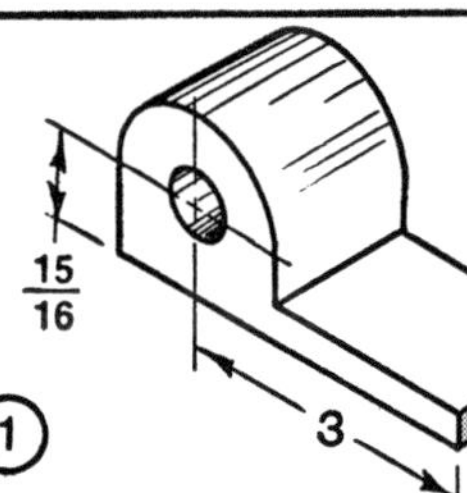

(1)

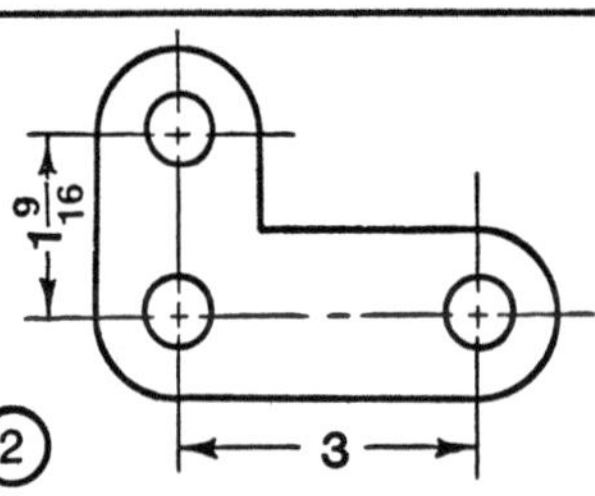

(2)

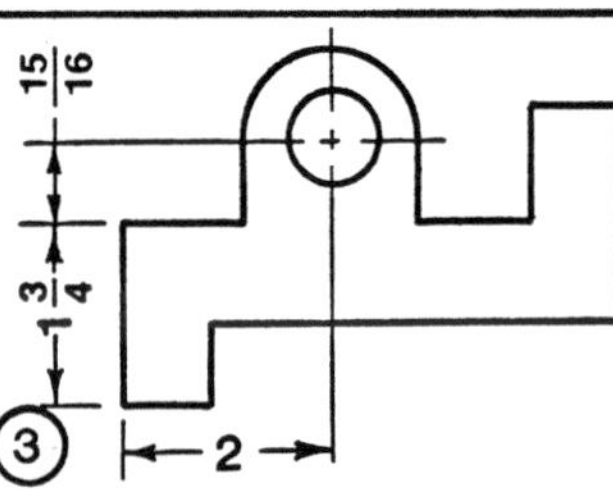

(3)

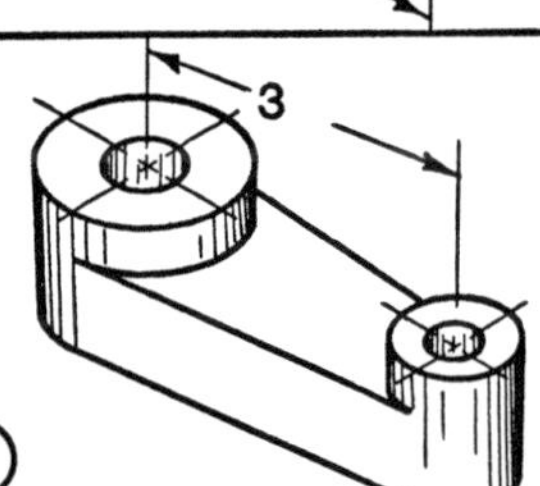

(4)

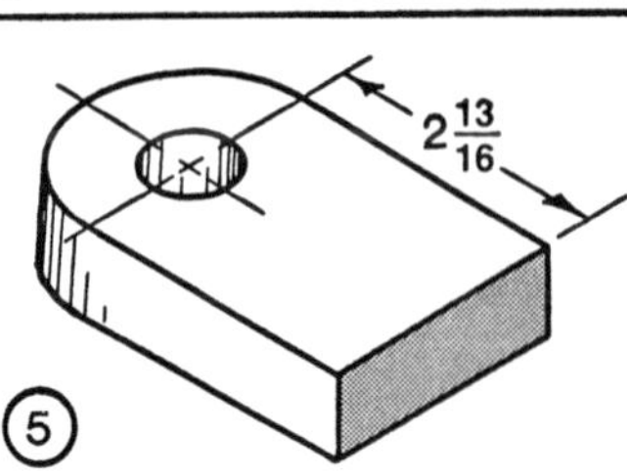

(5)

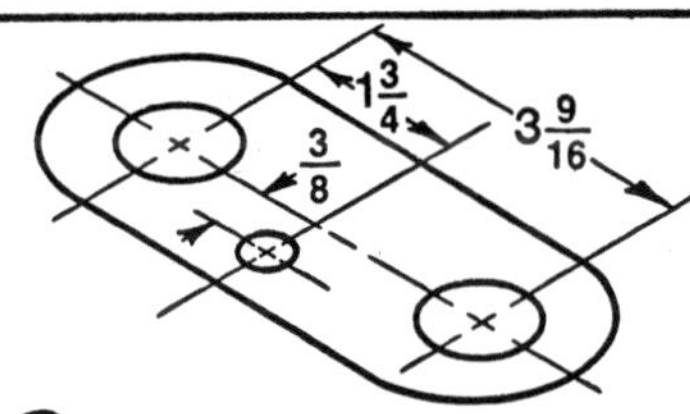

(6)

ASSIGNMENT

Make a working drawing of any object on this page.

THE DIMENSIONING OF ARCS (Size and Location)

The size of an arc is always indicated by designating the radius used.

A STUDY OF DIMENSIONING ARCS

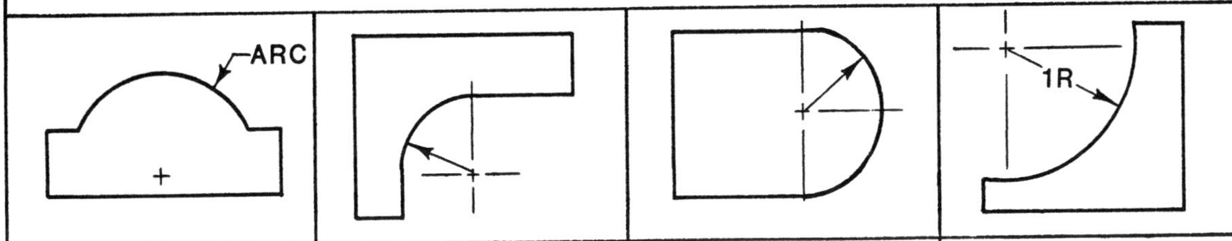

① An arc is a part of a circle.	② Whenever practical, show the center used for the arc.	③ The arrowhead must touch the arc.	④ Use the letter "R" to denote the radius used.

WAYS OF INDICATING THE RADIUS (Note how the "R" is shown.)

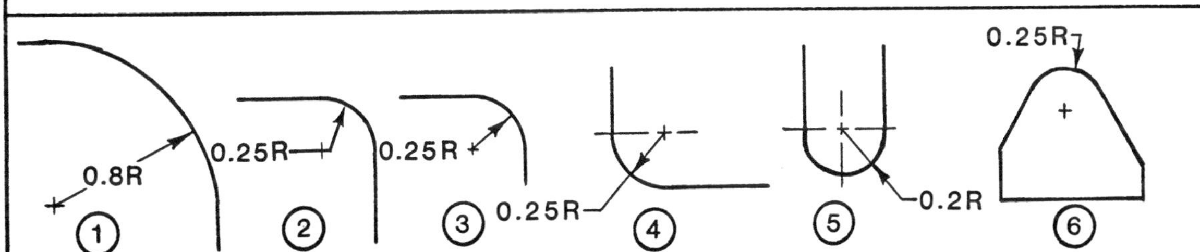

When the arc is large enough, dimension it as shown at ① .
Small arcs may be dimensioned as shown in ② , ③ , ④ , ⑤ , and ⑥ .
Always use the letter "R" to denote that the dimension is a radius.
The "R" is always placed by the leader (arrow shaft). It is in line with the numerator when a fraction is used.

NOTICE HOW THE CIRCLES AND ARCS ARE DIMENSIONED

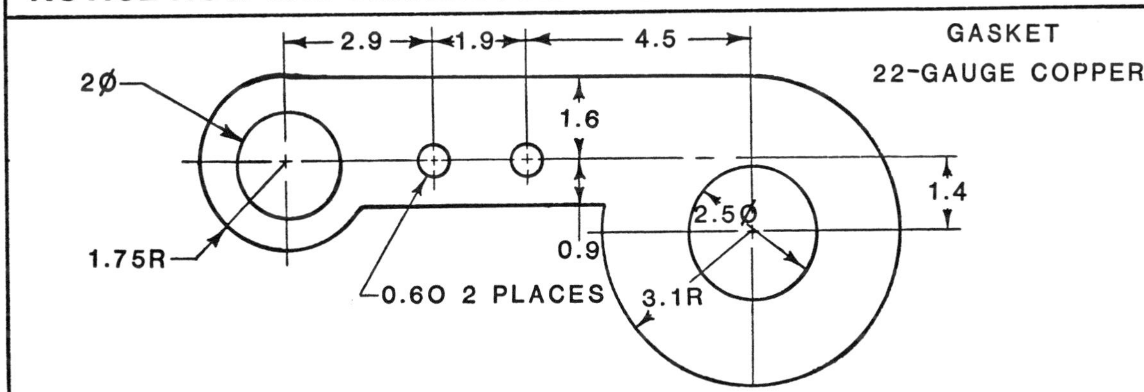

A SPECIAL SITUATION IN DIMENSIONING ARCS

Sometimes it is not necessary to show by dimensions the location of the center for an arc. Skilled mechanics will locate the center for themselves.

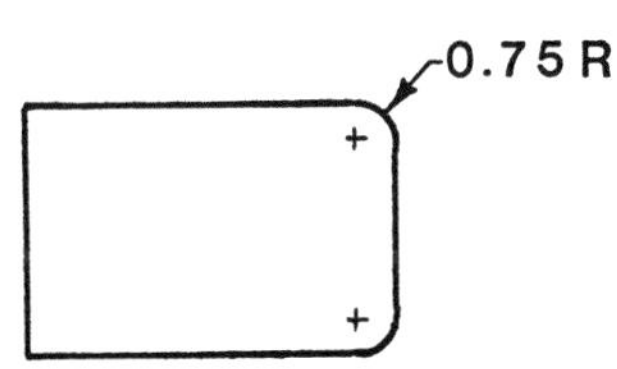

OBJECTS WITH ROUNDED ENDS OR ROUNDED CORNERS

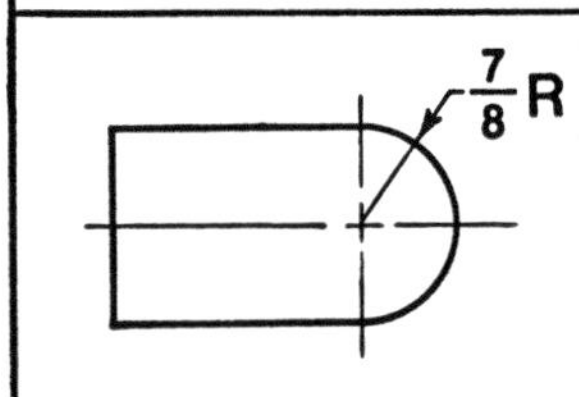

① The rounded end is an arc. Dimension it as an arc.

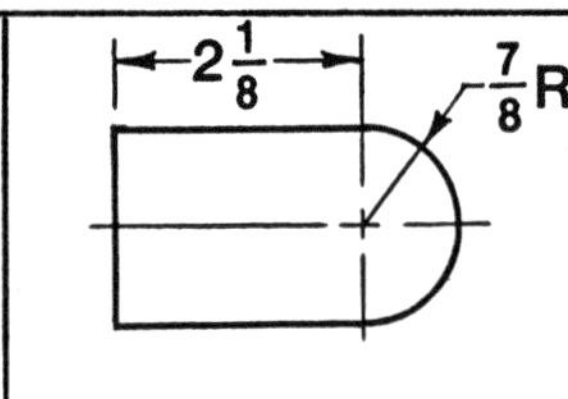

② Give the size of the radius. Locate the center.

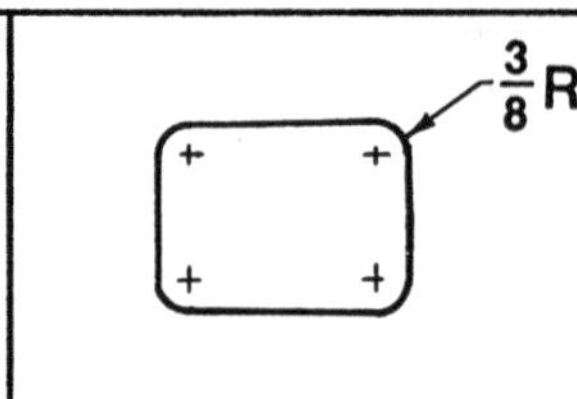

③ Only one corner need be dimensioned.

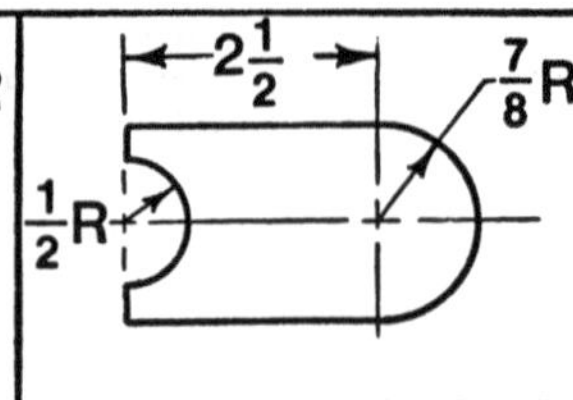

④ Give the radius of each arc and the distance between the centers.

LOCATION DIMENSIONS

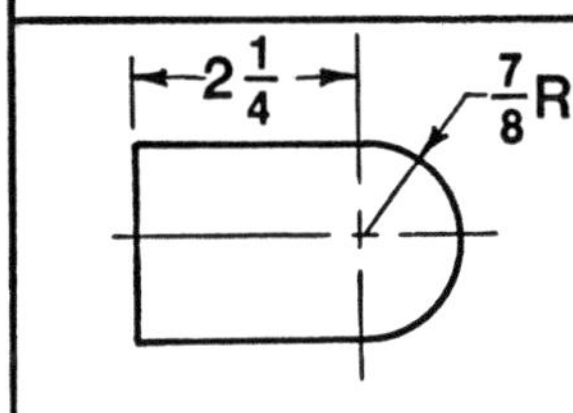

① Location: center to end.

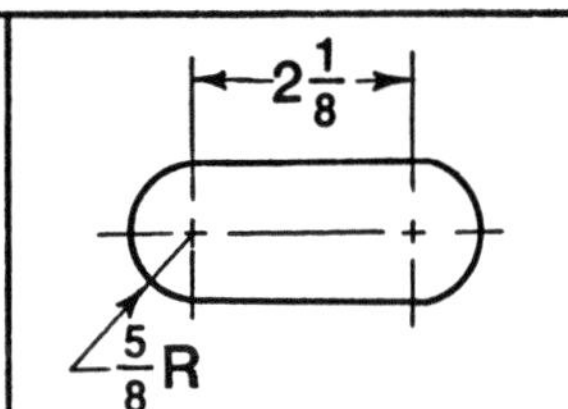

② Location: center to center.

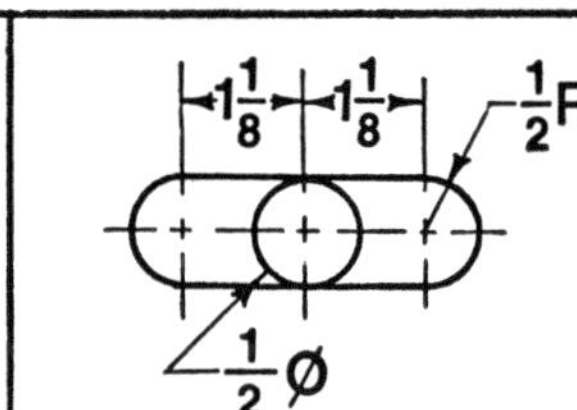

③ Location: center to center to center.

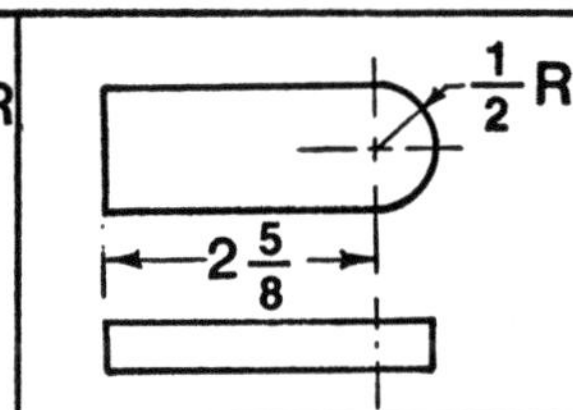

④ Always dimension an arc on the circular view.

IF TWO PARTS ARE OBVIOUSLY IDENTICAL, DIMENSION ONLY ONE

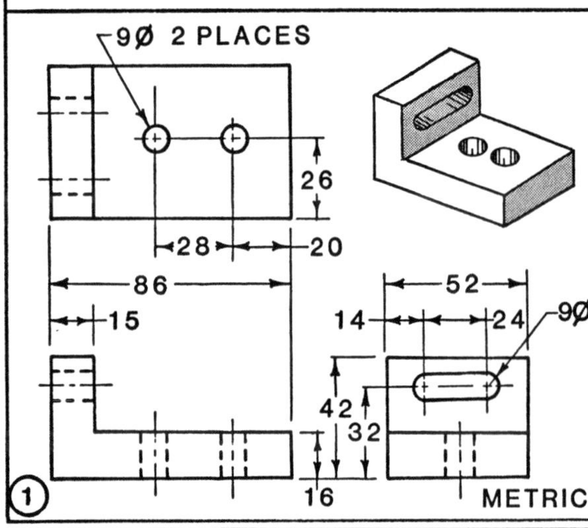

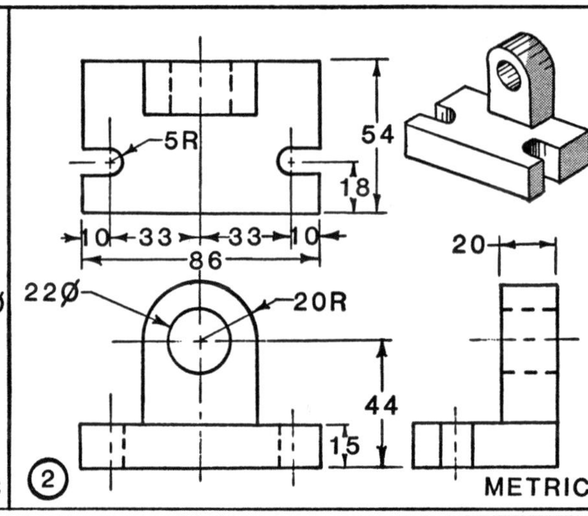

ASSIGNMENT

Make a working drawing of one of these objects. (Metric)

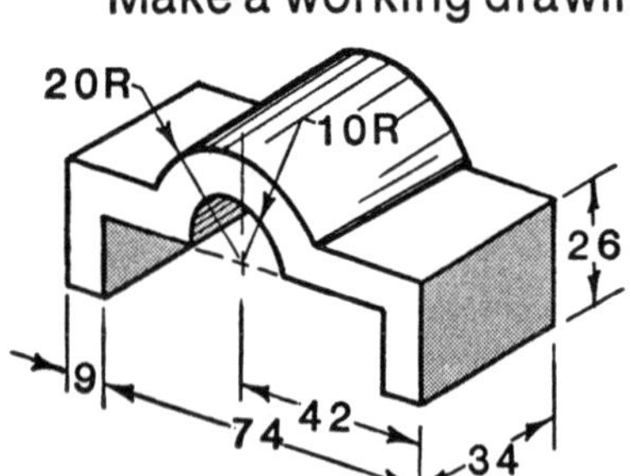

① BEARING CLAMP

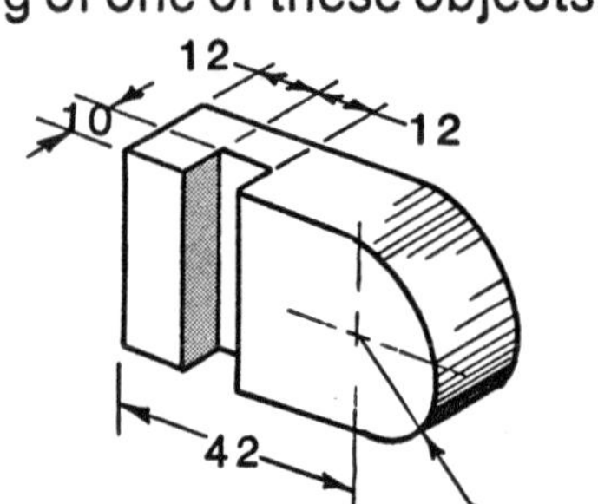

② NOTCHED GUIDE

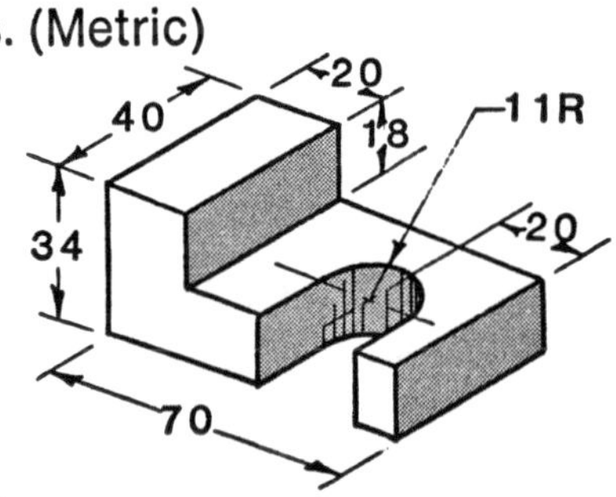

③ ADJUSTABLE BASE

DIMENSIONING ANGLES

An angle is a space between two converging lines. Angles are measured in degrees.

HOW TO DIMENSION AN ANGLE

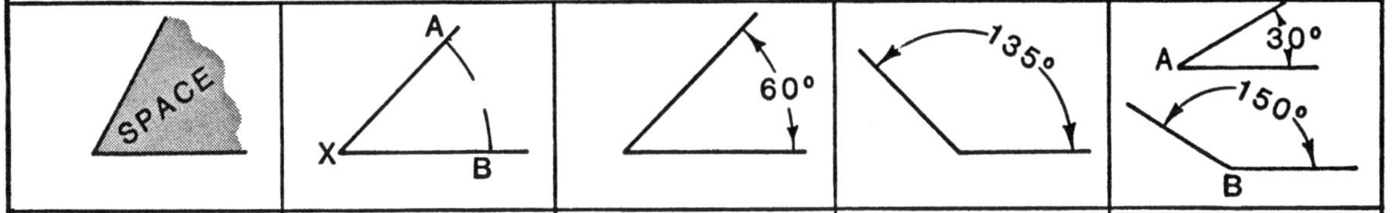

(1) An angle is a space between two converging lines.	(2) Place your compass needle at "X" and swing arc "AB."	(3) Draw the dimension in a horizontal position. Add arrowheads.	(4) The dimension can curve with the arrow on large angles.	(5) The dimension can be horizontal (A) or follow the dimension line (B).

FURTHER HELP ON DIMENSIONING ANGLES

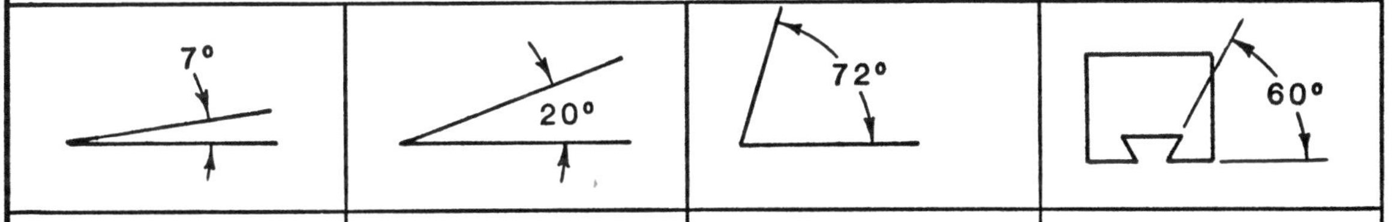

(1) On small angles, place the numeral and the arrows on the outside.	(2) When possible, place the numeral on the inside and the arrows outside.	(3) When space permits, place the numeral and the arrows inside.	(4) An angle may be extended in order to dimension it.

NOTICE HOW THESE ANGLES ARE DIMENSIONED

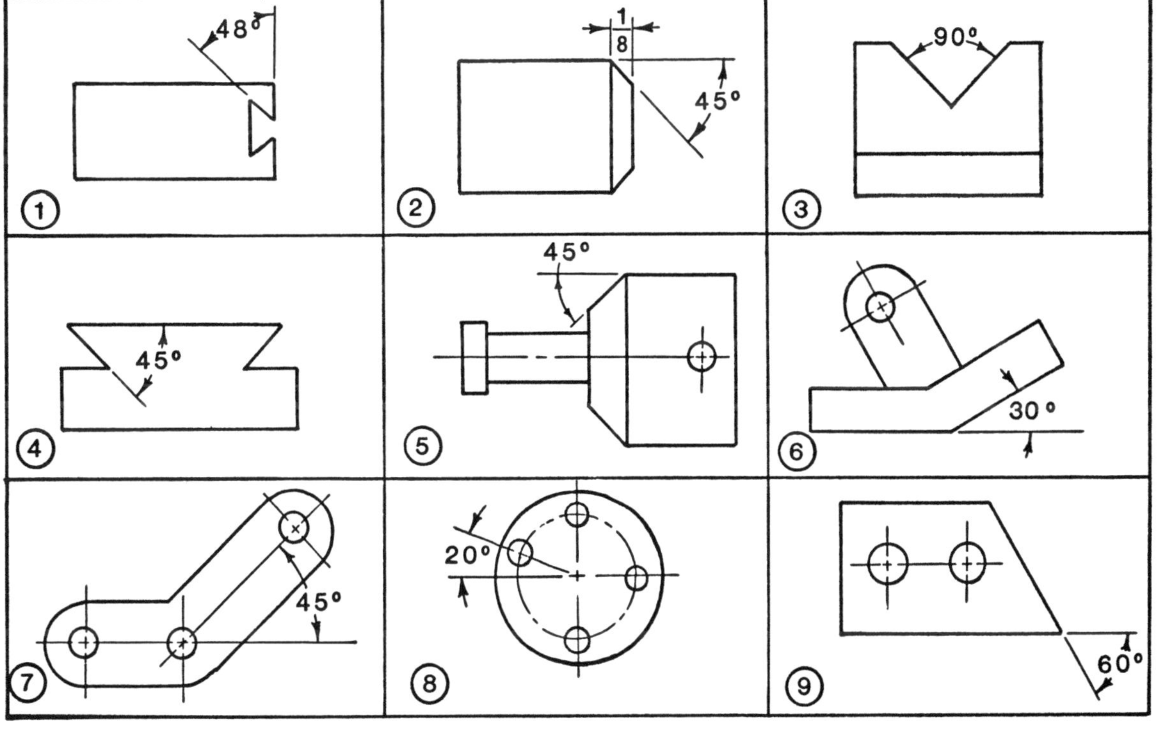

DIMENSIONING BY NOTE

At times, in a drawing, you will have to use notes rather than the methods previously described and illustrated. In such situations, the notes are lettered near the view concerned.

HOW AND WHERE NOTES ARE LETTERED

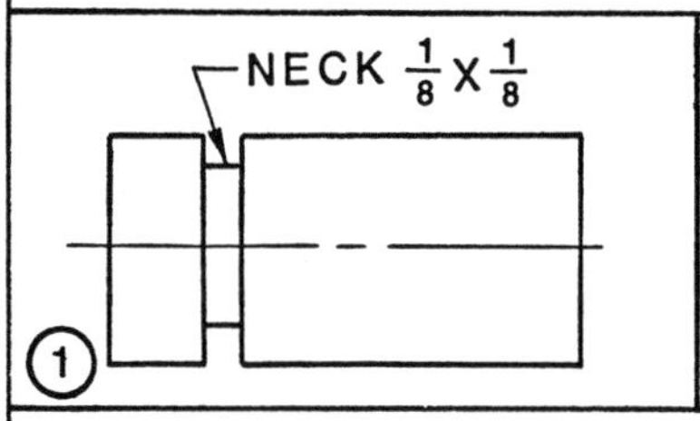

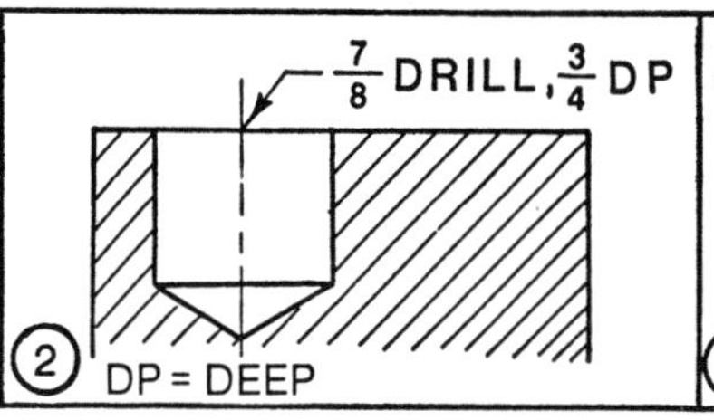

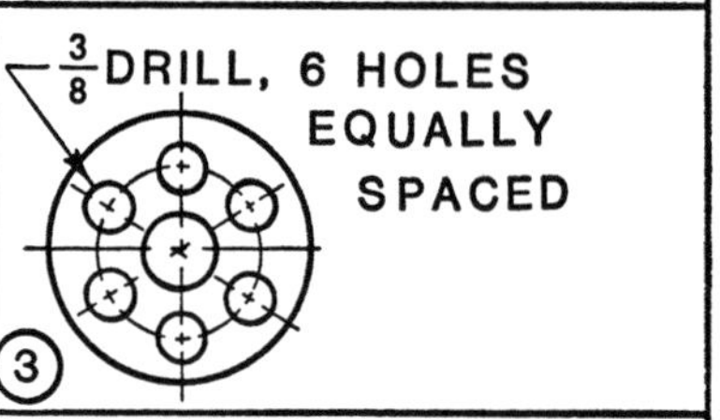

1. Use a leader between the note and the object described.
2. Use only horizontal lines of lettering (1/8" high).
3. Direct the leader at the center line in such cases as (2) and (3)

STUDY THE USEFULNESS OF THE NOTES ON THESE DRAWINGS

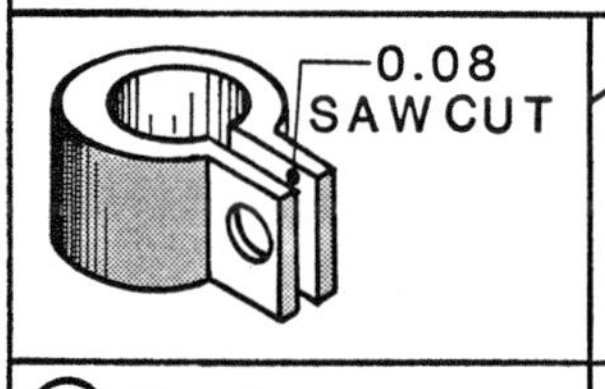

(1) Showing a mechanical operation and size.

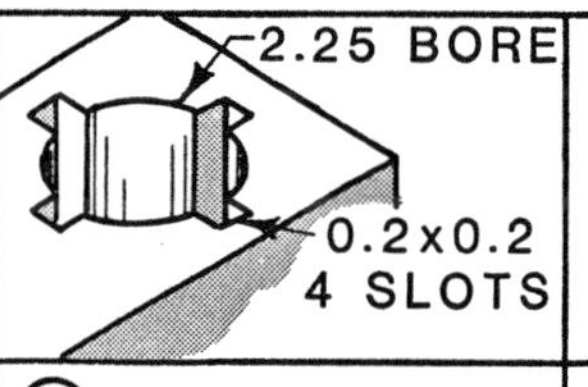

(2) Size and mechanical operations.

(3) Dimensioning a small detail.

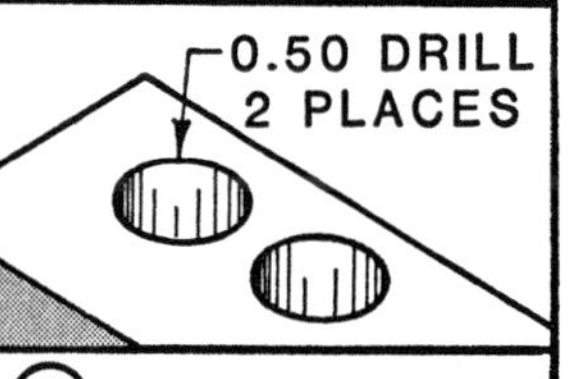

(4) Size and number of holes.

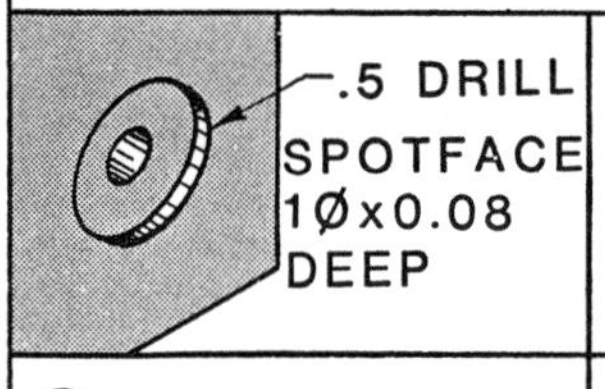

(5) Operation and size.

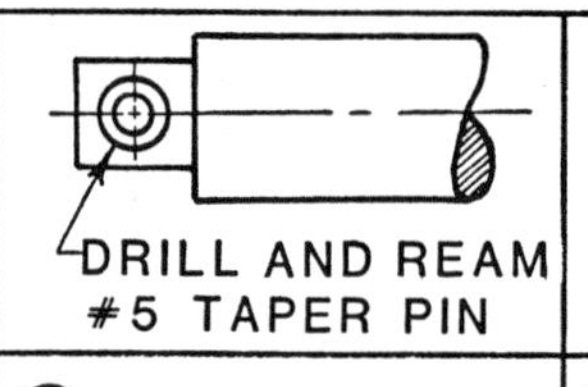

(6) Operation and size.

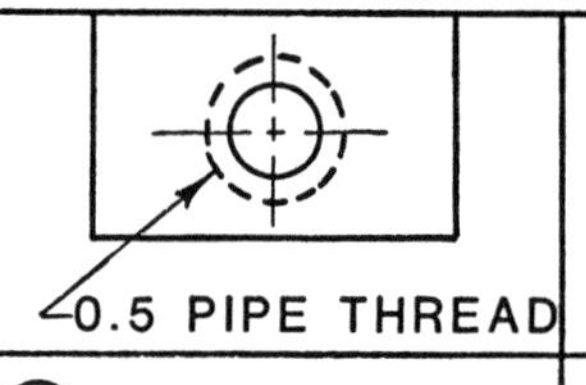

(7) 1/2" pipe threads.

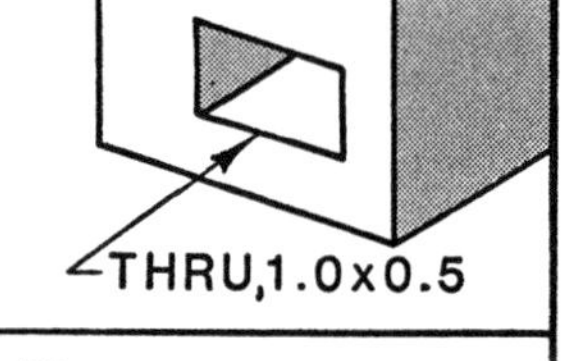

(8) Hole all the way through.

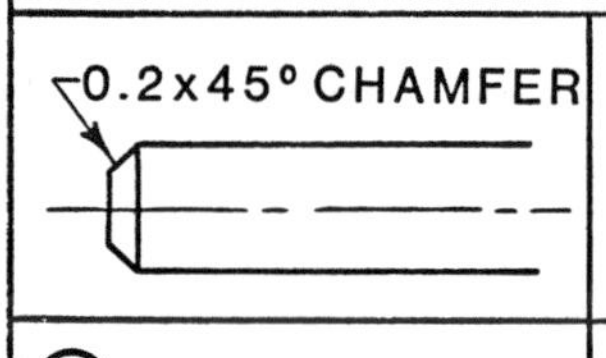

(9) Size and operation.

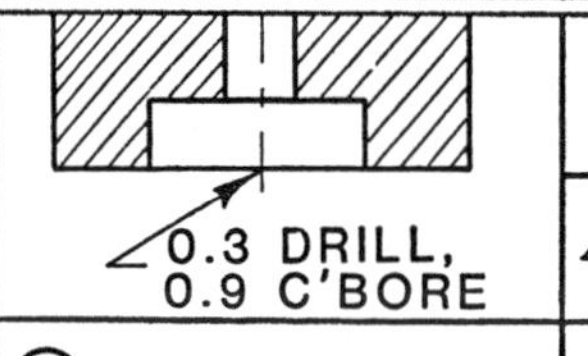

(10) 3/4" counterbore (c'bore).

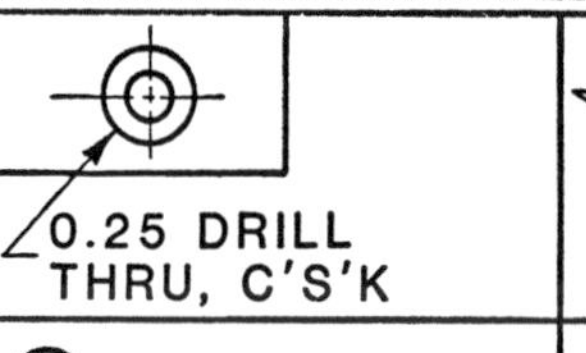

(11) Drill and counter-sink (c's'k).

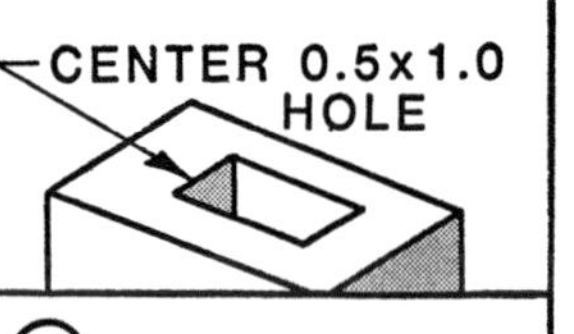

(12) Location and size.

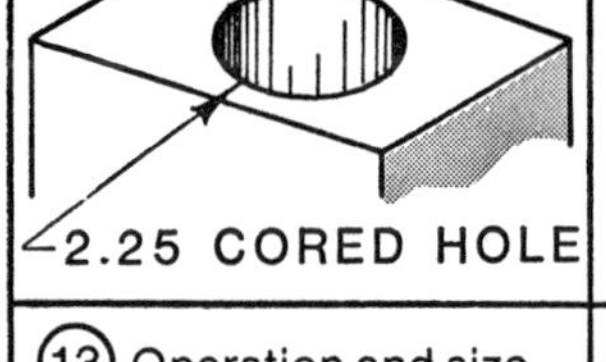

(13) Operation and size.

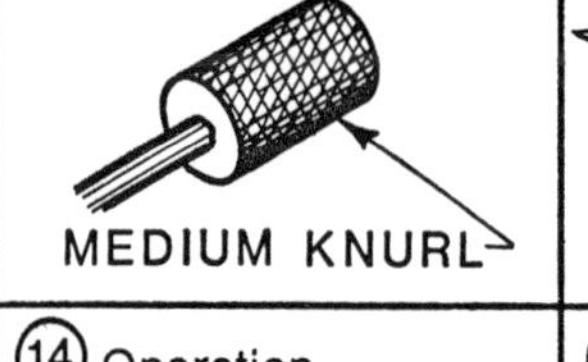

(14) Operation.

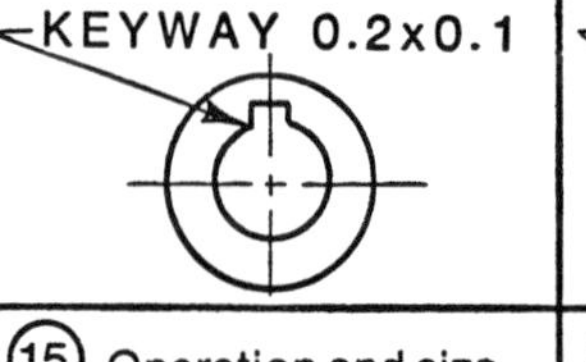

(15) Operation and size.

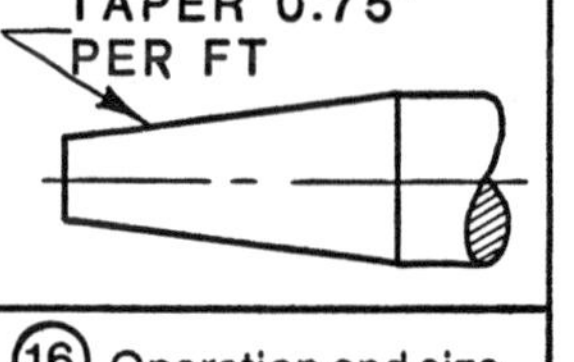

(16) Operation and size.

LIMITING DIMENSIONS

When accurate machining is required, the drafter indicates this to the fabricator by means of limiting dimensions.

The drafter indicates limiting dimensions by using decimal fractions instead of common fractions. In giving the dimension, the drafter allows a maximum limit and a minimum limit to the finished piece. These dimensions are written as decimals and are usually carried to three places.

TOLERANCE (THE ALLOWABLE AMOUNT OF VARIATION)

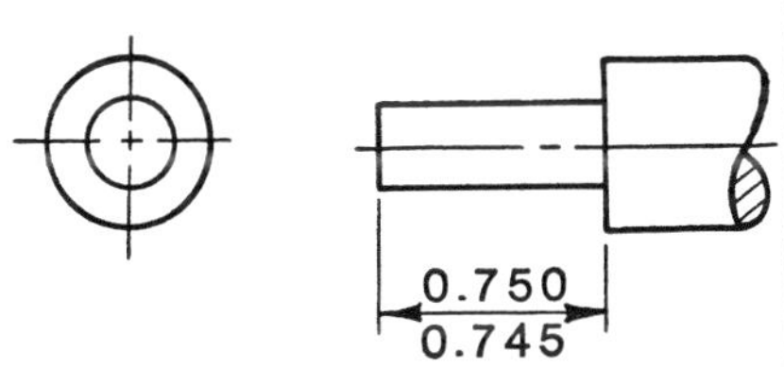

(1) This shaft may not be made longer than .750 of an inch, or may not be shorter than .745 of an inch. This variation is known as **tolerance.**

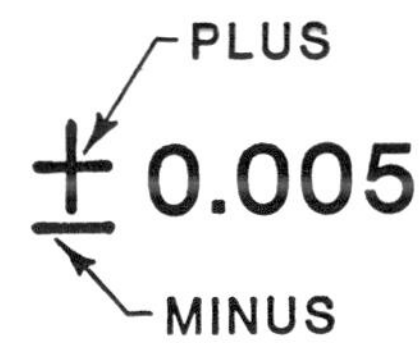

(2) This sign indicates that the piece may be .005" larger or smaller than the indicated dimension.

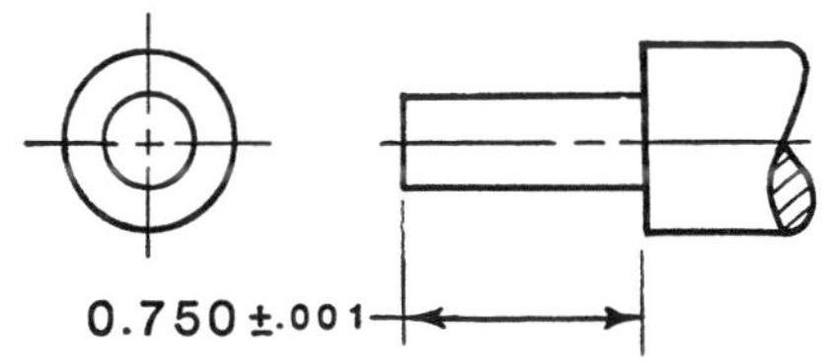

(3) This dimension means: Try to make the shaft .750" long. But it may be .751" (+ .001), or it may be .749" (- .001). No more, no less.

A PRINTED NOTE ON THE DRAWING

Many drawings contain the note shown on the right.

The range between the plus (+) and the minus (-) is the tolerance.

Unless otherwise specified:
Limits on decimal dimensions: ± .005
Limits on fractional dimensions: ± 1/64"
Limits on angular dimensions: ± 1/2°

NOTICE THESE LIMITING DIMENSIONS

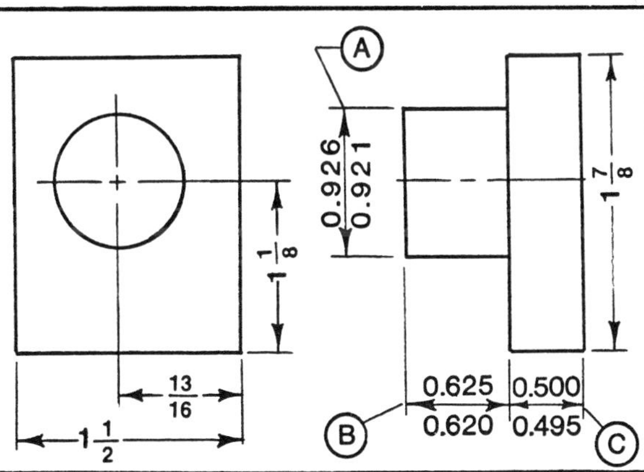

(1) Dimensions (A), (B), and (C) require accurate machining.

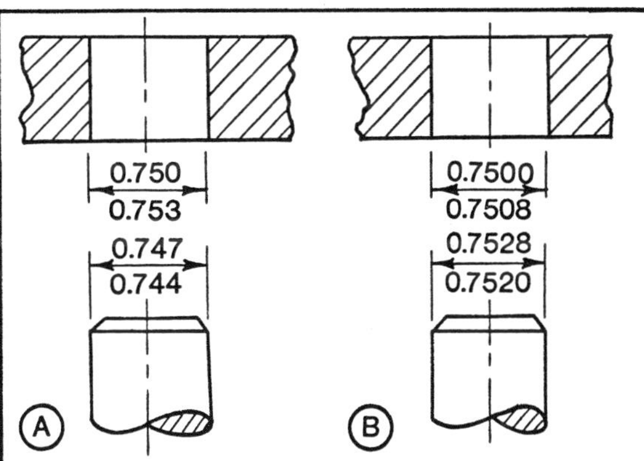

(2) (A) is a clearance fit.
(B) is an interference or force fit.

CLASSES OF FITS

1. **Clearance fit:** When two parts are machined so that one may be placed inside of the other (with a little room to spare), this is known as a clearance fit.

2. **Interference or force fit:** When two parts are machined so that one may be placed inside the other only with the use of force, this is known as an interference or force fit.

TABLE OF DECIMAL EQUIVALENTS OF FRACTIONS

A Table of Decimal Equivalents of Fractions may be found on p. 170.

ASSIGNMENT

Make a working drawing of one of these objects. (See Units 10 and 11.)

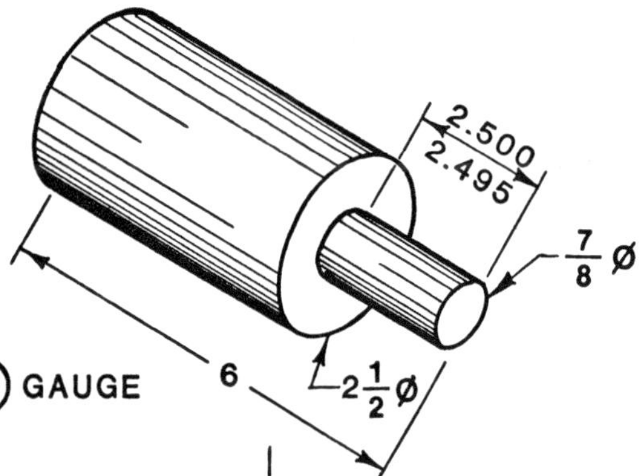

(1) GAUGE

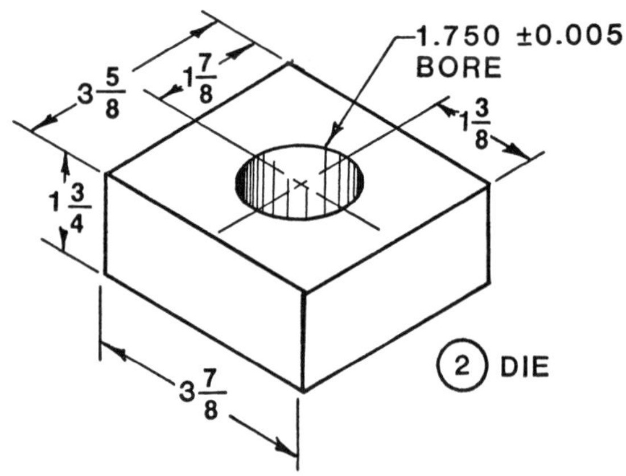

(2) DIE

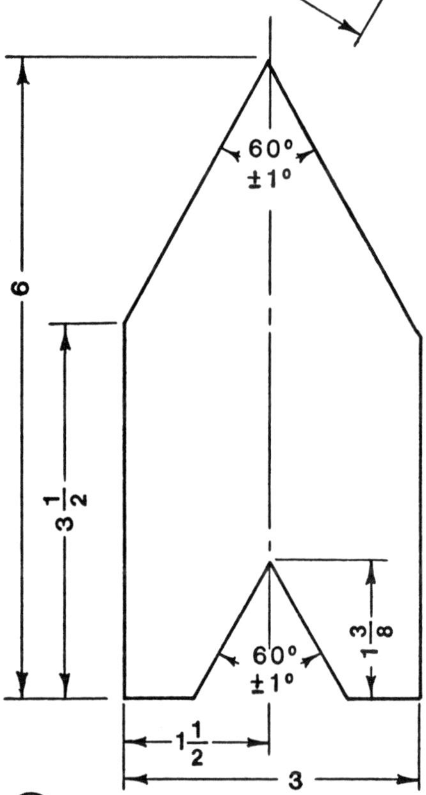

(3) GAUGE 0.04 STEEL

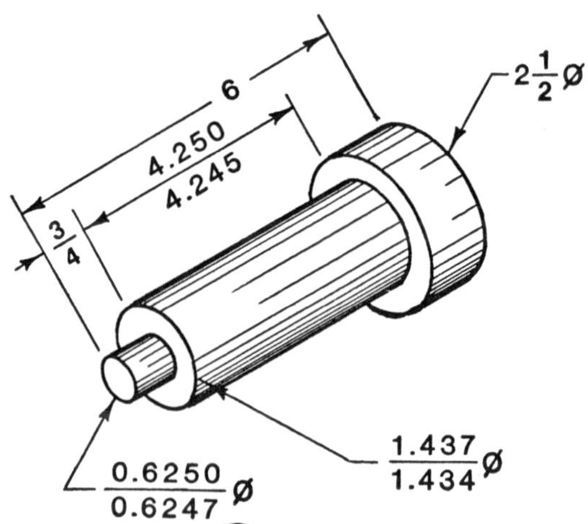

(4) PLUG

BASELINE DIMENSIONING

Baseline dimensioning (or co-ordinate dimensioning) is a system of dimensioning which uses a common reference point or extension of a point (a line) from which measurements are indicated. Baseline dimensioning is a common method of dimensioning used in some industries.

TWO METHODS OF DIMENSIONING

SUCCESSIVE DIMENSIONING	BASELINE DIMENSIONING
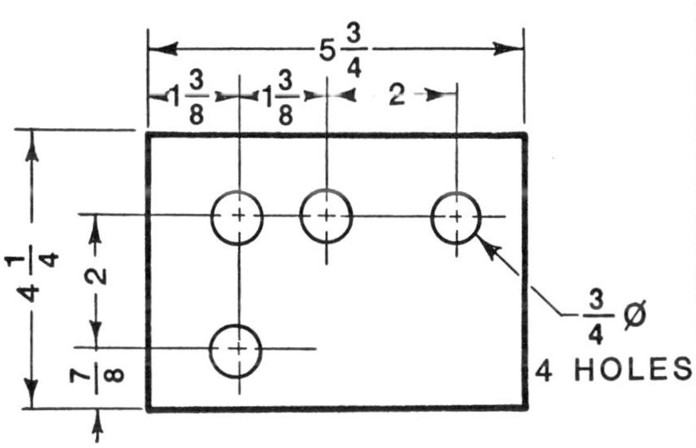	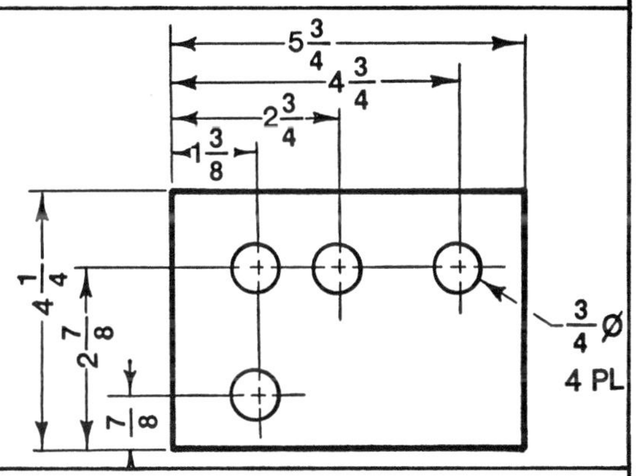
In successive dimensioning, each dimension is based on the one immediately preceding it.	In baseline dimensioning, all dimensions in a series refer to a common line or reference.

STUDY THE ARRANGEMENT OF BASELINE DIMENSIONING SHOWN BELOW

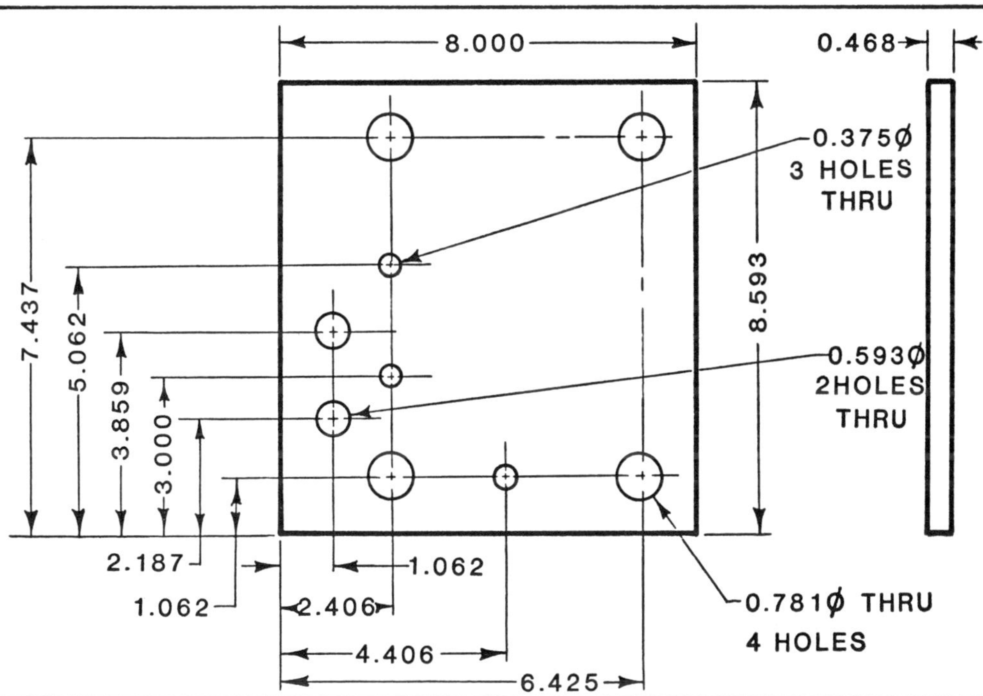

The notes indicate that the holes are drilled through. The holes do not have to be indicated in the end view. If they were indicated they would be so close together that the hidden edge lines would be confusing.

ASSIGNMENT

Make a working drawing of one of these drawings. (Use baseline dimensioning.)

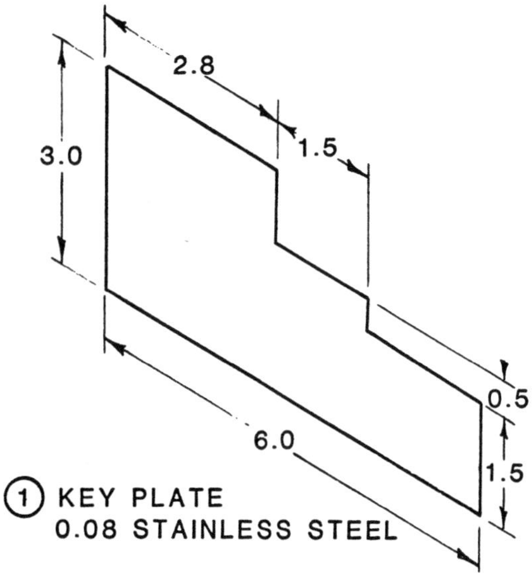

① KEY PLATE
0.08 STAINLESS STEEL

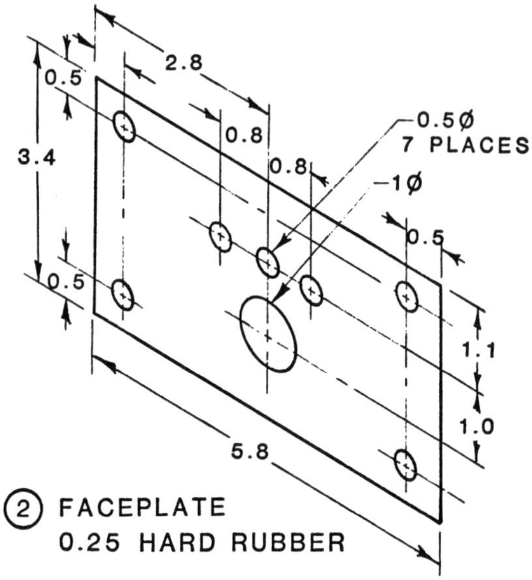

② FACEPLATE
0.25 HARD RUBBER

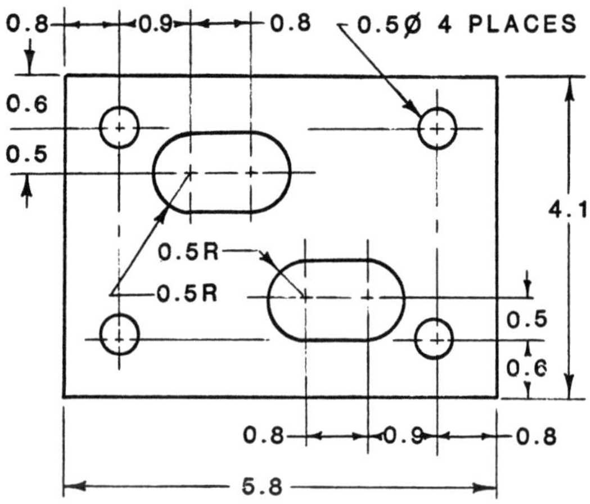

③ SWITCH PLATE 0.125 BRASS

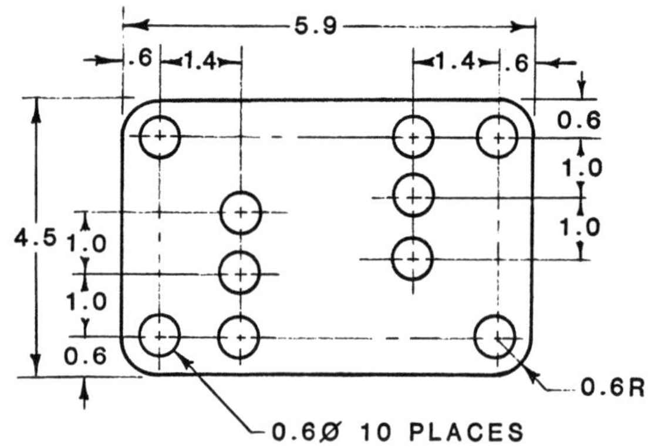

④ PANEL 0.125 BAKELITE

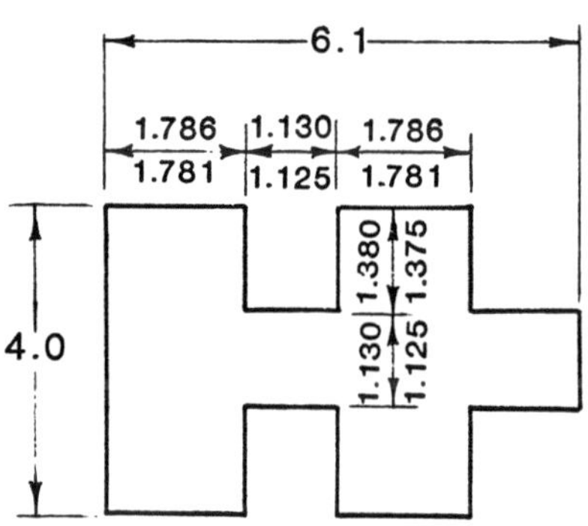

⑤ GAUGE 0.04 STEEL

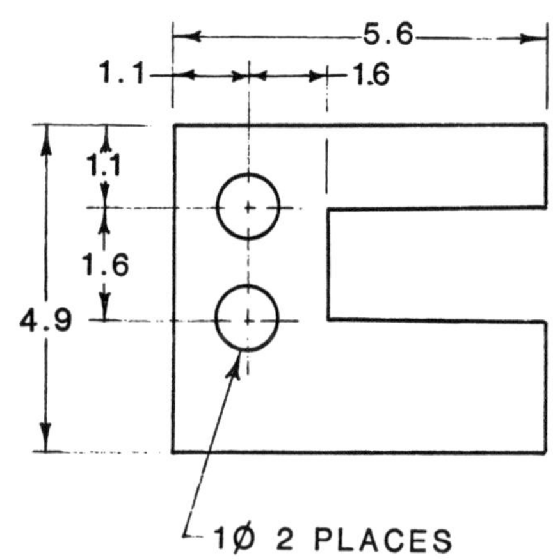

⑥ LAMINATION 0.08 SWEDISH STEEL

PICTORIAL DIMENSIONING

The method for dimensioning pictorial drawings is basically the same as the method used to dimension orthographic projections, namely:

1. Dimensions must read from the bottom or the right-hand side.
2. Smaller dimensions are placed closer to the drawing than larger dimensions but never closer than 1/4″.
3. Dimension lines must be drawn parallel to the surface dimensioned.
4. When possible, dimensions should be kept aligned.

VERTICAL DIMENSIONING OF VERTICAL SURFACES

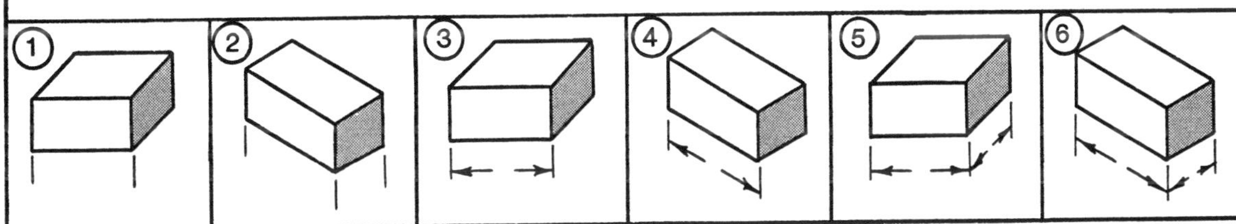

A. When dimensioning a vertical surface project vertical extension lines downward (see (1) and (2)).
B. Draw dimension lines parallel with the surface dimensioned (see (3), (4), (5), and (6)).
C. When possible, align the dimension lines (see (5) and (6)).

HORIZONTAL DIMENSIONING OF HORIZONTAL SURFACES

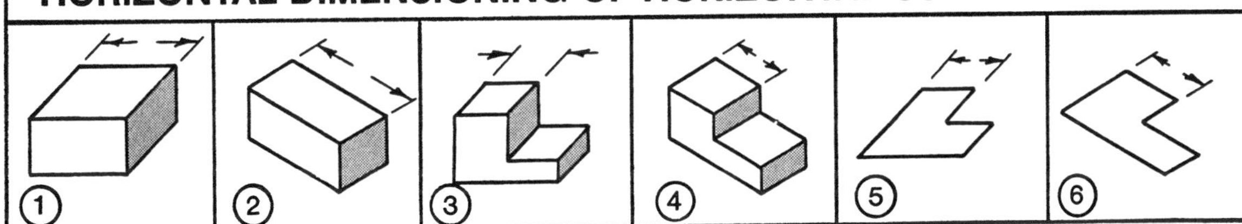

A. When dimensioning a horizontal surface, draw dimension lines in a horizontal plane (see (1), (2), (3), (4), (5), and (6)).
B. Draw dimension lines parallel to the surface dimensioned.

USE SLOPING DIMENSION LINES FOR SLOPING SURFACES

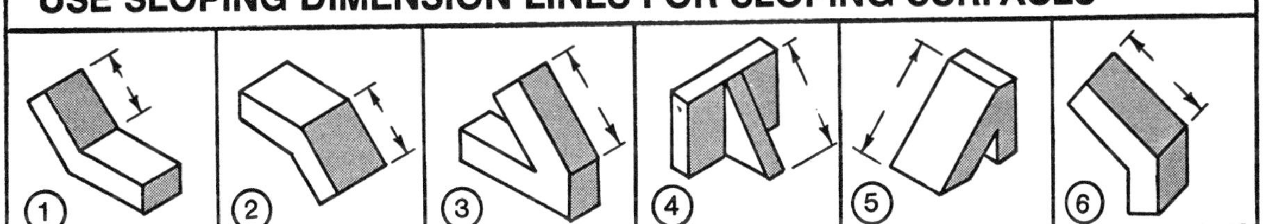

Follow the rule: A dimension line must be drawn parallel to the surface dimensioned.

A suggestion: Study the methods used in dimensioning every pictorial drawing from page 47 to this page. Concentrate on how the extension lines, dimension lines, and leaders have been drawn. Mentally compare the dimensioning of pictorial drawings with the dimensioning of orthographic projections.

NUMERALS IN PICTORIAL DIMENSIONING

The simple device shown below will aid you in determining how to place the numerals when dimensioning a pictorial drawing.

Imagine that the dimension 1½″ has been lettered on a piece of clear plastic, or on a piece of cardboard.

THE NUMBERING DEVICE (for dimensioning pictorial drawings)

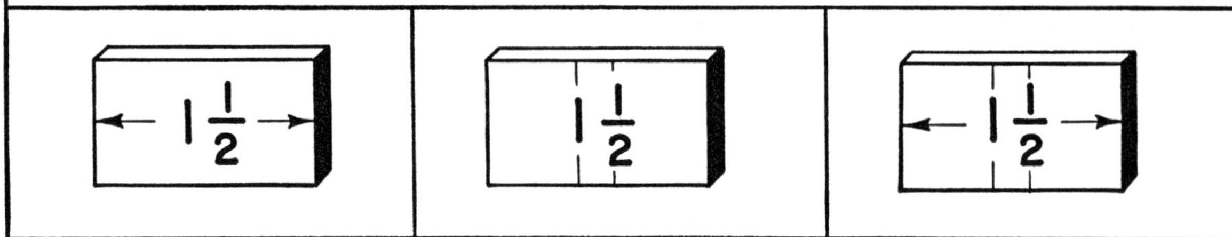

(1) The dimension 1½″ would look like this when lettered on a piece of plastic or on cardboard.	(2) The vertical lines show the position of the numerals.	(3) The horizontal line shows the position of the fraction line and the dimension line.

THE NUMBERING DEVICE SHOWN IN VARIOUS POSITIONS

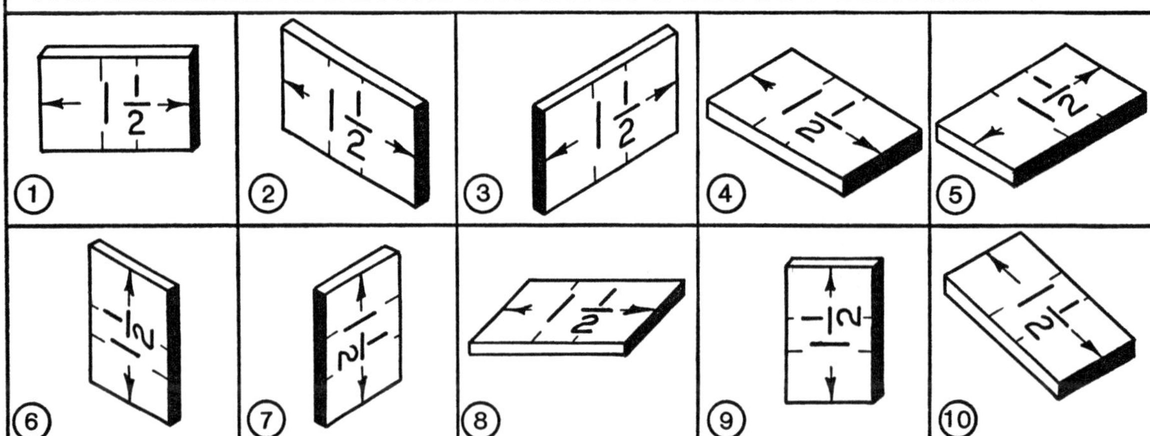

USING THE DEVICE AS AN AID IN DIMENSIONING

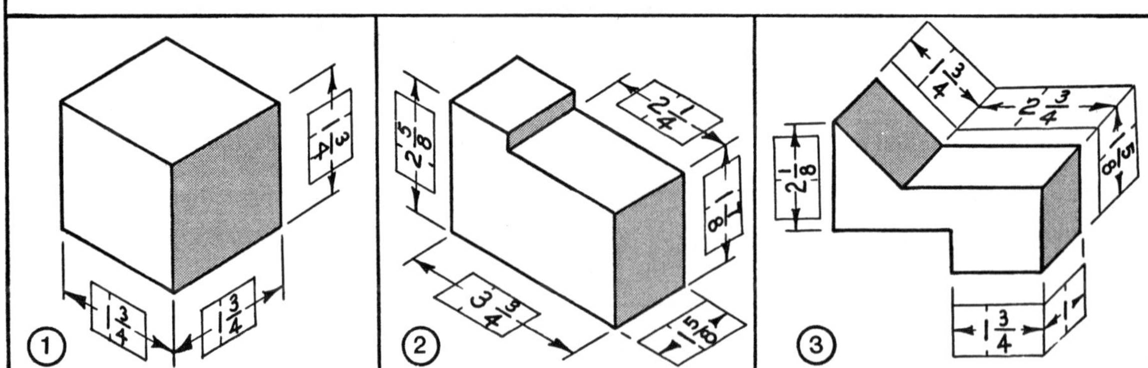

See (2), (3), and (7) above.	See (4), (5), (6), (7), and (10)	See (1), (3), (7), (8), (9), and (10).

Note that the backs of all the pictorial arrowheads are parallel to the extension line they touch.

STUDY THE PLACING OF DIMENSIONS ON THESE PICTORIAL DRAWINGS

A careful study of the dimensioning situations shown below will be well worth the effort.

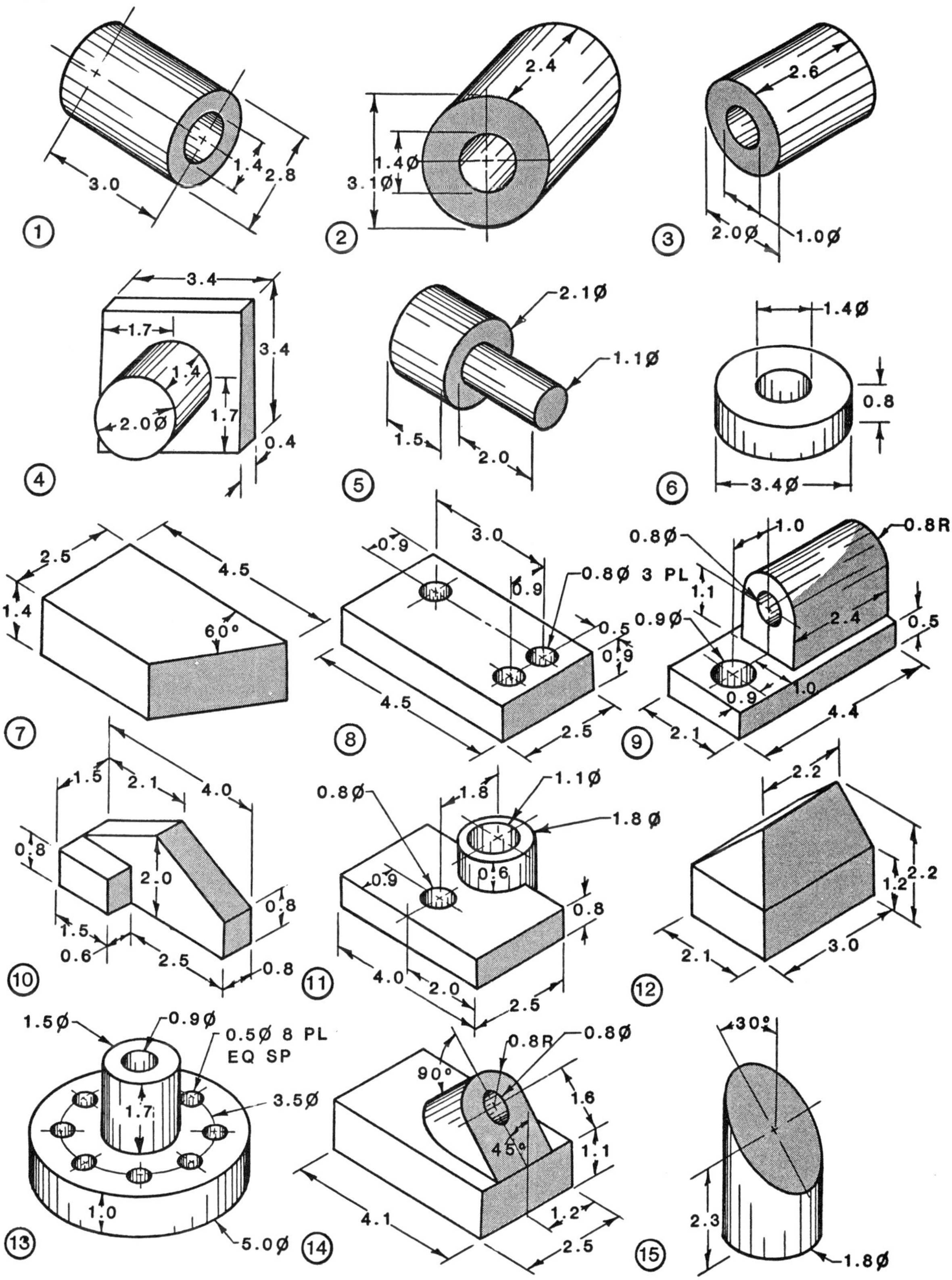

ASSIGNMENT

Draw and dimension the pictorial drawings.

UNIT 10

ORTHOGRAPHIC

An object may be described by means of drawings and symbols. This is the basic method of description which forms the language of drafting.

The most complex drawing consists of an arrangement of simple forms. The student must agree that it would be wise to obtain a thorough basic training in visualizing and drawing simple forms before attempting to make more complex drawings. Such training will permit the beginner to concentrate on one thing at a time. This leads to a better understanding and a sense of confidence that permits rapid advancement in the field.

The work covered in this unit provides a basis for all future work in drafting. **It is absolutely essential that you realize that the rectangle you draw is not an arrangement of four lines, but it is, rather, a drawing, a picture, or a view of one surface of an actual object.**

You must train yourself to visualize how the drawing of a surface will look. You should also learn to visualize one surface of a real object by studying a drawing of it.

A clear understanding of this mental process is the root of mechanical drawing.

A drawing for an object to be manufactured can have one or more views. Drawings with two or more views are called multiview or orthographic working drawings. A rule to follow is to draw the minimum number of views needed to manufacture the object. Most objects will require three views (front, top, and right-hand side). Note in the example how many views can be drawn of a die.

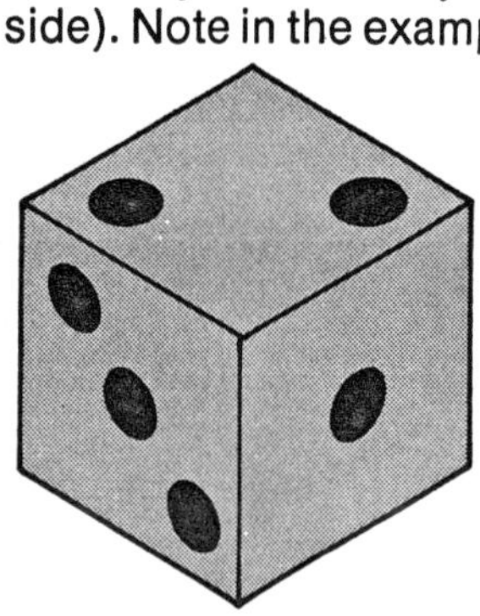

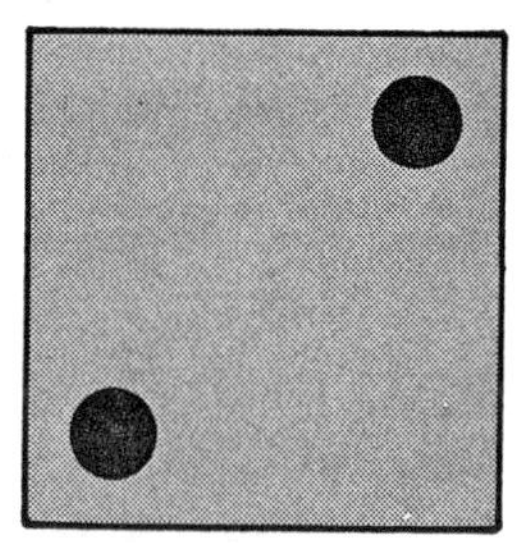

TOP VIEW

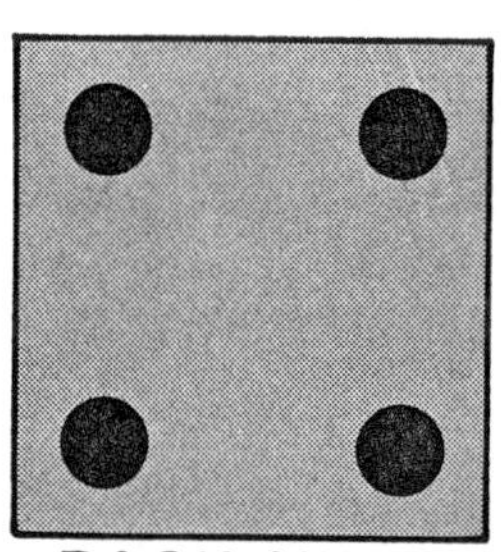

BACK VIEW

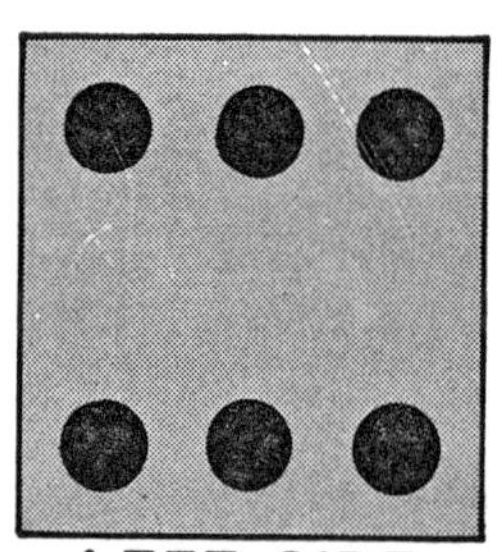

LEFT SIDE VIEW

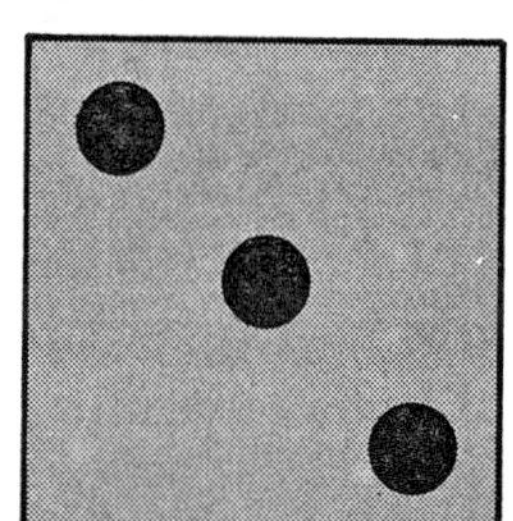

FRONT VIEW

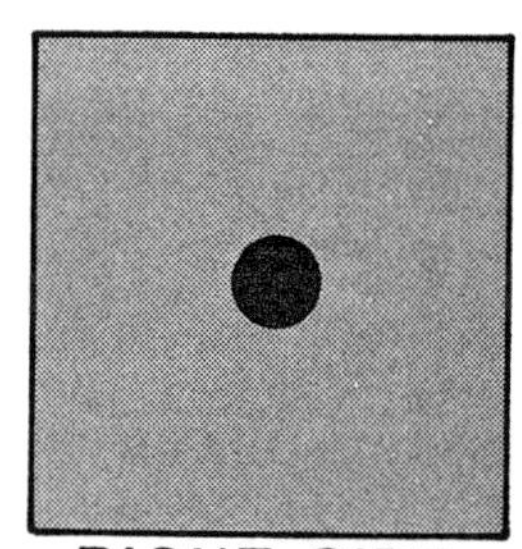

RIGHT SIDE VIEW

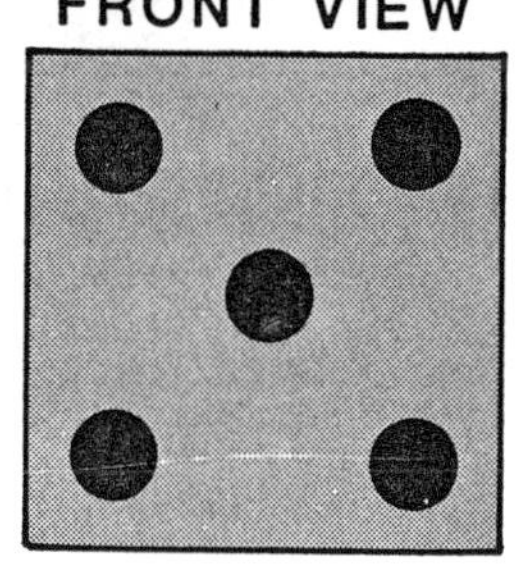

BOTTOM VIEW

DRAWING A NOTCHED PLANE

A plane is a flat surface having only length and height.

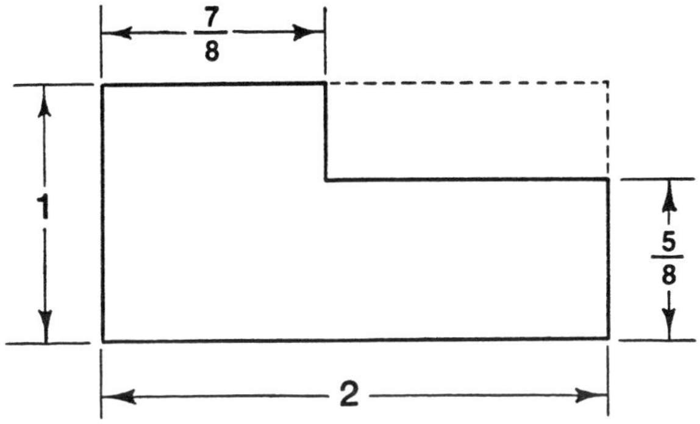

1. The basic shape is a rectangle which is 2″ long and 1″ wide.
2. The notch is in the upper right-hand corner. It is 7/8″ from the upper left-hand corner and 5/8″ from the lower right-hand corner.

DRAWING THE NOTCHED PLANE SHOWN ABOVE, STEP BY STEP

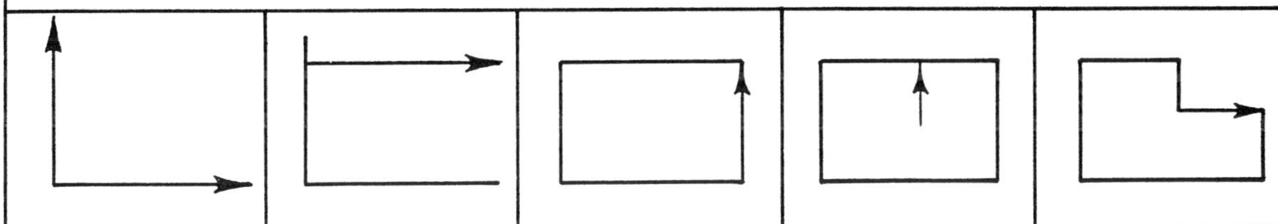

① Draw a horizontal line and a vertical line. (Use a T-square and triangle.)	② Measure the height (1″) and draw the top edge of the basic shape.	③ Measure and draw the length of the basic shape (2″).	④ Measure and draw the left side of the notch (7/8″).	⑤ Draw the depth of the notch. Darken the completed drawing (5/8″).

MORE PRACTICE, STEP BY STEP

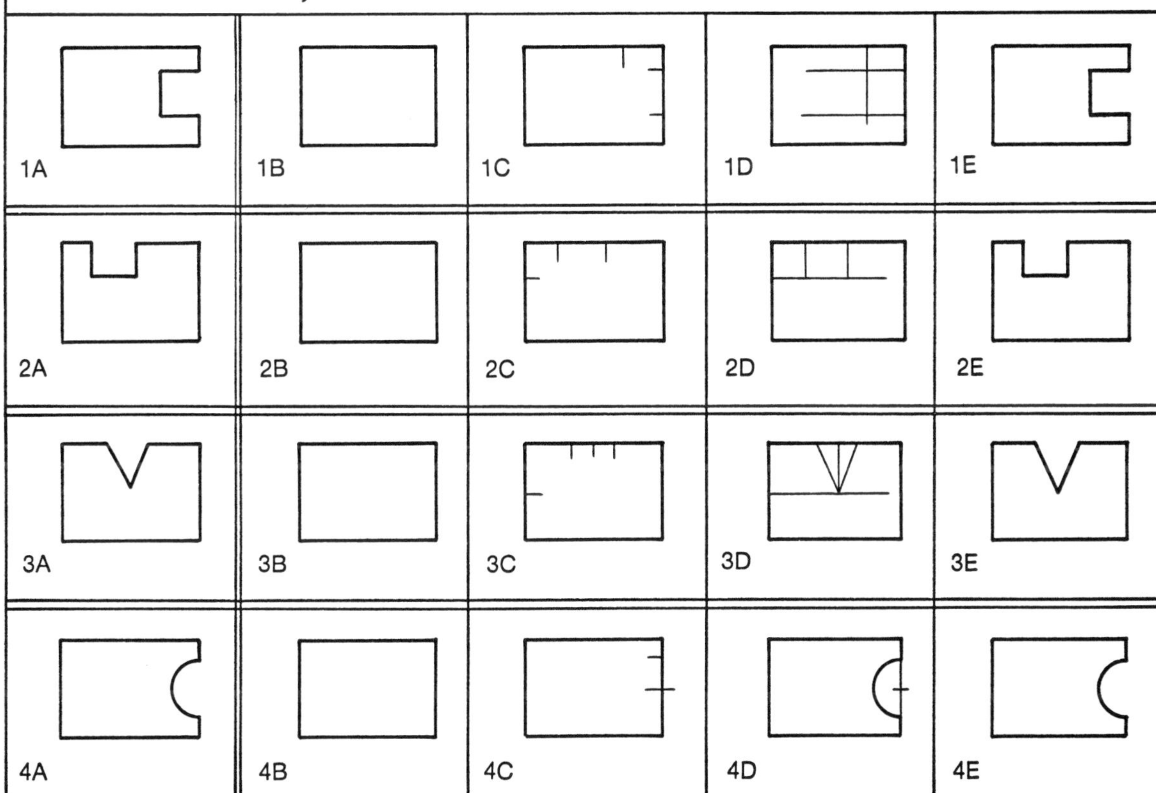

HOW TO "SEE" THE FRONT VIEW

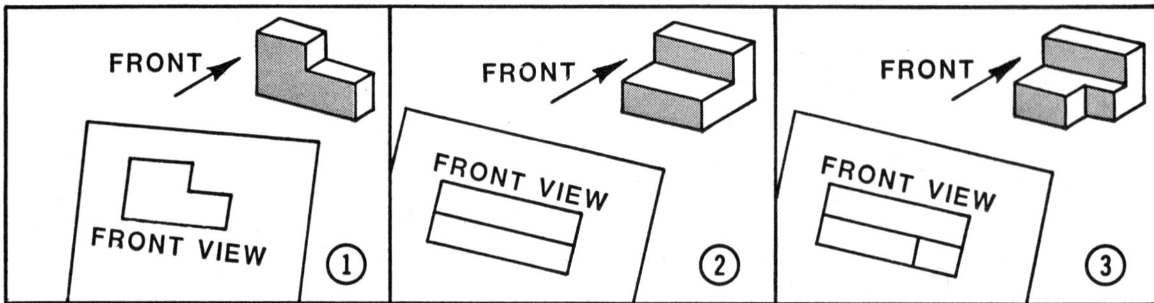

1. Stand directly in front of the object.
2. "See" only the front-view surfaces.
3. Imagine that you cannot "see" any of the top surface.
4. Imagine that you cannot "see" any of the end surface.
5. Study the drawing of each front view, and you will learn to "see" the front view, just as a draftsperson sees it.

PRACTICE IN "SEEING" THE FRONT VIEW

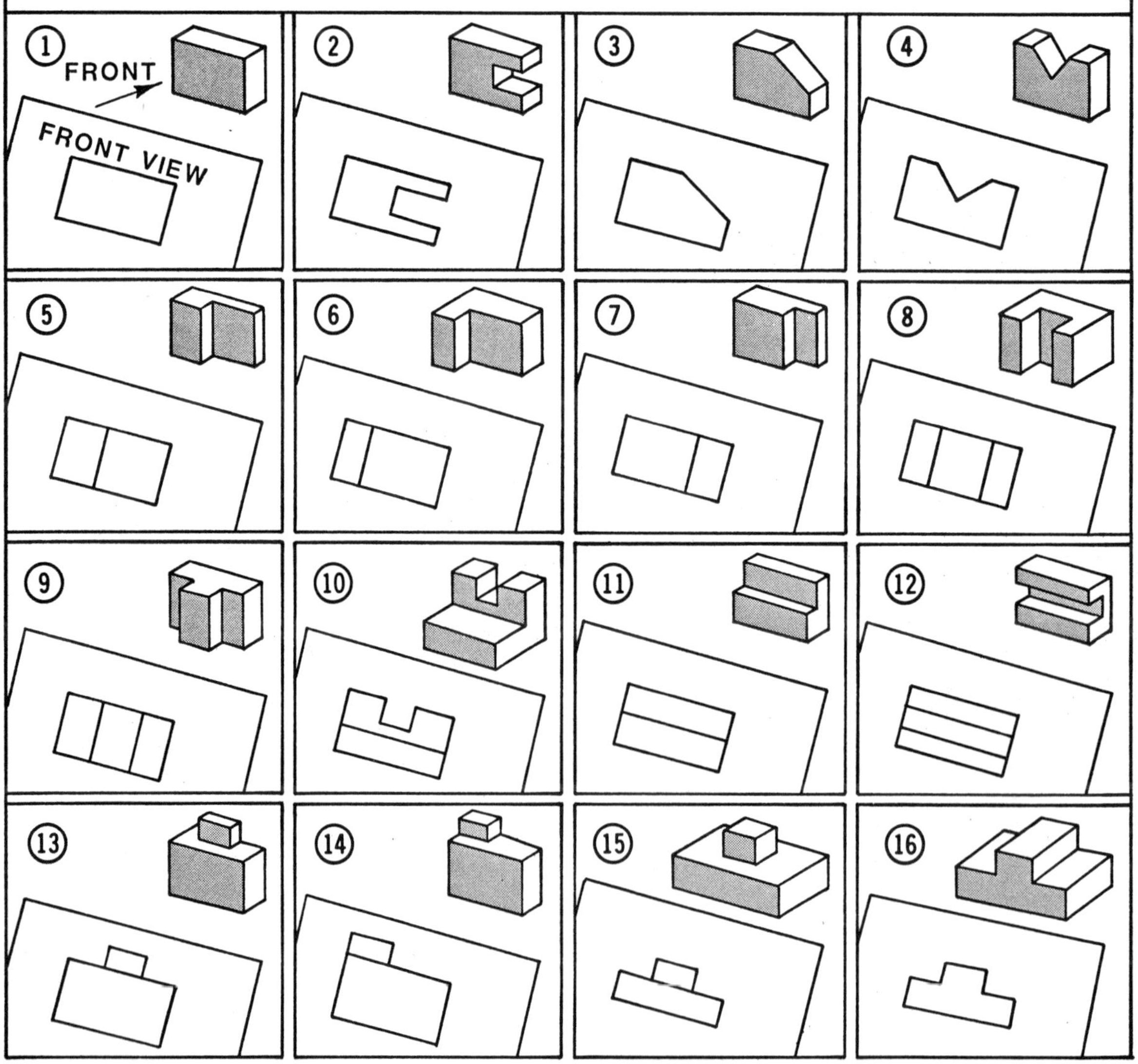

DRAWING THE FRONT VIEW

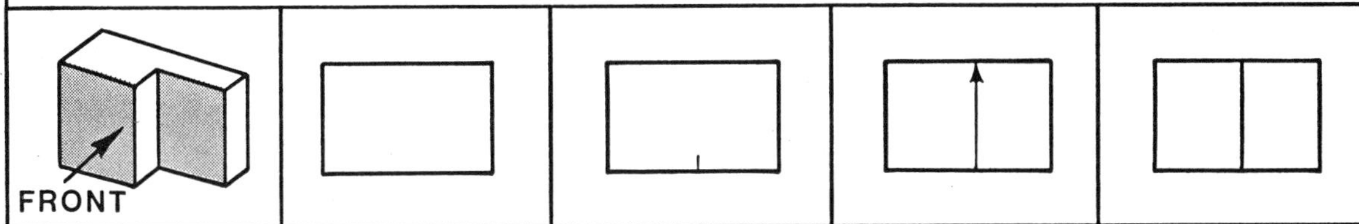

① Study the basic shape of the front view.	② Measure and draw the basic shape.	③ Measure for the necessary details.	④ Draw the details.	⑤ Darken the completed drawing.

PRACTICE IN DRAWING THE FRONT VIEW

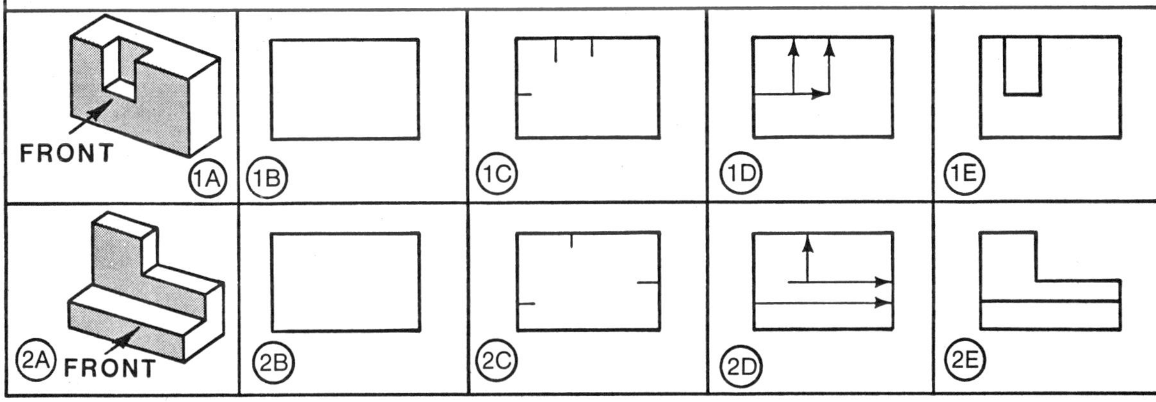

ASSIGNMENT

TITLE: FRONT VIEW

Draw the front view.

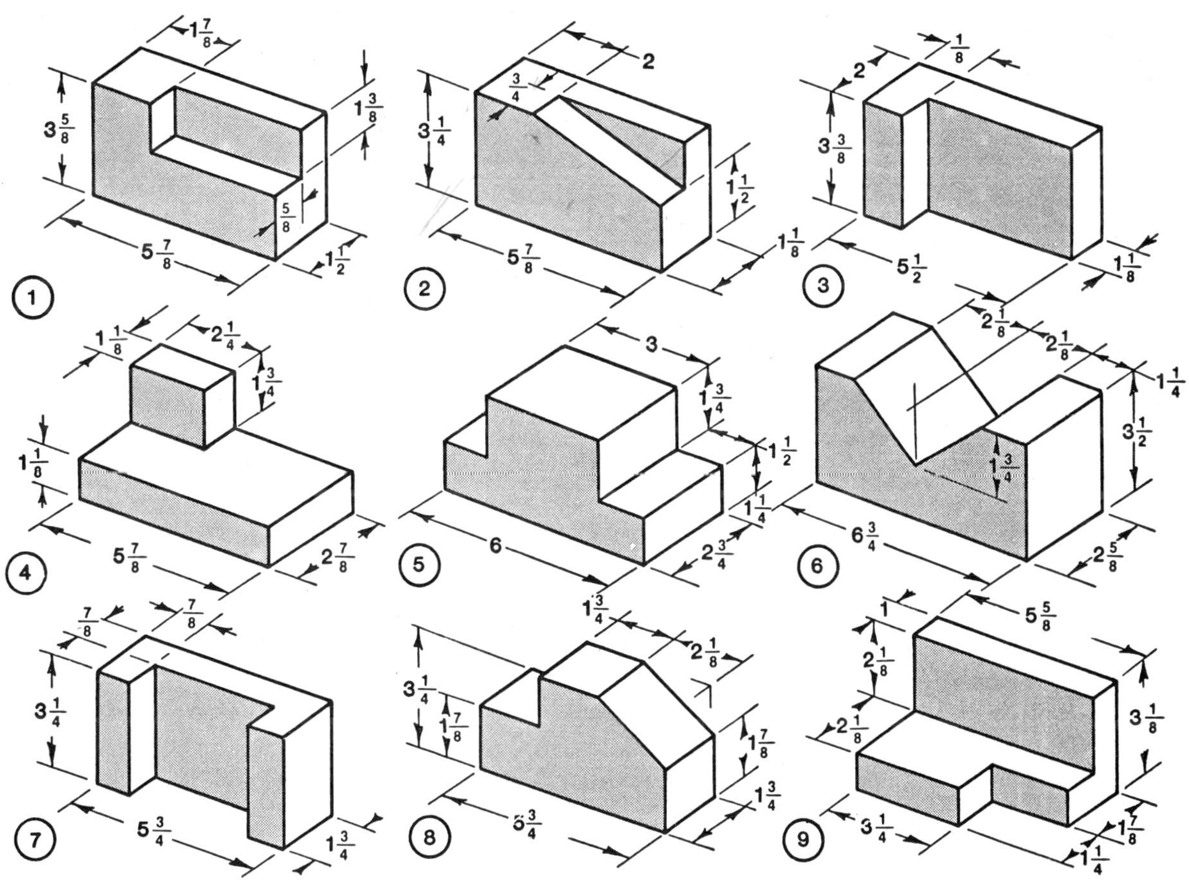

TRAINING IN VISUALIZATION

You must train yourself to visualize the true nature of the surface of an object. Develop the ability to recognize the fact that one surface may be nearer to you than another. You must be able to visualize that a surface may slope toward or away from you.

TRAINING IN VISUALIZATION

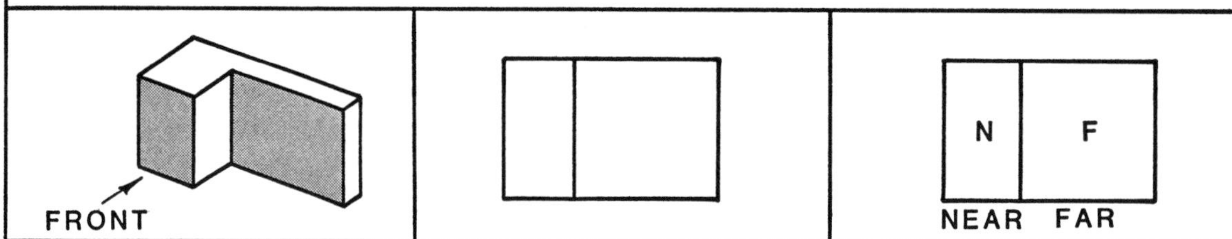

① Study this object and determine what the front view will look like.

② You should have formed a mental picture that looked like this.

③ You must also realize that the surface marked "N" is near you. The surface marked "F" is farther away.

PRACTICE IN VISUALIZATION

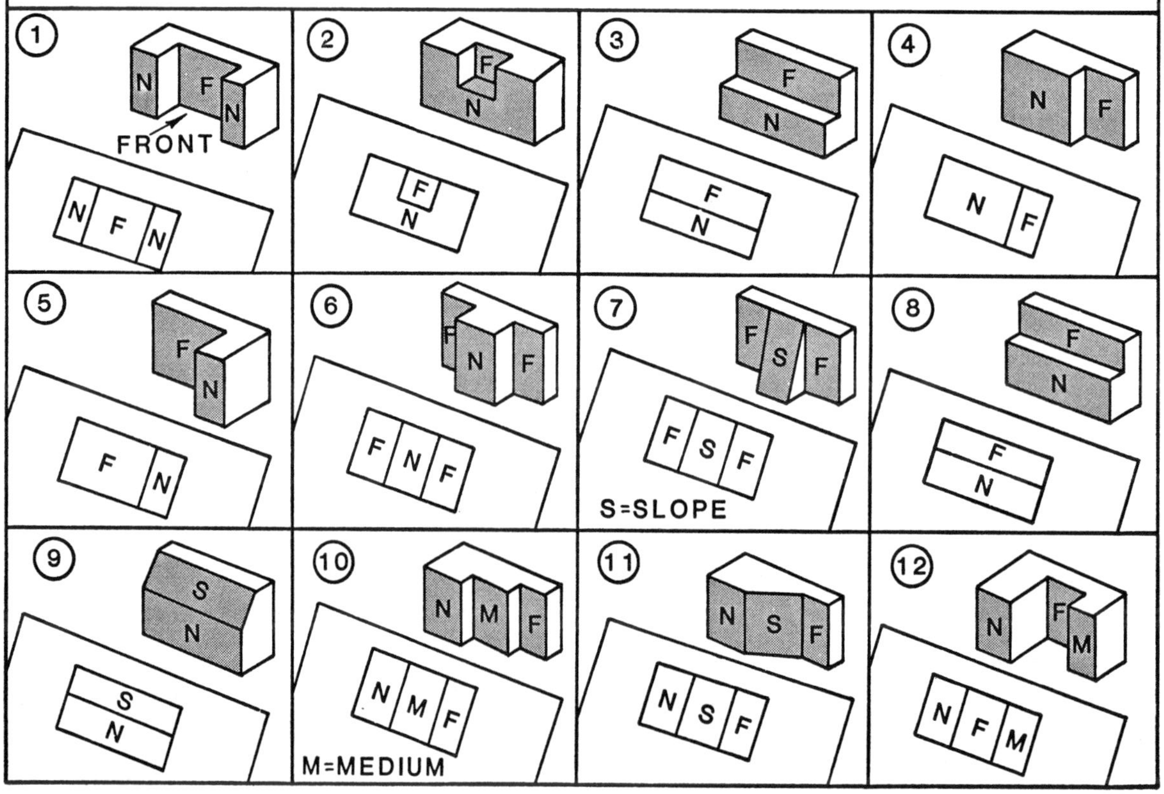

ANALYSIS OF THE ABOVE EXPERIENCES

You should now realize that you must visualize a front view with deeper understanding in order to form a more intelligent picture of it.

It is interesting to compare the front views of 4 and 5, 1 and 6, 6 and 7 and 10, 11, and 12.

Train yourself to visualize with a complete understanding of surfaces.

THE RESULT OF TRAINING IN VISUALIZATION

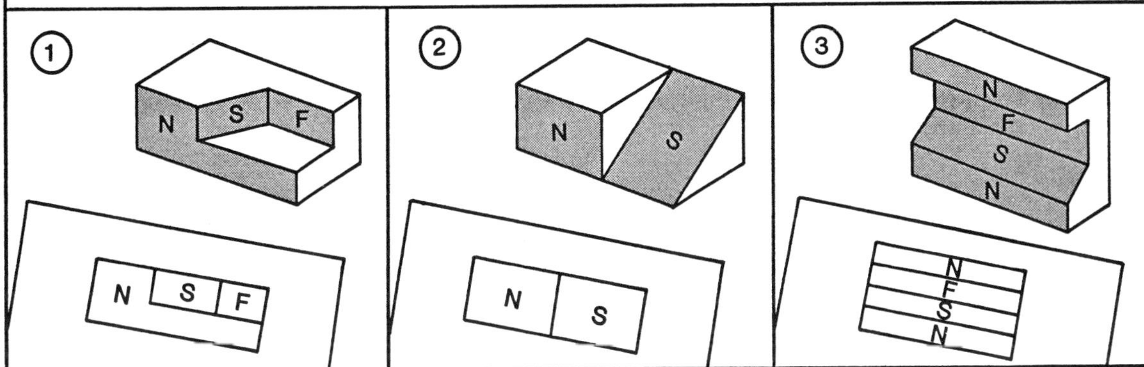

It is doubtful whether you would have been able to visualize the true nature of the front view of these objects without having studied the drawings on the preceding pages.

ASSIGNMENT

Draw the front view of one of these objects.

TITLE: FRONT VIEW

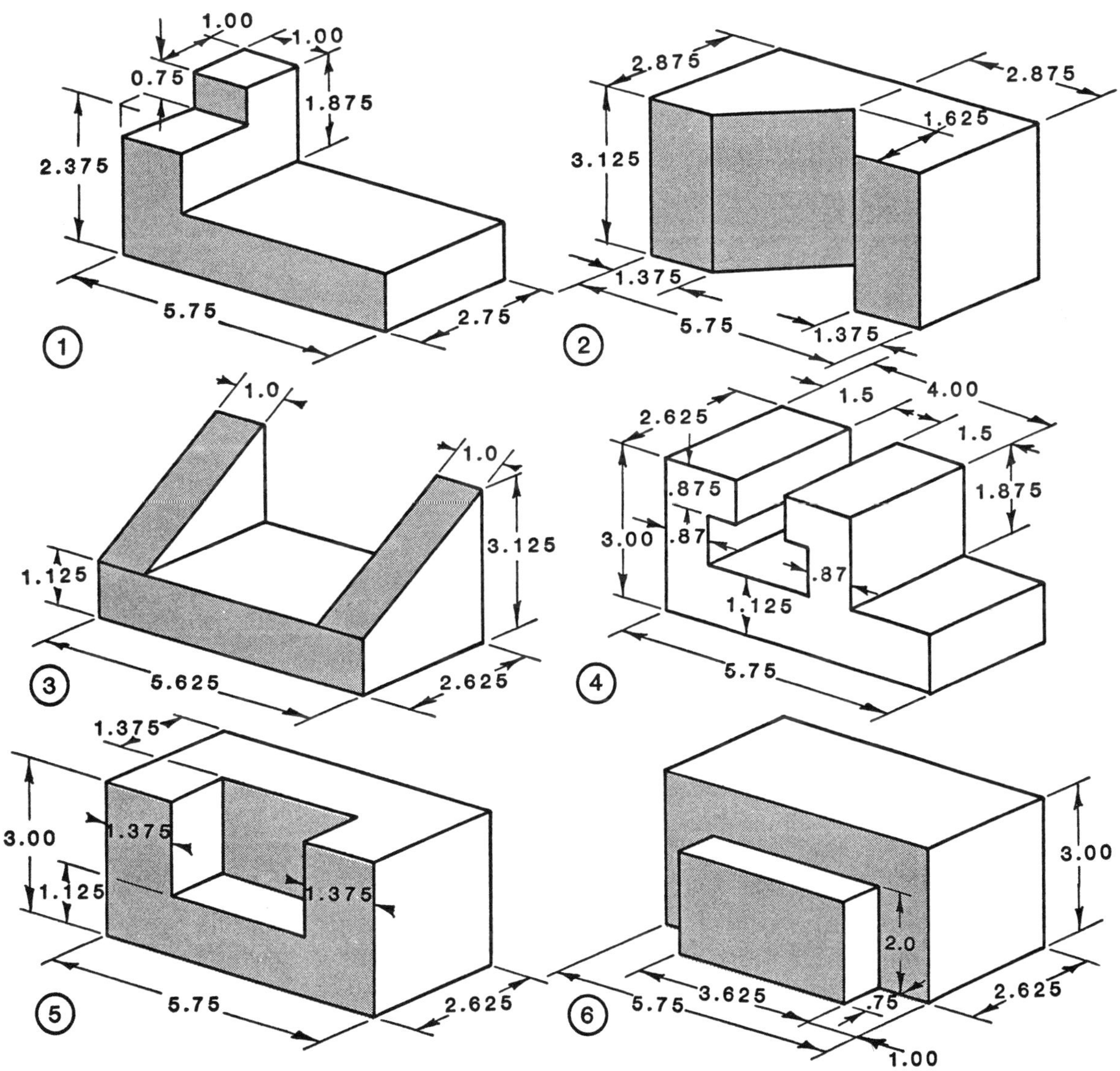

HOW TO "SEE" THE TOP VIEW

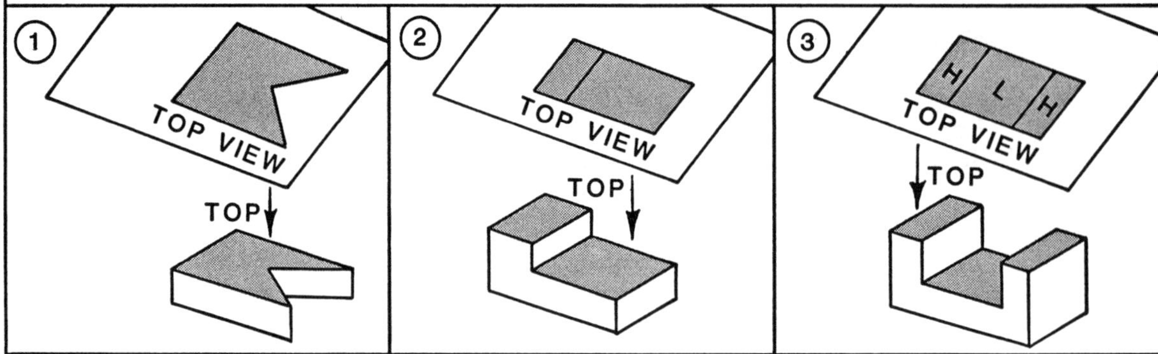

1. Stand directly in front of the object and directly over it.
2. "See" only the surfaces that can be seen from the top.
3. Imagine that you cannot "see" any of the front or end surfaces.
4. Study the three drawings of the top views.

PRACTICE IN "SEEING" THE TOP VIEW

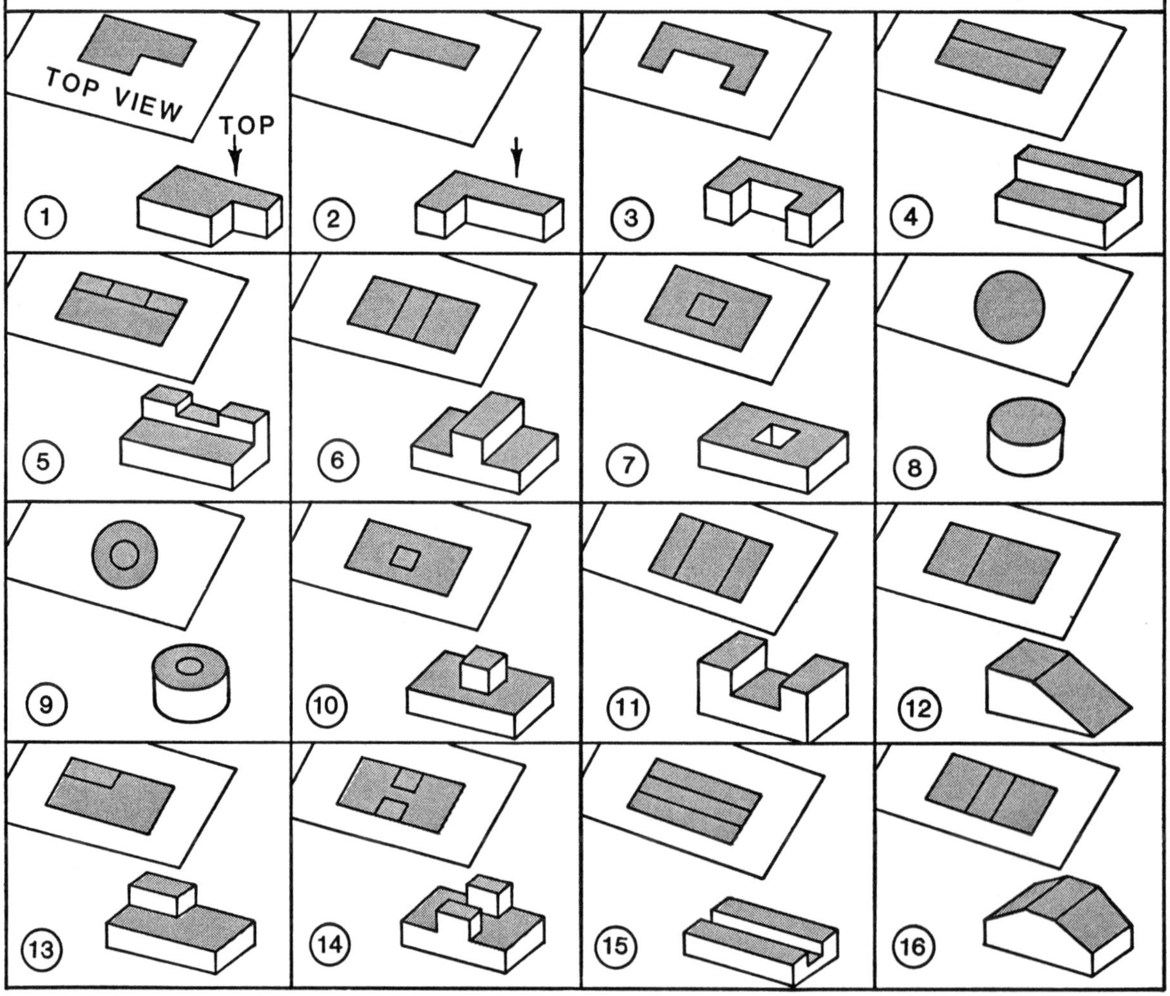

It is extremely important to recognize high, low, and sloping surfaces.

It is important to remember that you are standing directly in front of, and directly above, an object when drawing a top view.

DRAWING THE TOP VIEW

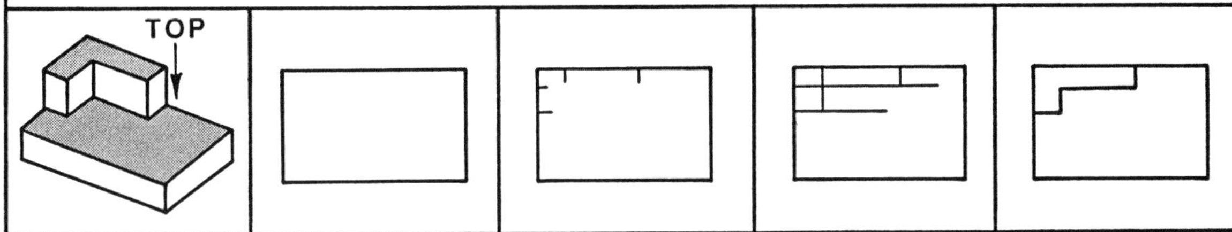

1. Study the basic shape and the shape of the details.
2. Measure and draw the basic shape lightly.
3. Measure and lightly lay out the details.
4. Draw the lines for the details lightly.
5. Darken and complete the top view.

PRACTICE IN VISUALIZATION

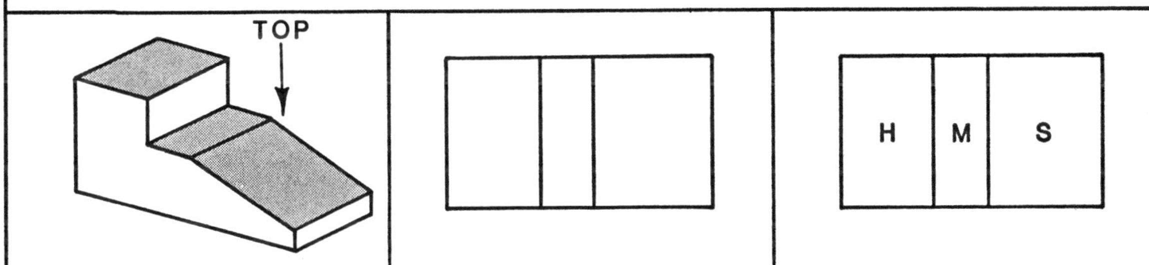

1. How does the top view of this object look?
2. Your mental picture of the top view should have looked like this.
3. The three surfaces are: high (H), medium (M), and sloping (S).

ASSIGNMENT

TITLE: TOP VIEW

Draw the top view of one of these in millimeters (mm).

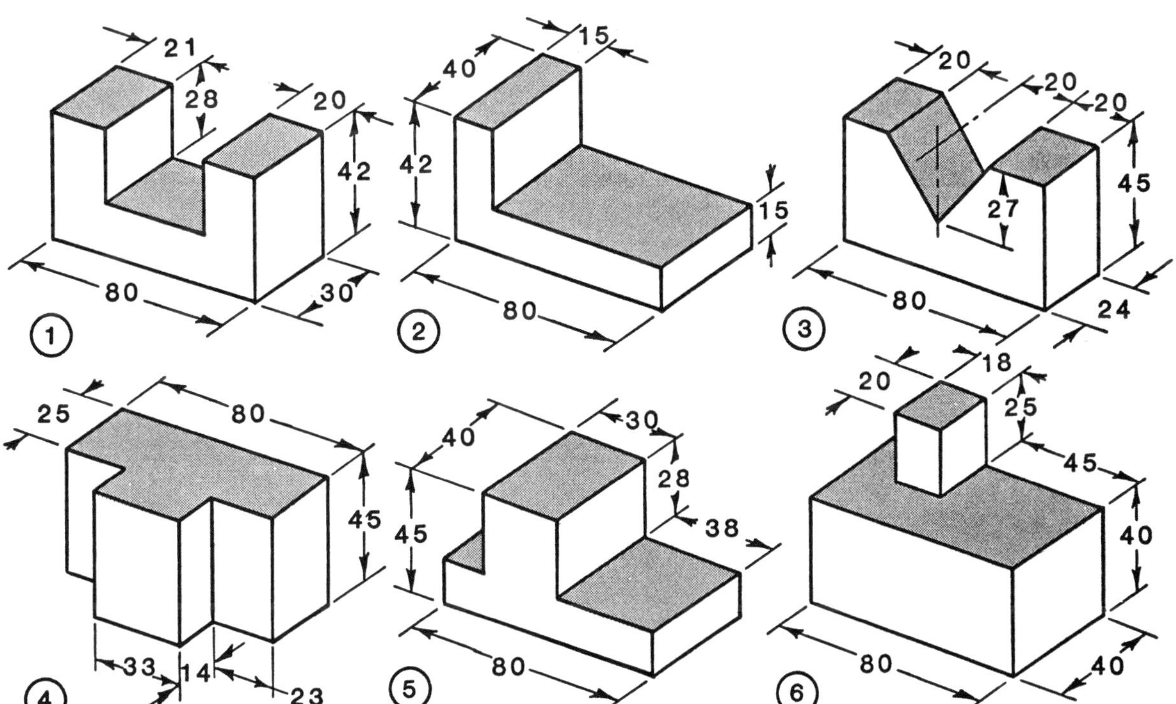

HOW TO "SEE" THE END VIEW

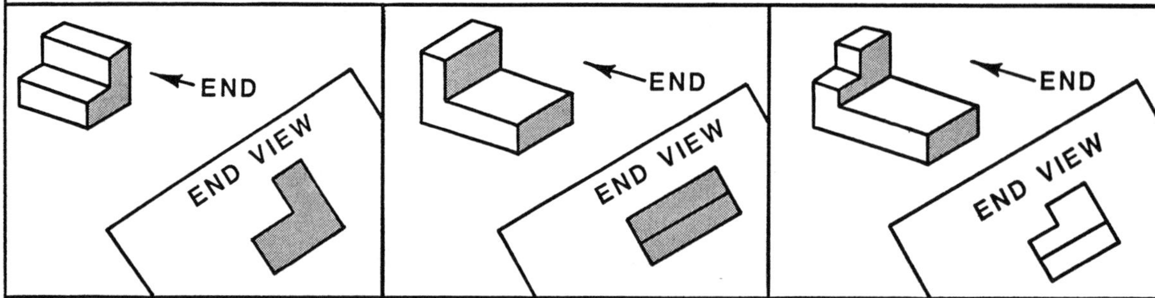

1. Walk in a quarter circle to the right until you are facing the end of the object.
2. Imagine that you cannot "see" any of the front surfaces.
3. Imagine that you cannot "see" any of the top surfaces.
4. "See" only the end surfaces.
5. Study the three drawings of the end views.

PRACTICE IN "SEEING" END VIEWS

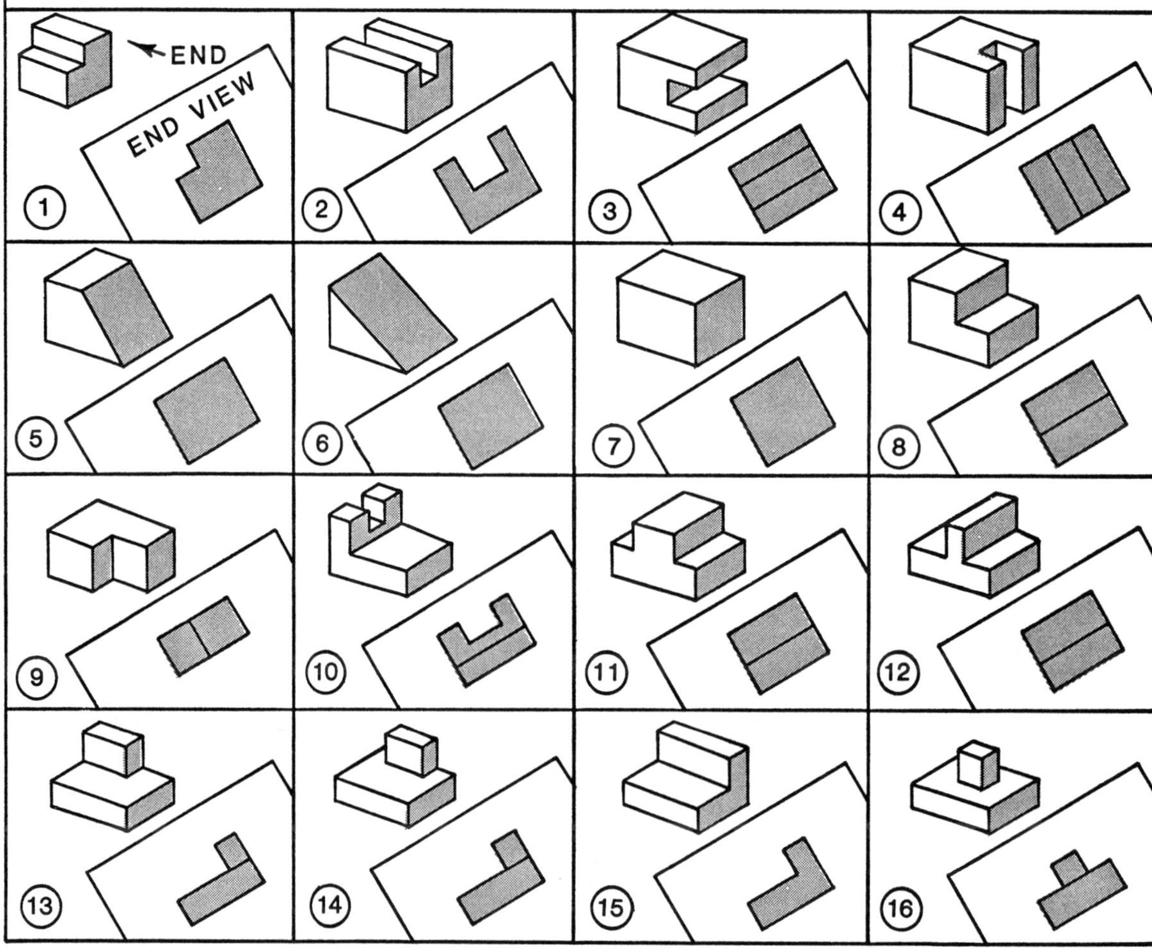

As you observe the end view of 5, 6, and 7, you realize that each square represents a different surface condition. When you understand why these views are drawn as they are, you realize that 5 represents a steep slope, 6 a gradual slope, and 7 a vertical surface.

DRAWING THE END VIEW

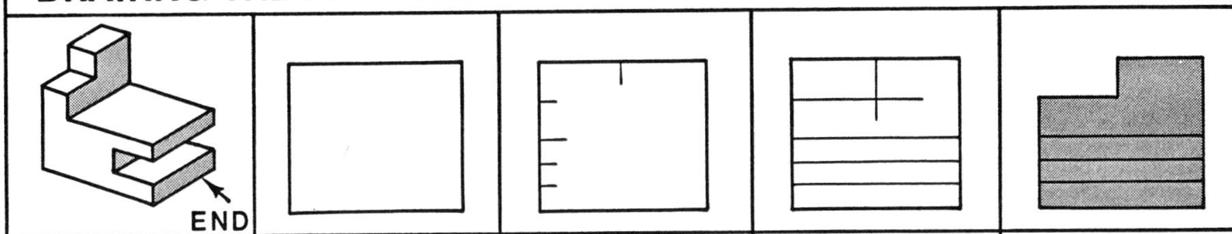

(1) Study the basic shape and the shape of each detail.	(2) Measure and lightly draw the basic shape.	(3) Measure and lightly lay out each of the details.	(4) Draw the lines lightly for the details.	(5) Darken the lines necessary for the end view.

PRACTICE IN VISUALIZATION

(1) Visualize the end view of this object.	(2) This is the end view. Visualize the two surfaces.	(3) "N" is near you. "F" is farther away.

ASSIGNMENT

TITLE: END VIEW

Draw the end view of one of these objects.

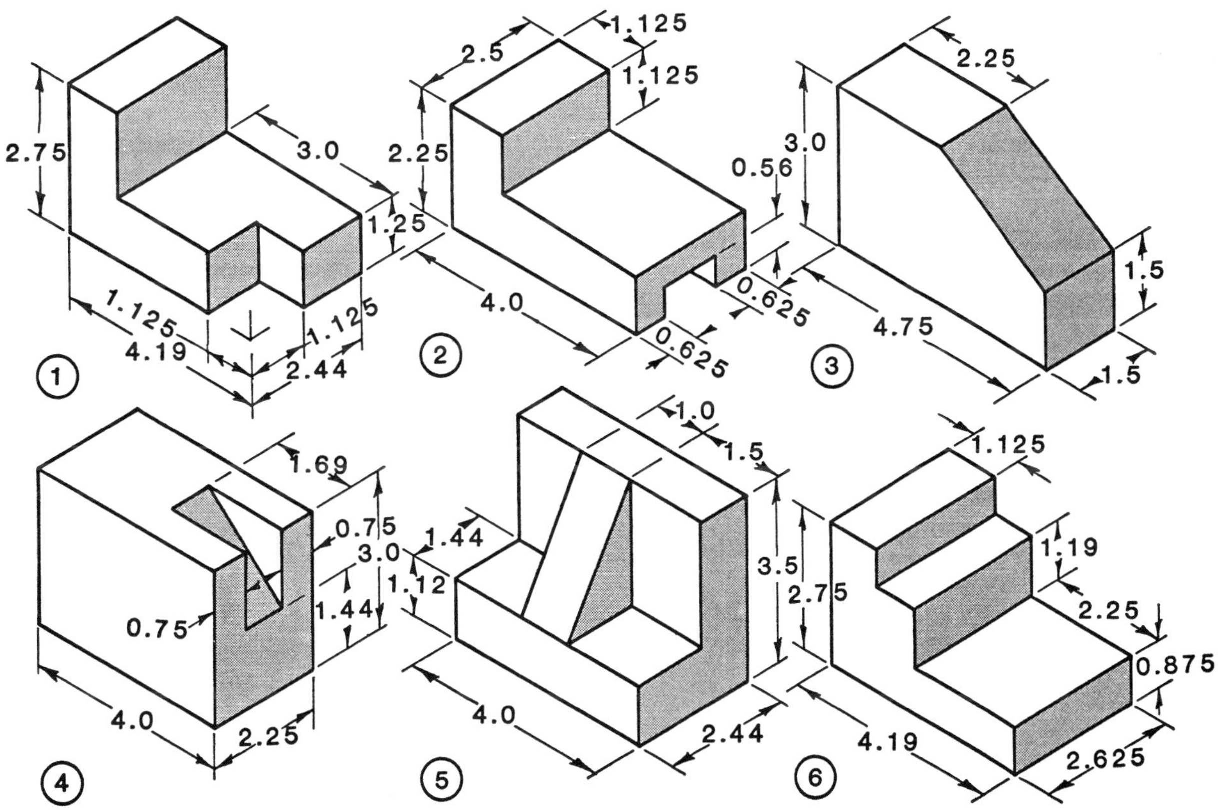

HOW TO "SEE" WITH AN "X-RAY EYE"

Imagine that you can look through a solid object in order to "see" a hidden edge.

In the front view of this object:

1. You can see the hole.
2. You can see its exact shape.
3. You can draw the shape of the hole on the front view.

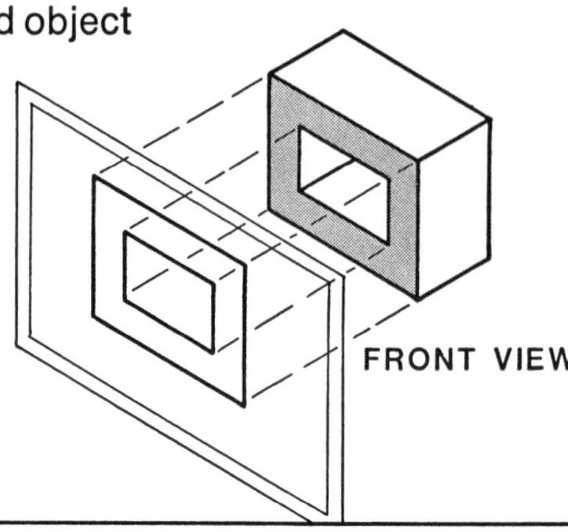

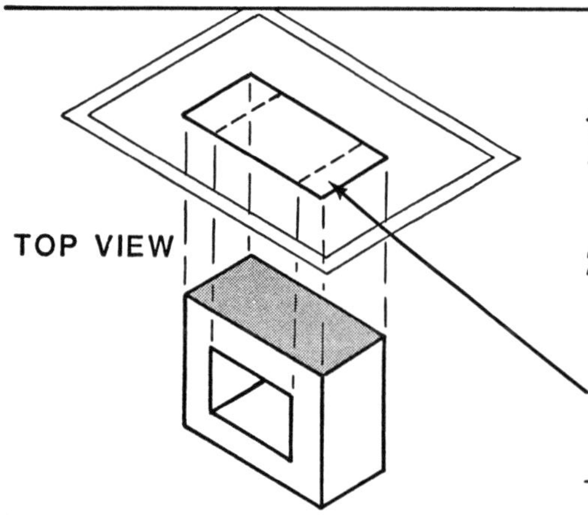

The hole is hidden when viewed from the top.

1. Try to see the hole as though you had an "X-ray eye."
2. Indicate the size and position of the hole by using two hidden edge lines.

This is a sample of a hidden line.

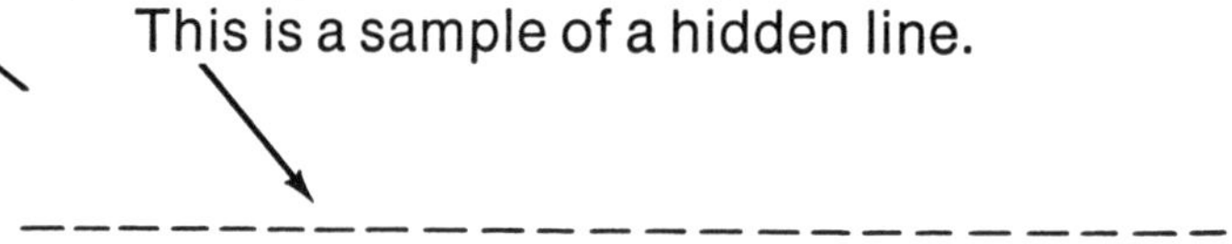

The hole is also hidden when viewed from the end-view position.

1. Try to "see" the hole as though you had an "X-ray eye."
2. Indicate the size and position of the hole in the end view by using two hidden edge lines.

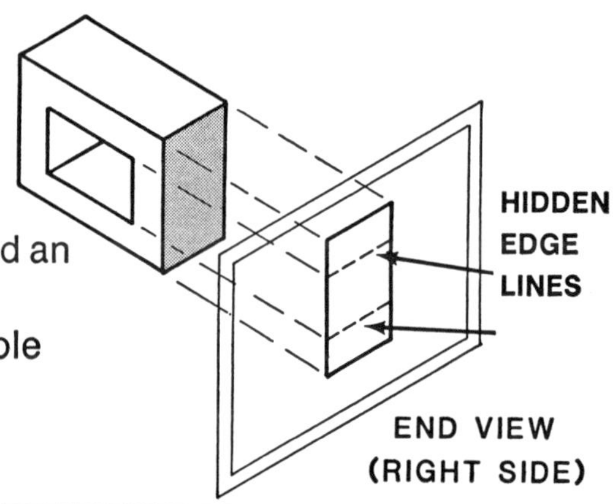

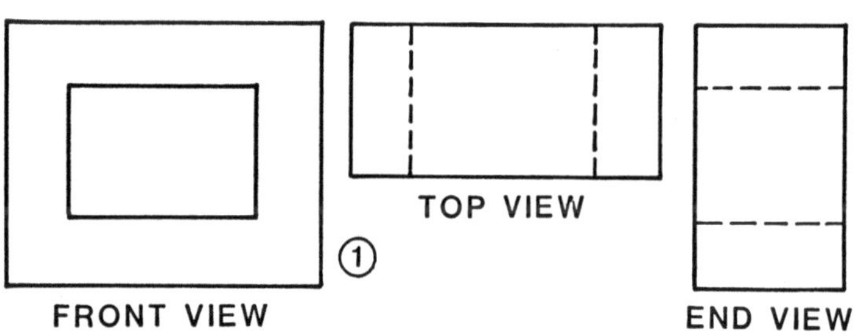

The three separate views as drawn above.

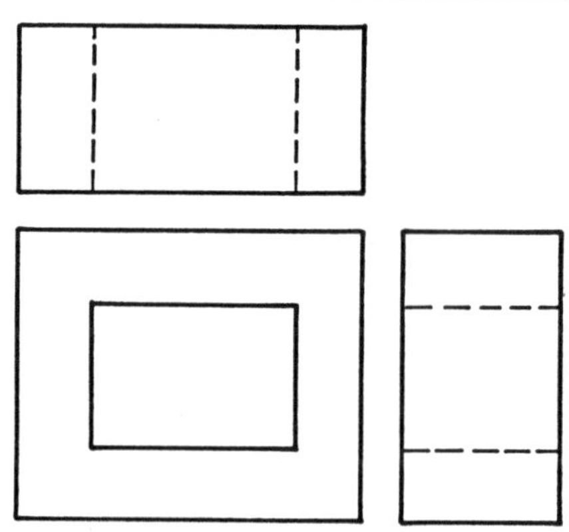

②ORTHOGRAPHIC PROJECTION

The hidden edges, showing the size and location of the hole in the top and end views, can be drawn by orthographic projection.

A hole may be projected from one view to another just as though it were a solid block. A hole must appear **in all three views** whether it is visible or hidden.

A STUDY OF THE CYLINDER

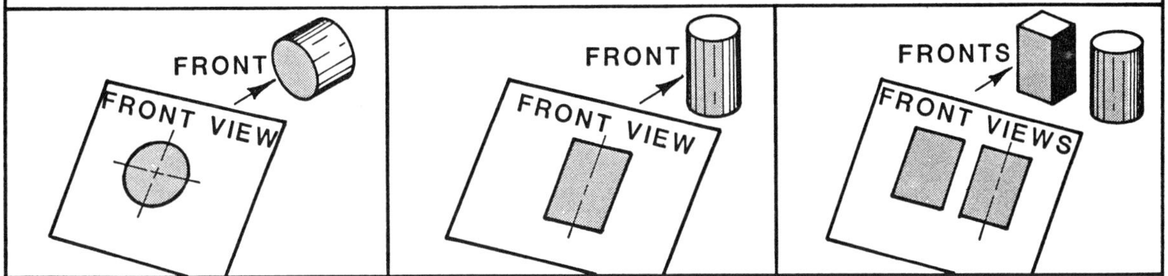

(1) The front view of a cylinder in this position must be drawn as a perfect circle.

(2) In this position, the top edge and the bottom edge appear as straight, horizontal lines.

(3) The center line must be used in order to show which of the two drawings represents the cylindrical object.

A STUDY OF THE CENTER LINE

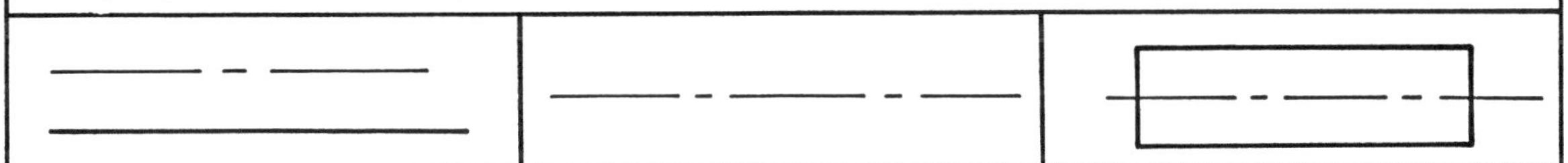

(1) The center line is only a symbol. It is drawn one-fourth as thick as an object line.

(2) Study the spacing used to create a center line. Always draw thin center lines.

(3) A center line must never appear to be part of the picture. It is only a symbol.

FACTS ABOUT CENTER LINES

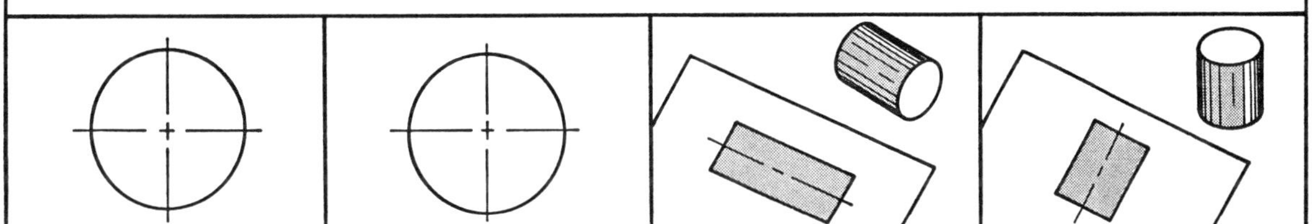

(1) A circle must have a horizontal and a vertical center line.

(2) Center lines extend only a short distance beyond the circle.

(3) A center line forms the axis of a cylinder.

(4) Center lines extend only a short distance beyond the end of the object.

DRAWING A CENTER LINE

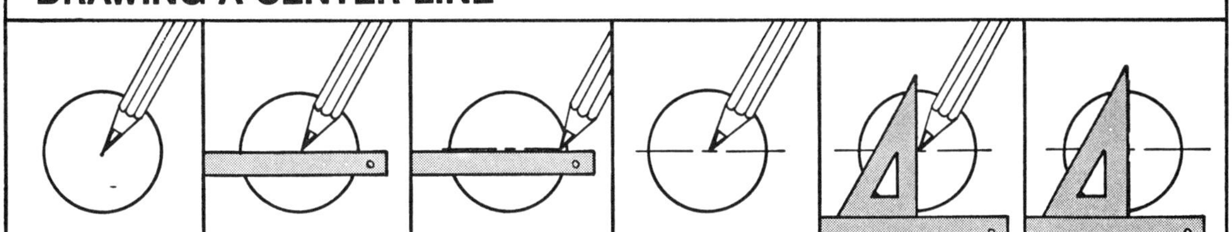

(1) Place the pencil on the center of the circle.

(2) Move the T-square up to the pencil.

(3) Draw the horizontal center line.

(4) Place the pencil on the center of the circle.

(5) Make the triangle touch the pencil.

(6) Draw the vertical center line.

MORE PRACTICE DRAWING CYLINDERS

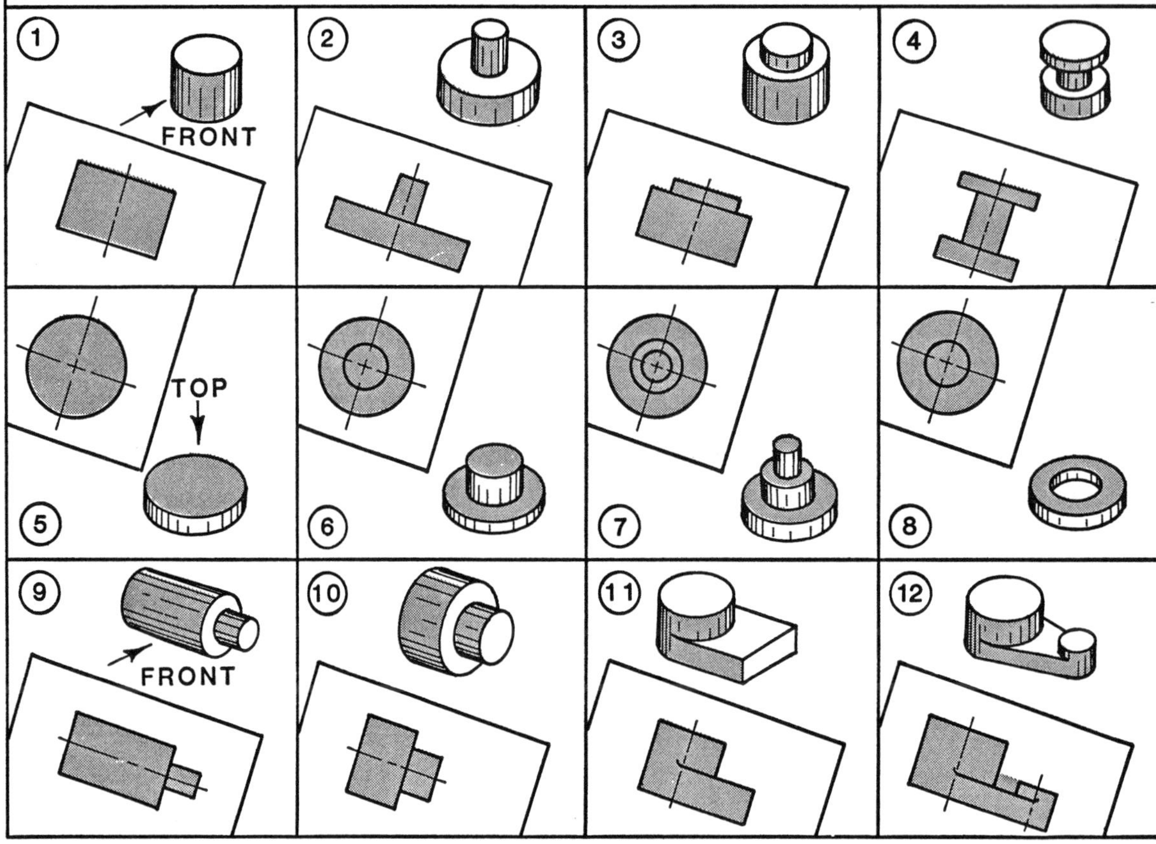

ASSIGNMENT

Make a drawing of one of these objects. Draw the view that best describes its shape. Dimensions should be in millimeters (mm).

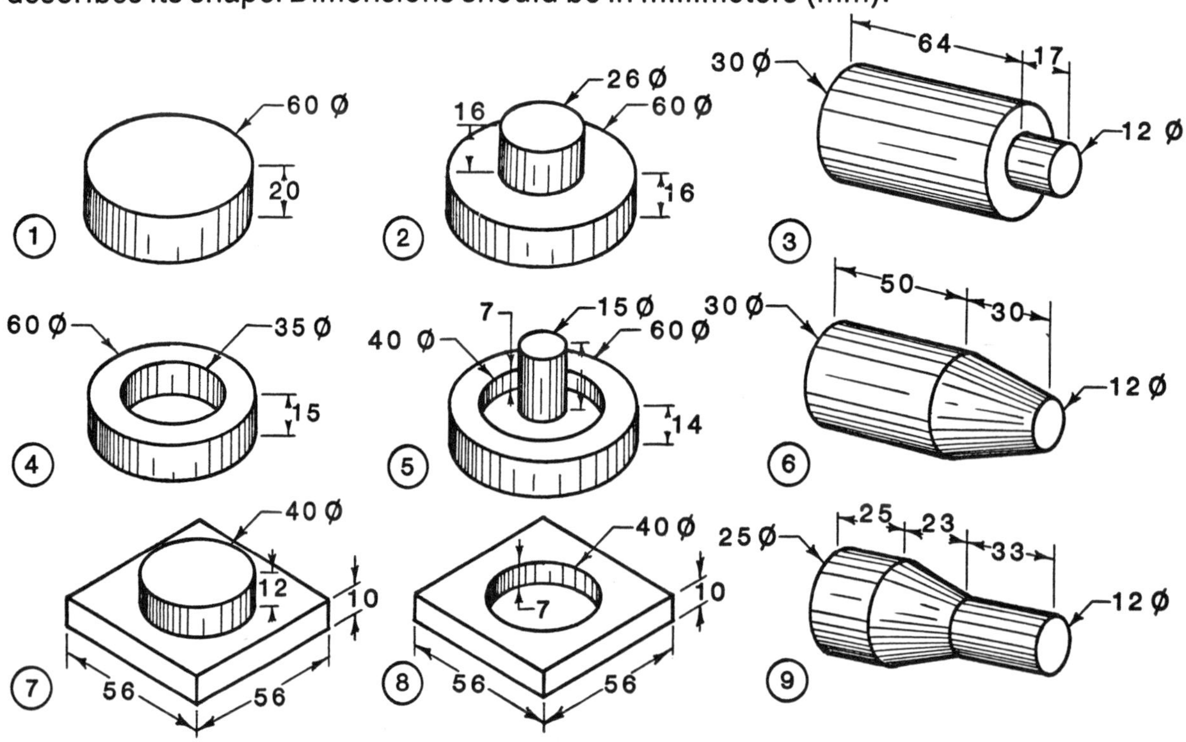

THE BASIC IDEA BEHIND ORTHOGRAPHIC PROJECTION

When the front surface is projected at right angles, the resulting drawing is called an orthographic projection.

When the drafter projects from view to view, the three views are called an orthographic projection.

COMBINING THREE VIEWS INTO AN ORTHOGRAPHIC PROJECTION

(1)

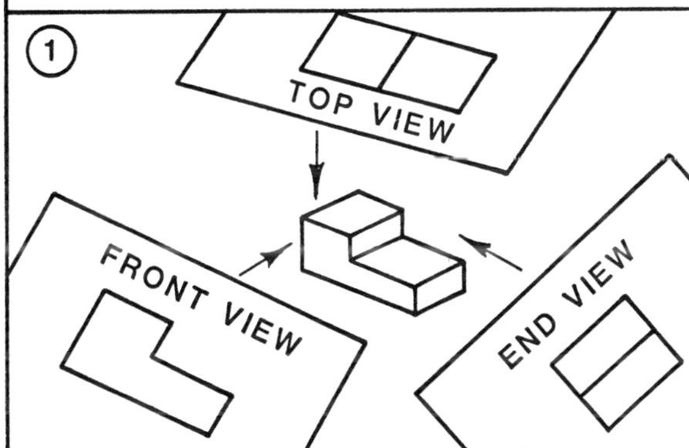

(2)

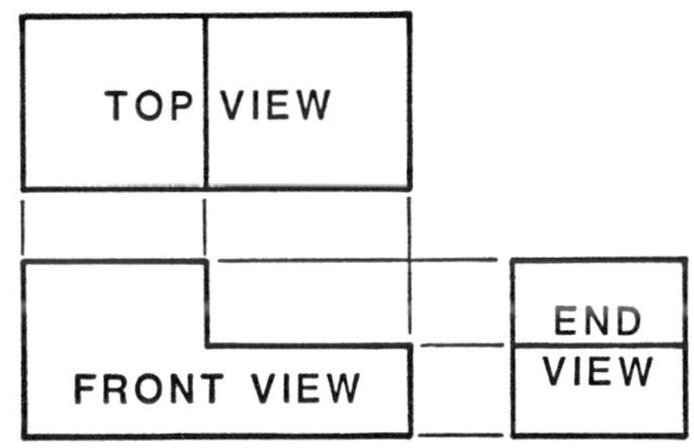

The front view, top view, and end view shown in (1) above are combined in order to create the three-view orthographic projection at (2).

THE MEANING OF THE TERM "ORTHOGRAPHIC"

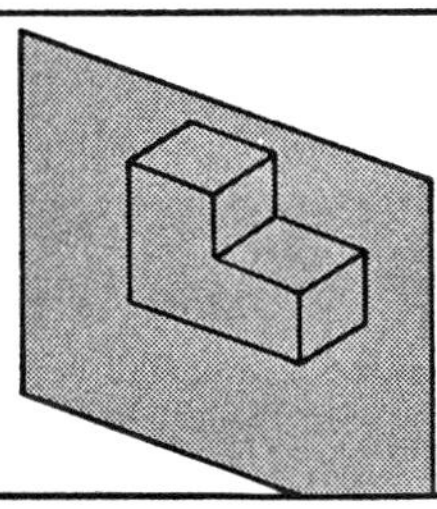

(1) An accurate front view may be obtained by tracing the front surface of this object.

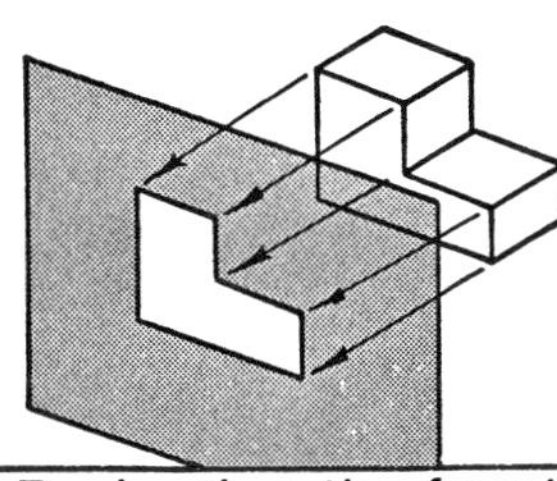

(2) Projecting the front surface of this object at right angles will result in an orthographic projection.

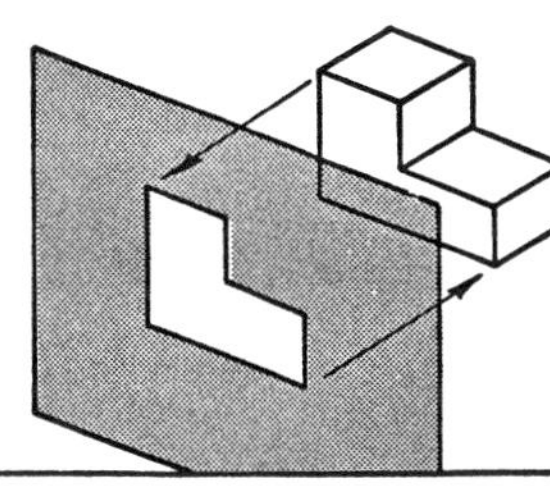

(3) Measuring and drawing the front view will result in an orthographic projection.

THE MEANING OF THE TERM "PROJECTION"

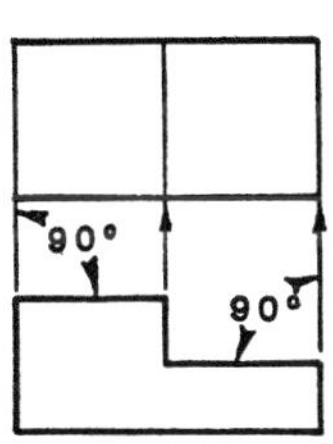

(1) The edges may be projected at 90° in order to draw the top view.

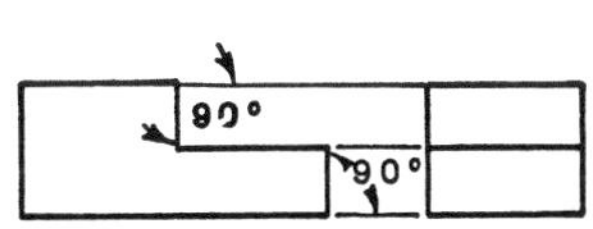

(2) The edges may be projected at 90° in order to draw the end view.

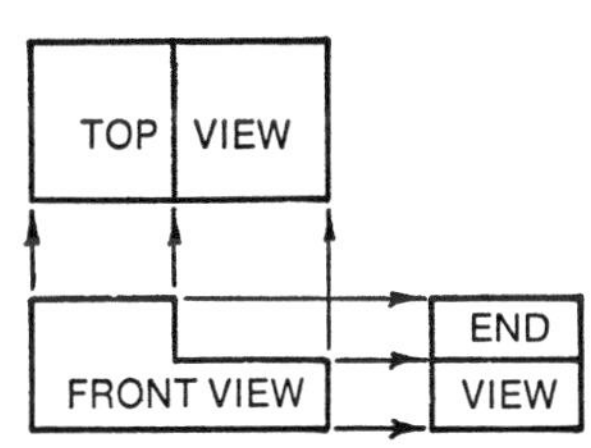

(3) Projecting at 90° upward or across produces an orthographic projection.

ORTHOGRAPHIC PROJECTION

This is a very important lesson in mechanical drawing.

(1)

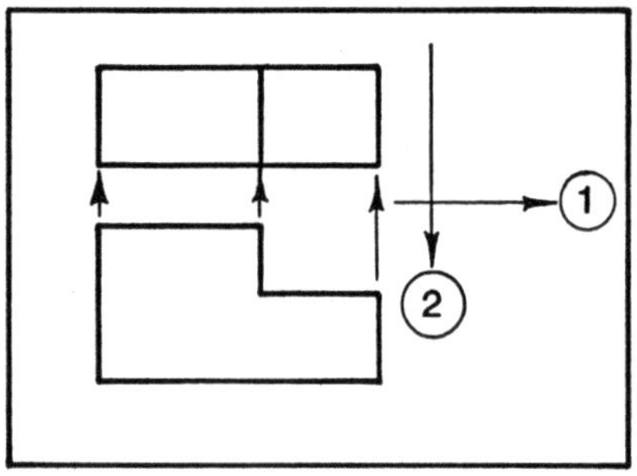

A. Draw the front view.
B. Project upward from the front view.
C. Measure and draw the top view. (Leave a 1" space between the front view and the top view.)
D. Draw a light horizontal line (1) about 1/2" below the top view. Extend this line to the right as shown.
E. Draw a light vertical line (2) about 1/2" to the right of the top view as shown.

(2)

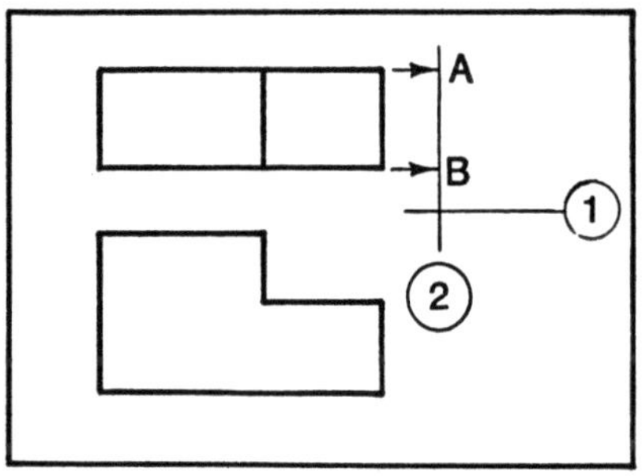

A. Extend the top horizontal edge of the top view until it touches line (2) at point "A."
B. Extend the bottom horizontal edge of the top view until it touches line (2) at point "B."

Basic Rule: Extend every horizontal line in the top view until it touches line (2).

(3)

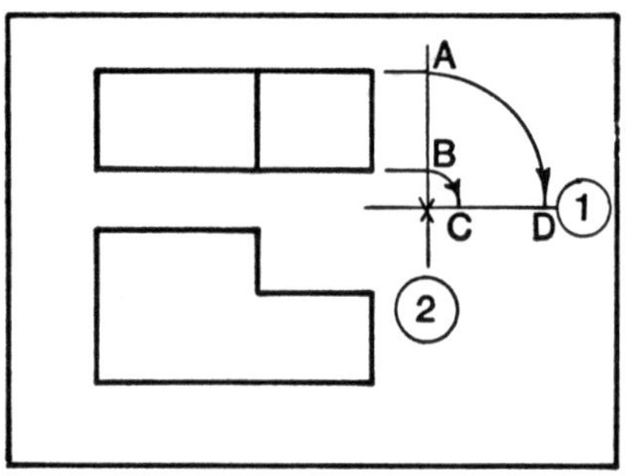

A. Put your compass needle at the point where lines (1) and (2) cross (marked "X").
B. Open the compass from "X" to "B."
C. Swing arc "BC" until it touches line (1) at "C."
D. Open the compass from "X" to "A." Swing "AD."

Basic Rule: Swing every line that touches line (2) downward until it touches line (1). (Always use "X" as the center.)

(4)

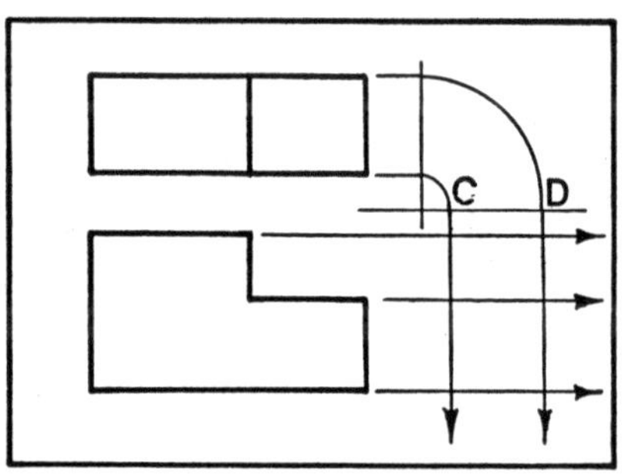

A. Drop a vertical line downward from "C."
B. Drop a vertical line downward from "D."
C. Extend the three horizontal edges of the front view toward the right as shown. (You can now see the end view.)

Basic Rule: Drop every line that touches line (1) vertically in order to help make the end view.

(5)

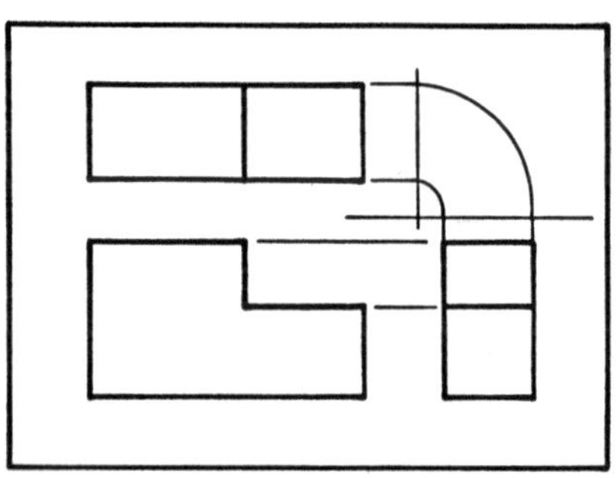

DARKEN THE END VIEW

This is the method used by drafters when they wish to make an orthographic projection.

Learn this method . . . thoroughly!

In a short time, this method of projection will become automatic with you.

TWO OTHER WAYS OF "MAKING THE TURN" FROM THE TOP VIEW

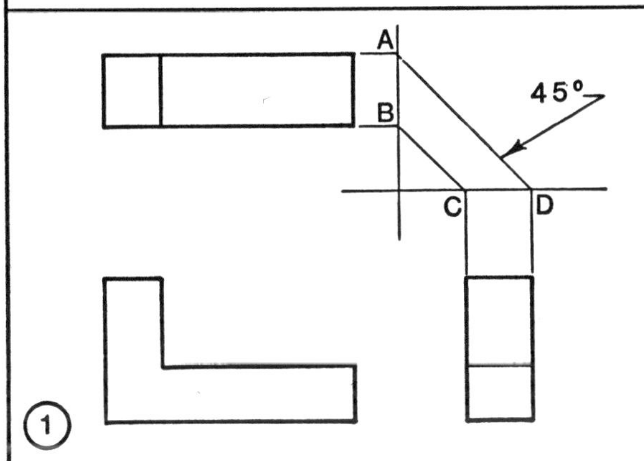

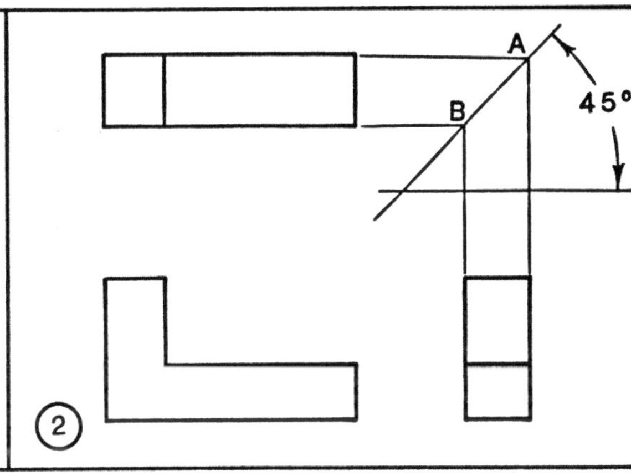

A 45° angle may be used in "making the turn" from the top view to the end view. The result will be the same as that obtained with the compass. Study methods (1) and (2) as shown above. You now have three ways of "making the turn" from the top view in an orthographic projection.

ASSIGNMENT

Make a three-view orthographic projection of one of these objects. (Leave a 1″ space between views.) The end views are shaded.

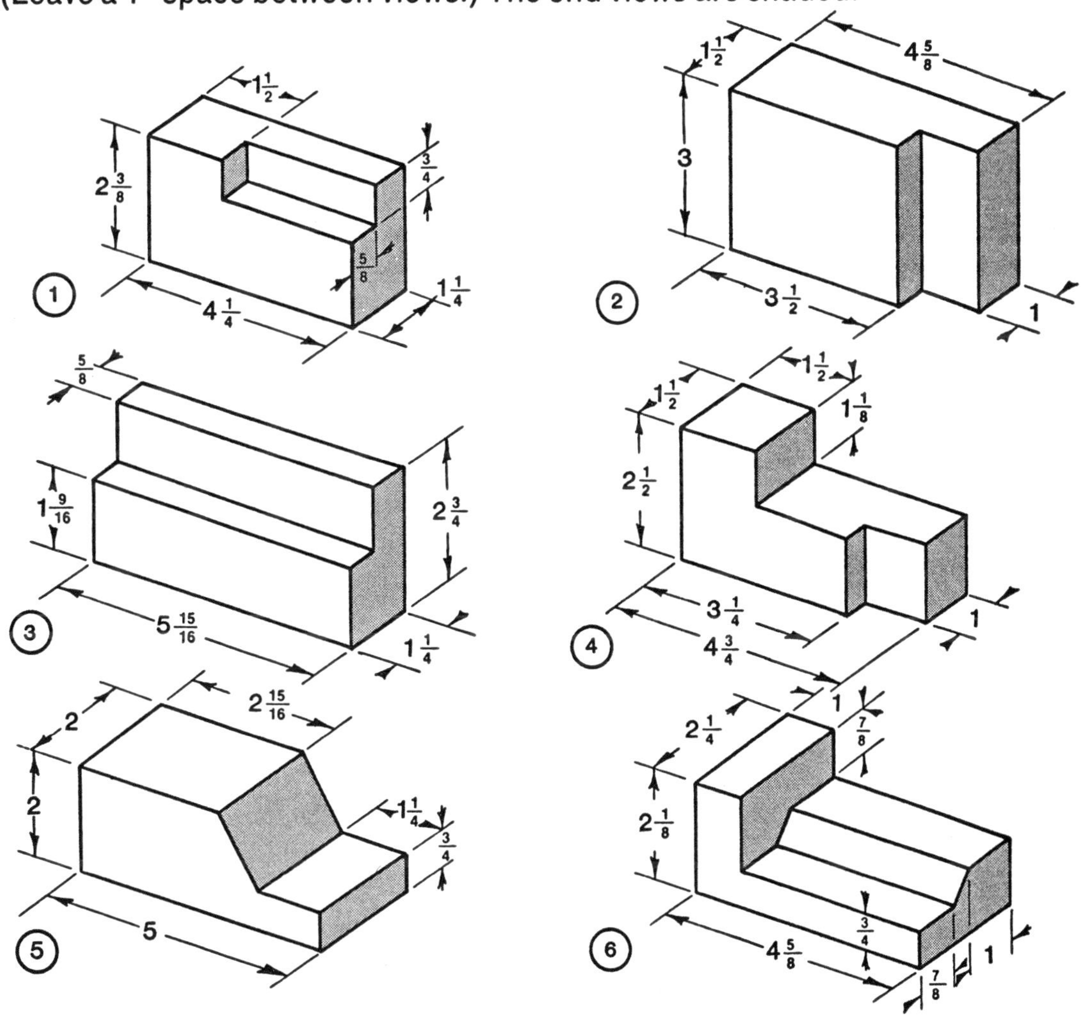

BE SYSTEMATIC IN DRAWING

Establish a definite way of drawing an orthographic projection. Use the same method unless there is a good reason for changing. Note that the side view is shaded on the pictorial drawings.

A BASIC METHOD FOR MAKING ORTHOGRAPHIC PROJECTIONS

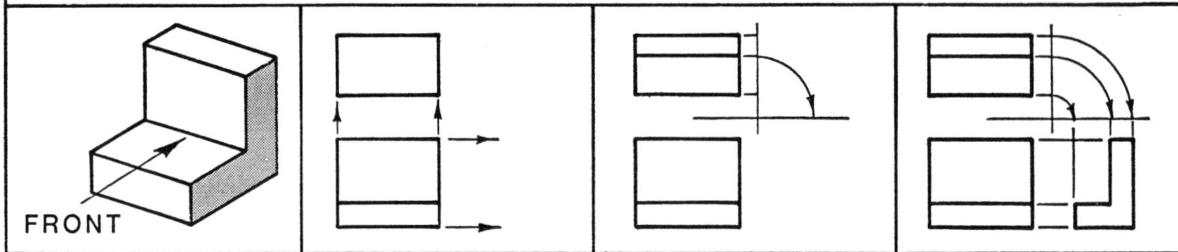

① Study the basic shape of the front view.	② Draw the front view and then the top view.	③ Make the "turn." Use the compass or the 45° triangle.	④ Complete the end view.

STUDY THESE ORTHOGRAPHIC PROJECTIONS

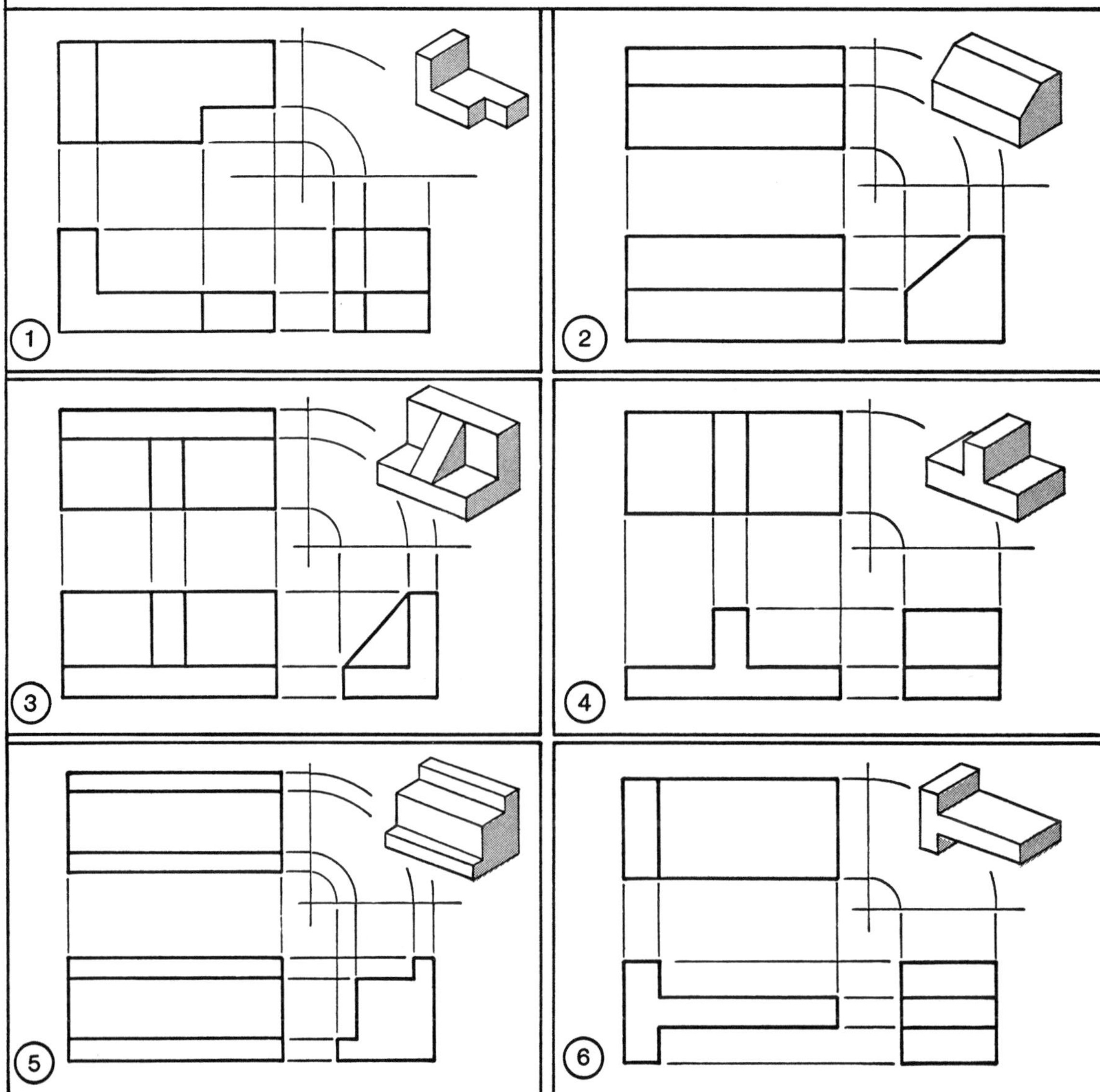

CREATING A 1″ SPACE BETWEEN VIEWS

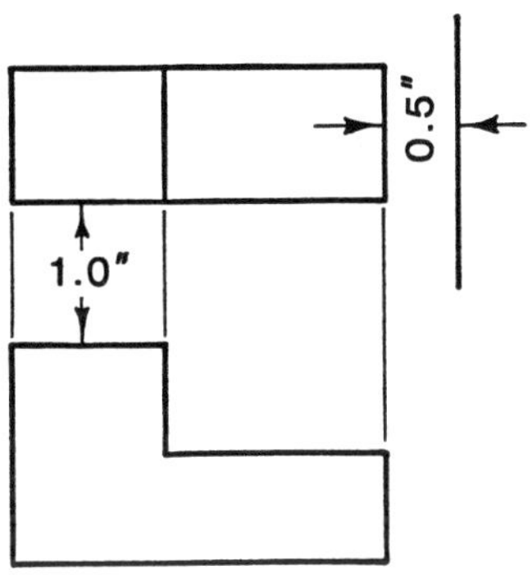

1

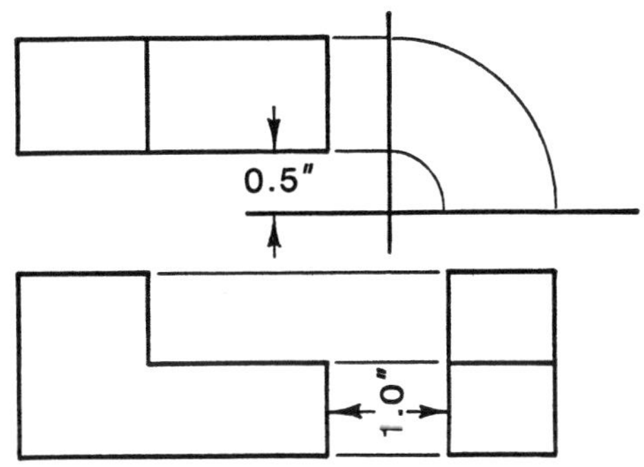

2

(1) Leave a 1″ space between the front view and the top view.

To obtain a 1″ space between the front view and the end view, measure one half of that distance (1/2″) to the right of the top view. Draw the vertical line used in "making the turn."

(2) Draw the horizontal line used in "making the turn," 1/2″ below the top view. Finish "turning" the top view. There will be a 1″ space between the front view and the end view.

ASSIGNMENT

TITLE: ORTHOGRAPHIC PROJECTION

Make a three-view orthographic projection of one of these objects. Note that the side view is shaded on the pictorial drawing.

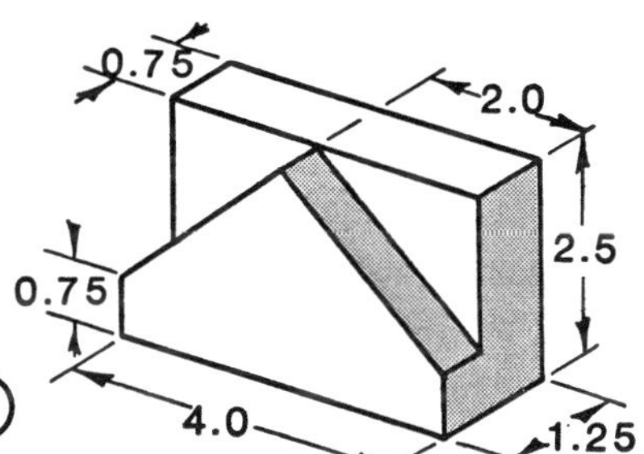

1

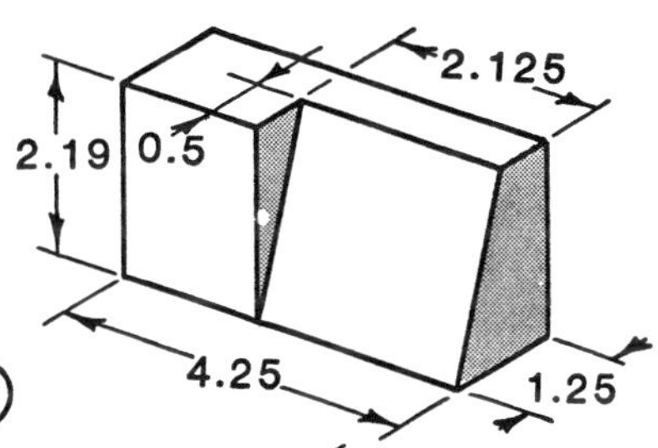

2

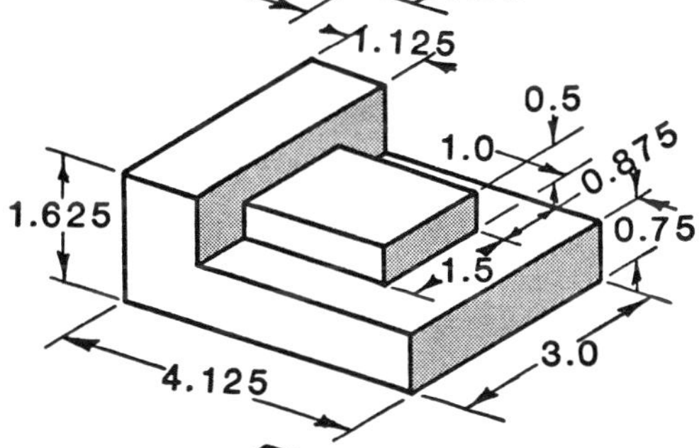

3

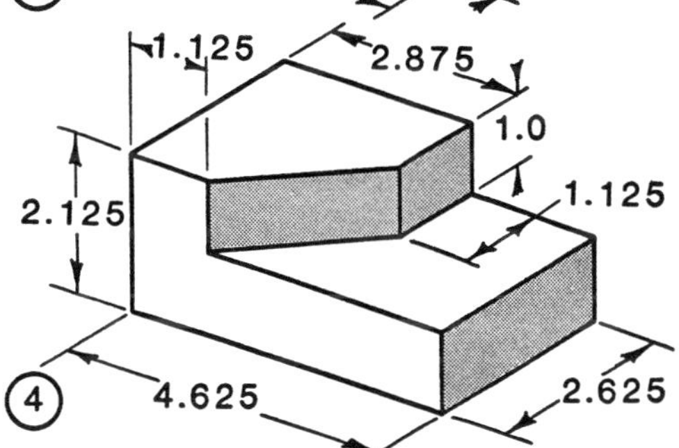

4

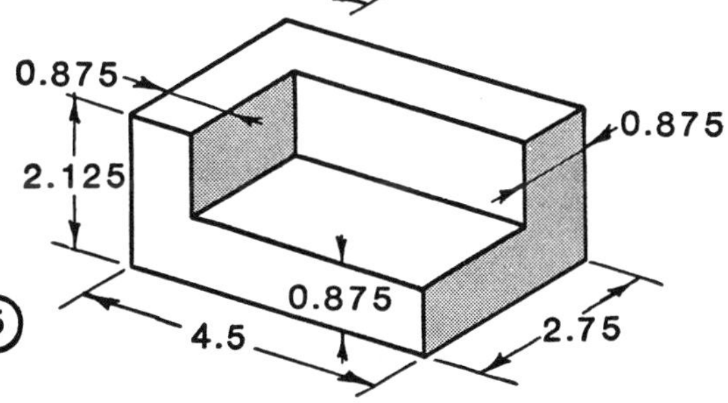

5

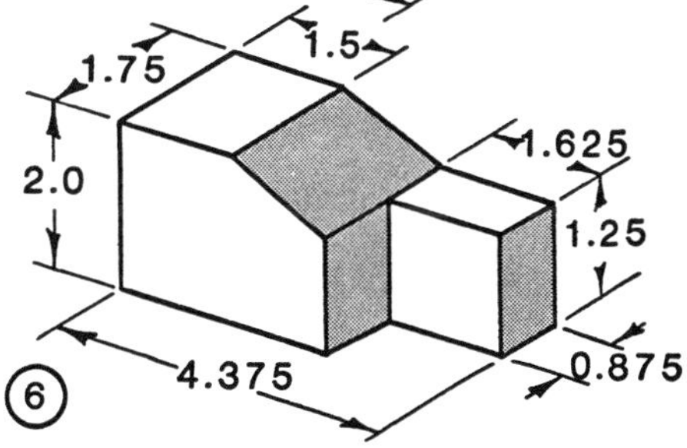

6

HOW TO CENTER A DRAWING

Your drawing should be centered to look balanced on the paper.

WHAT IS MEANT BY CENTERING

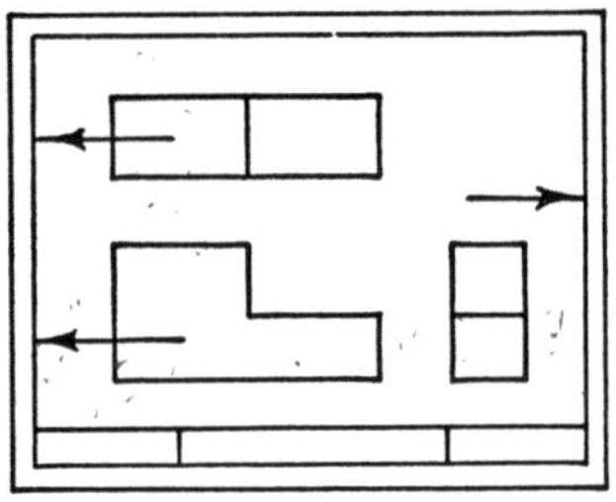

(1) Every drawing must be centered in relation to the side margins of the drawing area.

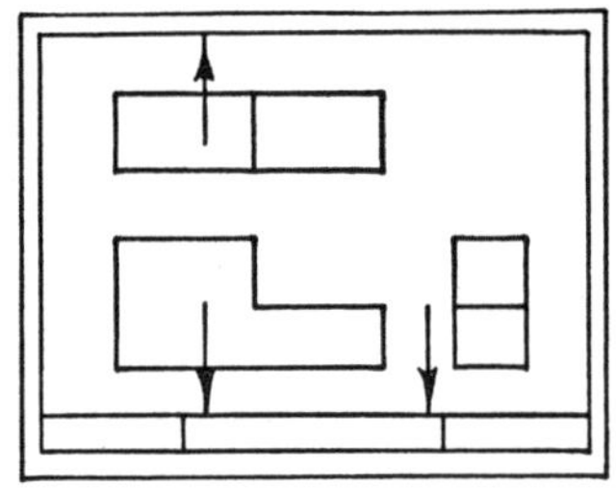

(2) Every drawing must be centered in relation to the top and bottom margins of the drawing area.

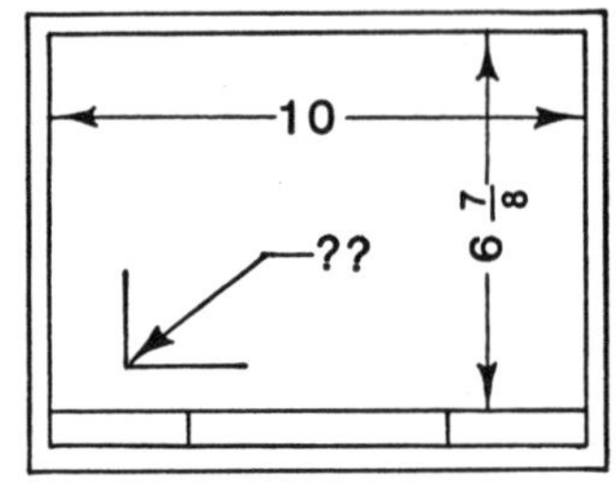

(3) The starting point is usually in the vicinity of the indicated point, but the exact location must be found.

VISUALIZE THE THREE VIEWS AND THE DIMENSIONS

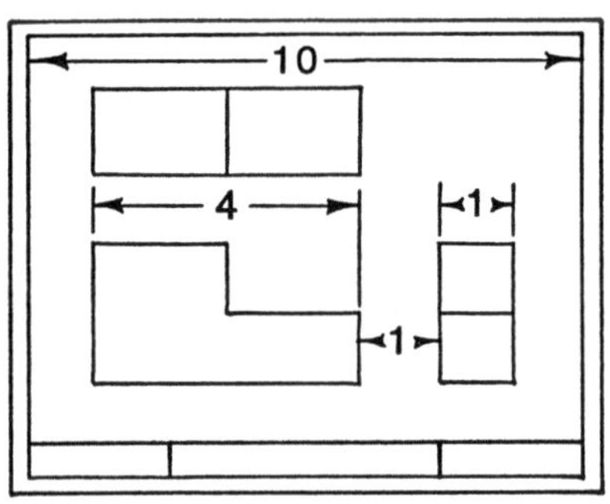

(1) When you visualize the three views, you will get an accurate idea of the space you will need (4″ + 1″ space + 1″ width = 6″ horizontal).

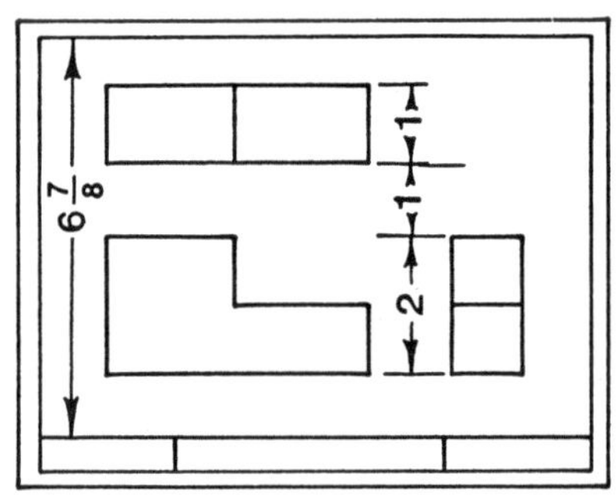

(2) Consider the height of the front view, and the width of the top view (2″ + 1″ space + 1″ width = 4″ vertical).

LOCATING THE STARTING POINT

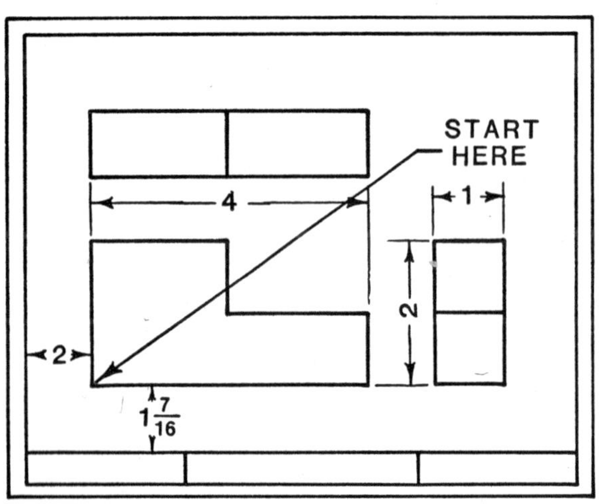

(1) The length of the finished drawing will be 6″ (4″ + 1″ + 1″).

(2) The length of the drawing area is 10″.

(3) The side margins will be 2″ (10″ - 6″ = 4″; 4″ ÷ 2 = 2″).

(4) The height of the finished drawing will be 4″ (2″ + 1″ + 1″).

(5) The width of the drawing area is 6⅞″.

(6) The top and bottom margins will be 1 7/16″ (6⅞″ - 4″ = 2⅞″; 2⅞″ ÷ = 1 7/16″).

DETERMINING THE "OVERALL" OR BASIC SHAPE

Always draw the "overall" or basic shape of the three views before attempting to draw any of the details.

DRAWING THE BASIC SHAPE

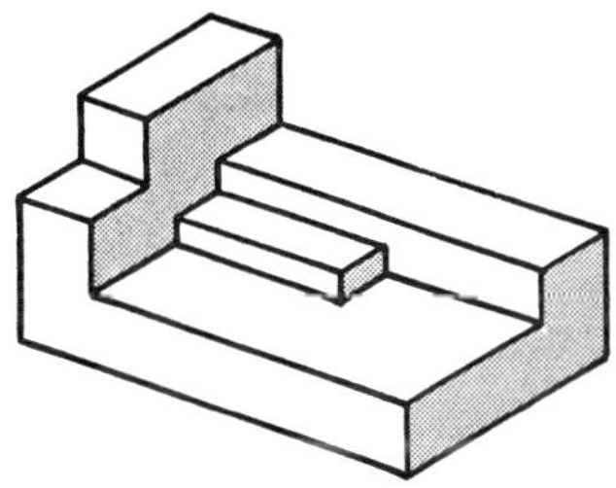

(1) Regardless of how complicated an object may appear, it has a basic shape. Always study an object before drawing it.

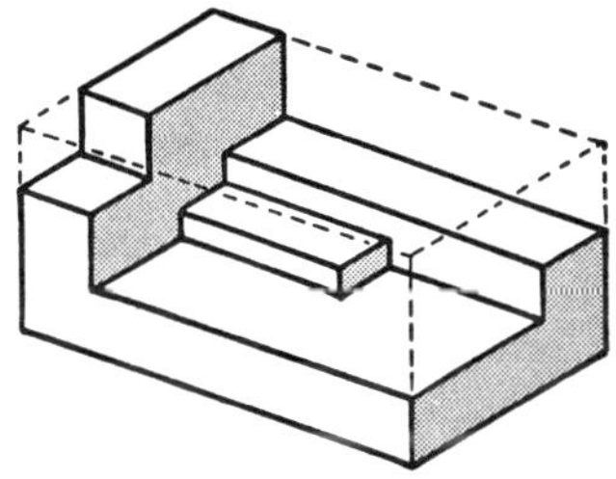

(2) The basic shape is often determined by the full length, the full width, and the full height.

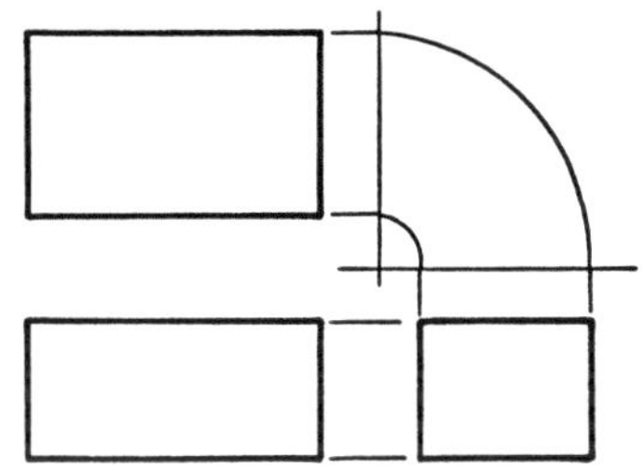

(3) Draw the basic shape of each of the three views before attempting to draw any of the details.

STUDY THE BASIC SHAPES OF THESE OBJECTS

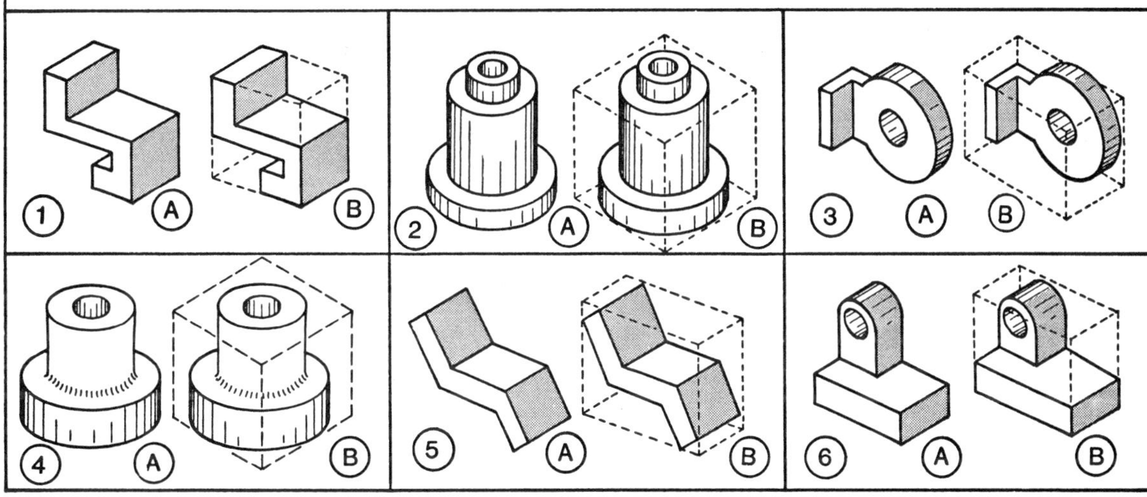

ASSIGNMENT

Make a three-view orthographic projection of one of these objects.

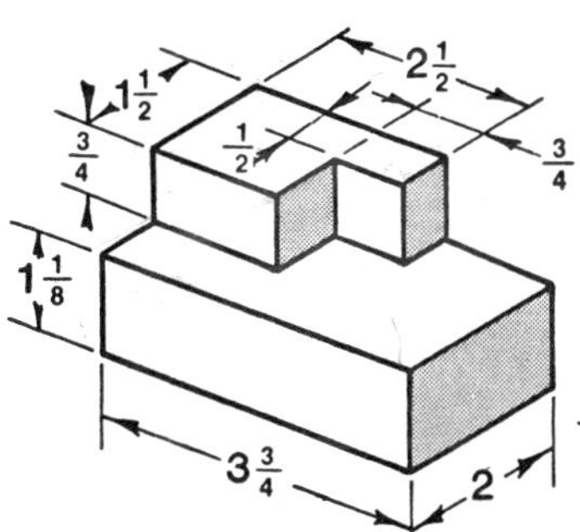

(1) GUIDE BLOCK

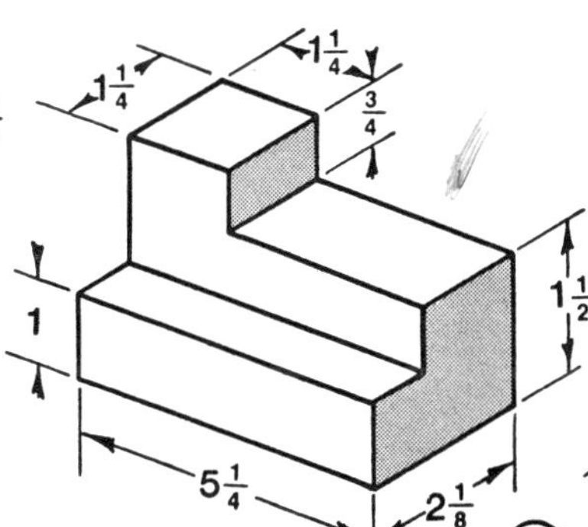

(2) STEPPED BLOCK

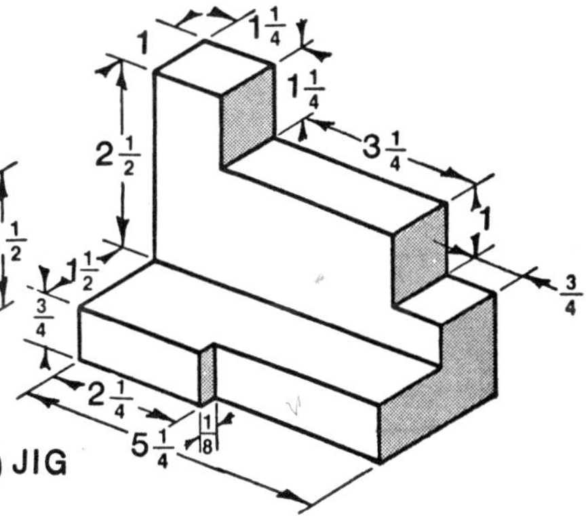

(3) JIG

A STUDY OF HIDDEN EDGES

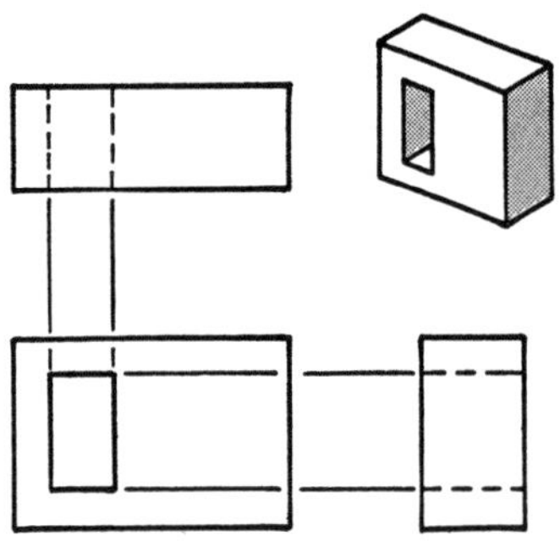

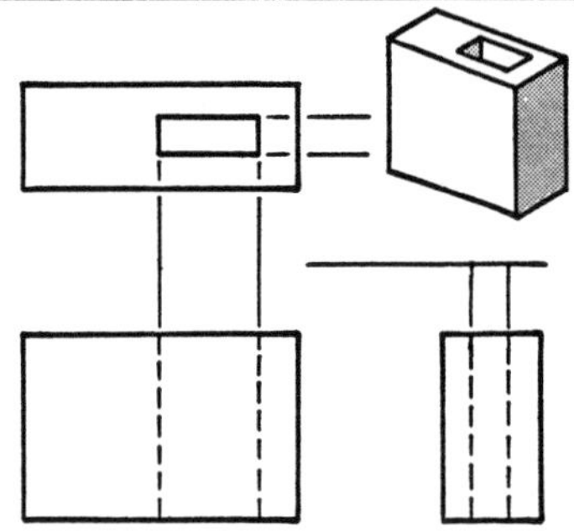

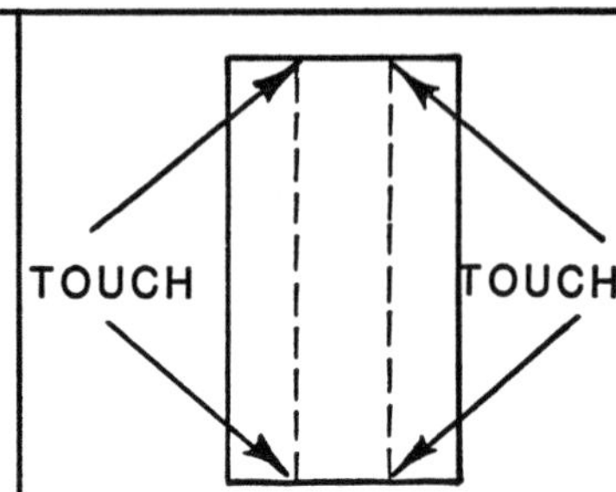

① A hole must be projected to the other two views.

② A hole must appear in each of the three views.

③ The hidden edge lines must touch the object lines.

STUDY THESE ILLUSTRATIONS OF ROUND HOLES

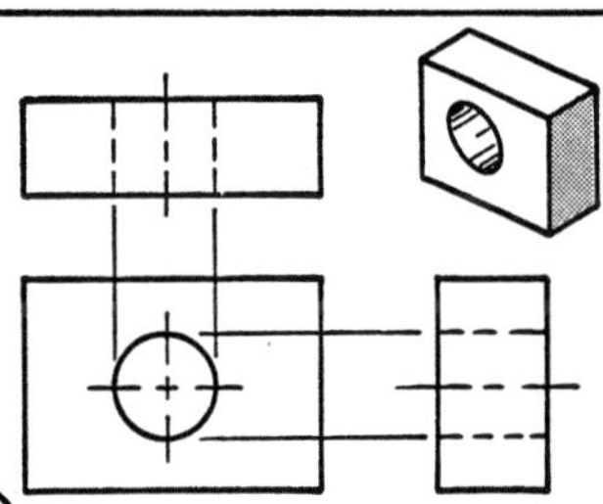
①

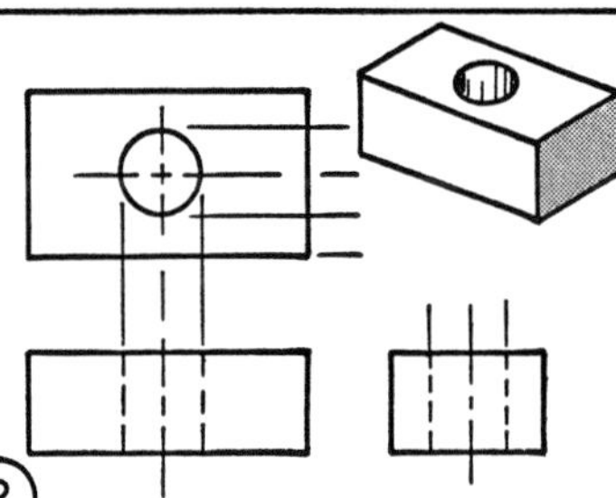
②

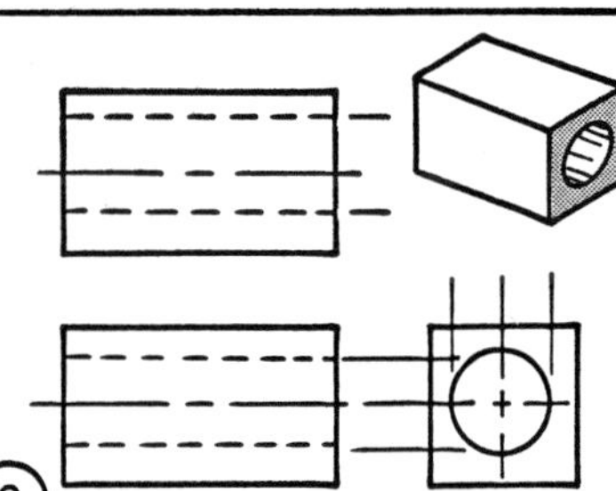
③

Study ①, ②, and ③ above. Learn and remember the following points:

Always draw the circular view of a circular hole first; then, project to the other views. (Always project the center line of a circular hole to the other views.)

Cylindrical (circular) holes are projected in the same manner as rectangular holes, with this exception: The center line must always be projected on circular holes.

ASSIGNMENT

Make a three-view orthographic drawing.

TITLE: ORTHOGRAPHIC PROJECTION

①
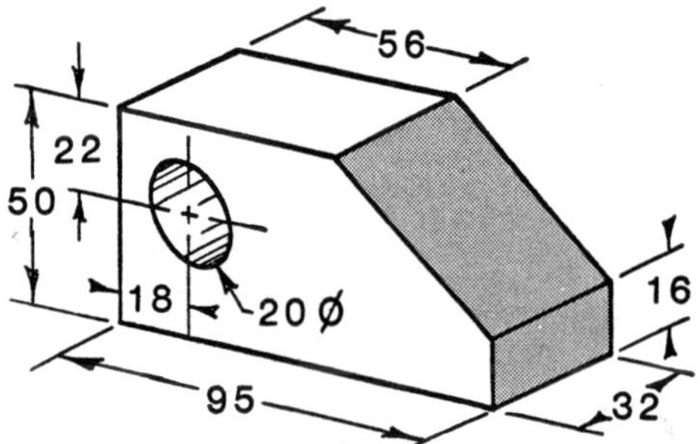

②
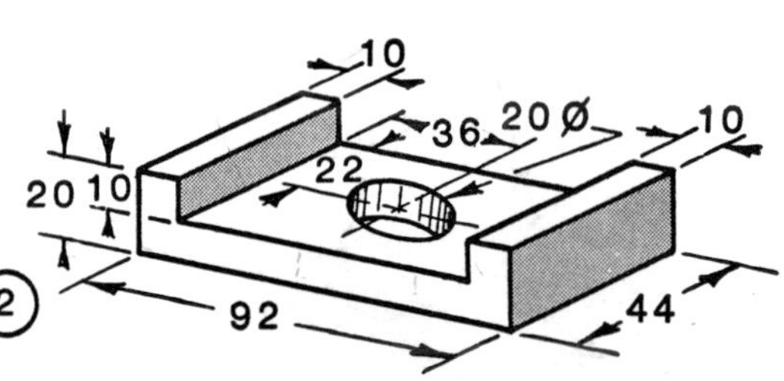

③
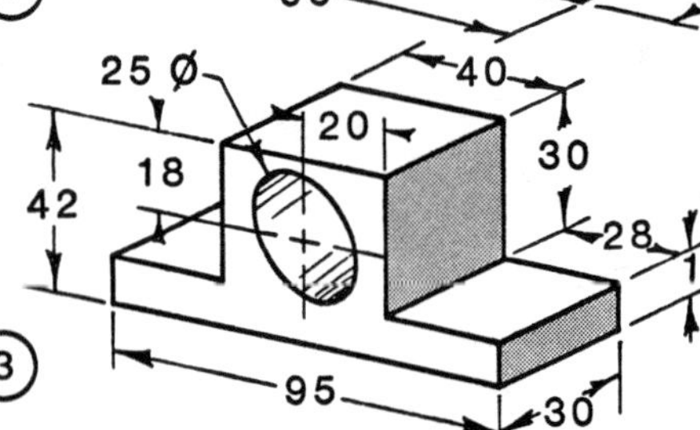

④
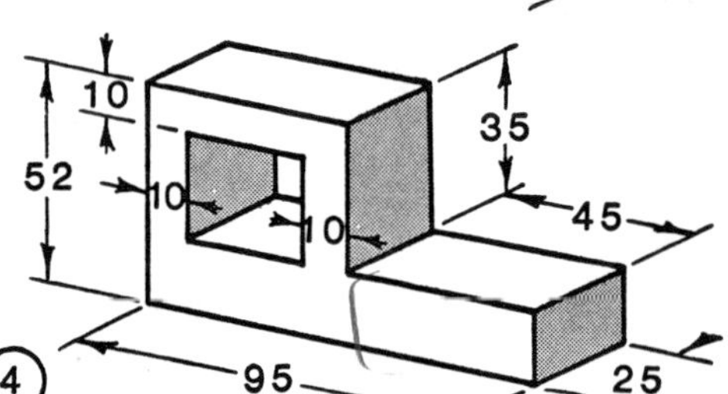

STUDY THE HIDDEN EDGES

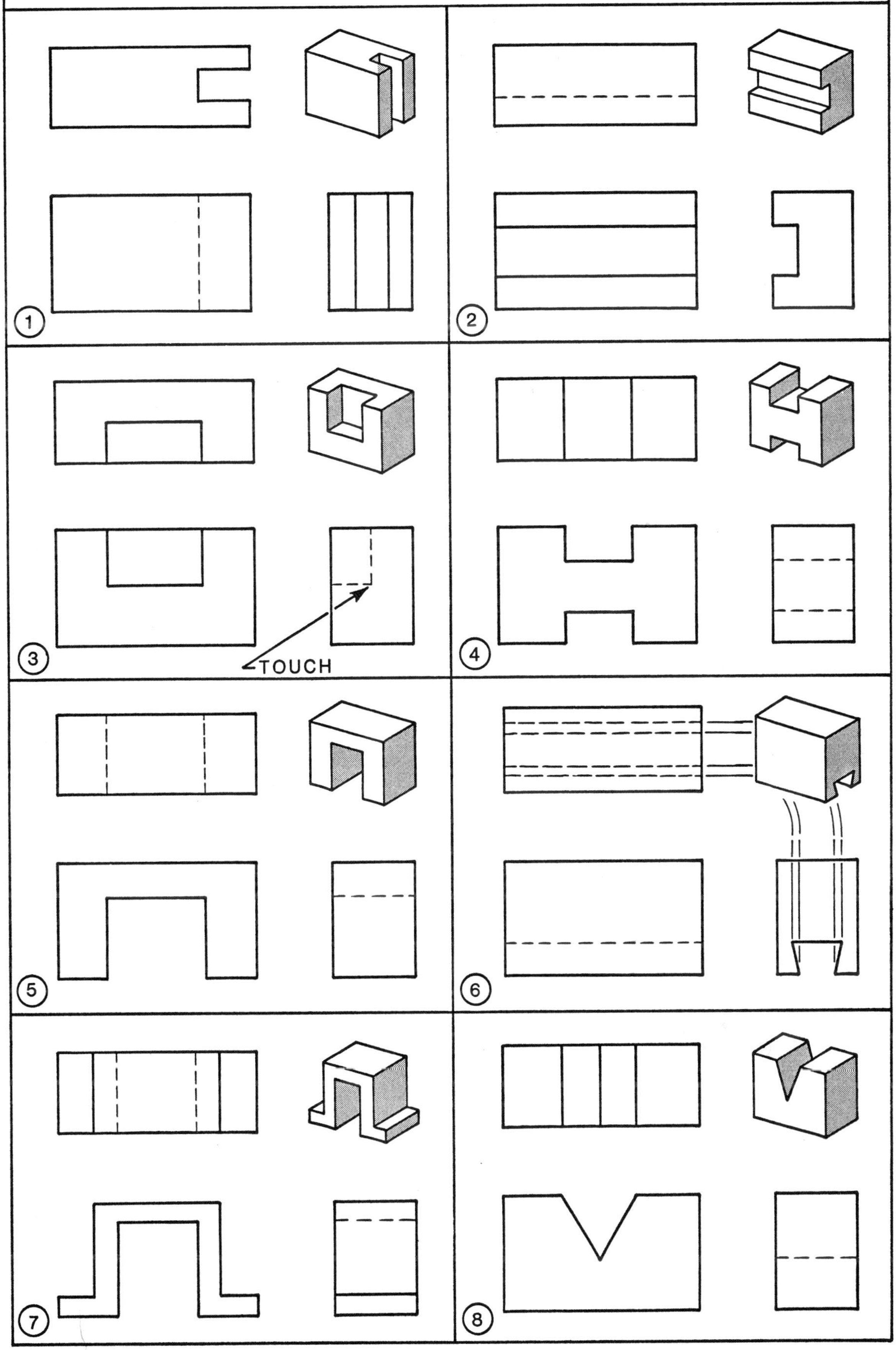

AVOID HIDDEN EDGES IF POSSIBLE

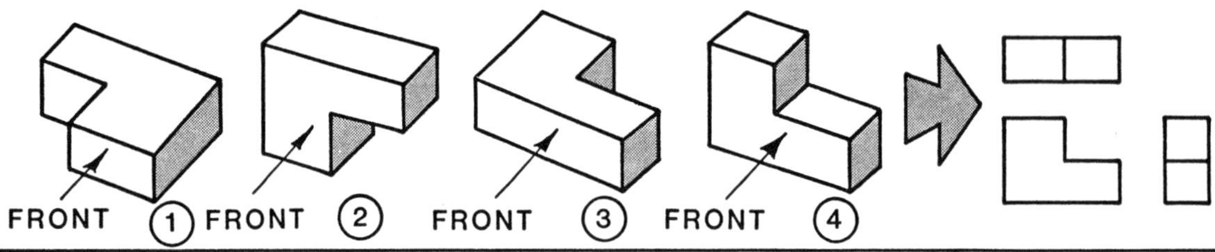

Do not hesitate to revolve an object in an effort to avoid hidden edges. This object has been revolved four times. Only the fourth position will result in an orthographic projection that has no hidden edges.

IN WHICH VIEW OR VIEWS WOULD THE HIDDEN EDGE OCCUR?

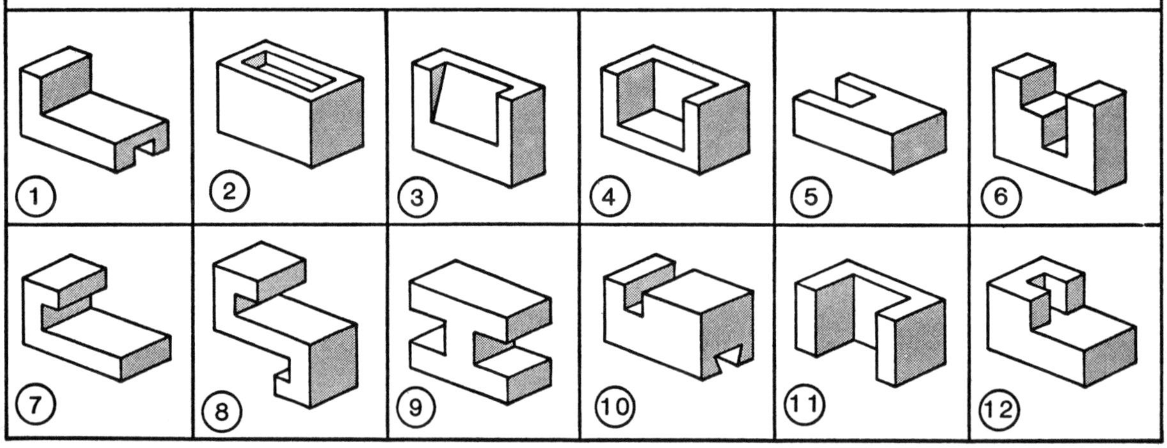

ASSIGNMENT

Draw a three-view orthographic projection.

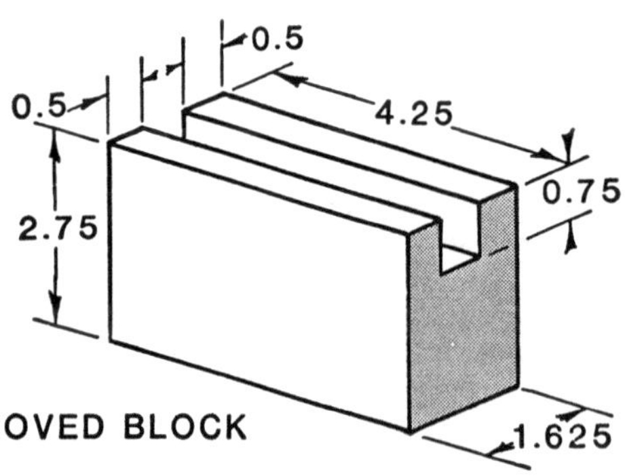

(1) GROOVED BLOCK

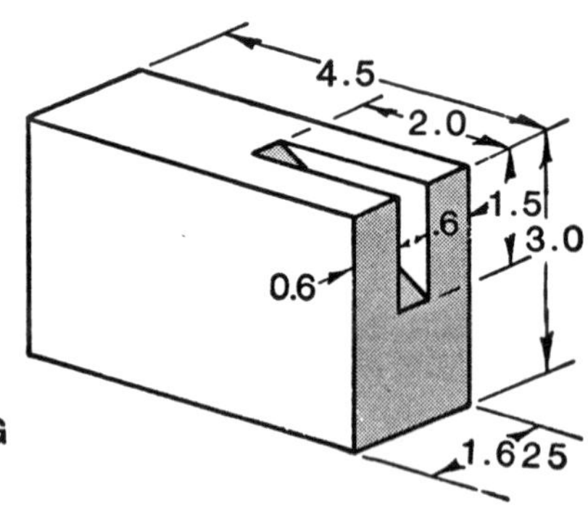

(2) JIG

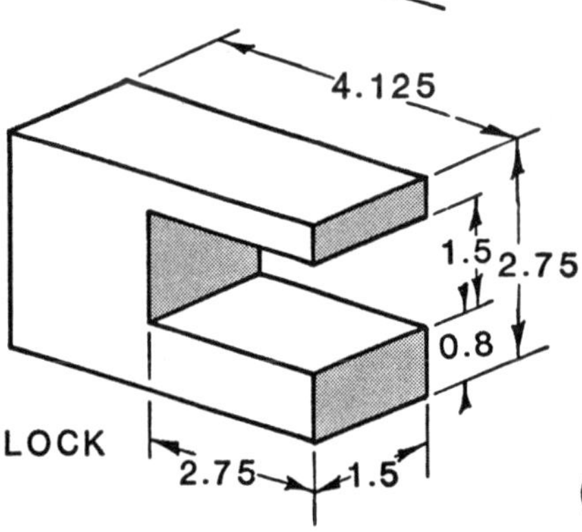

(3) FILLER BLOCK

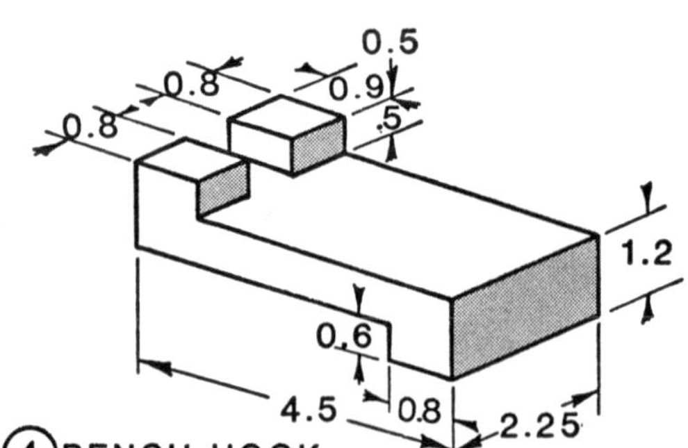

(4) BENCH HOOK

STUDY THESE HIDDEN EDGES VERY CAREFULLY

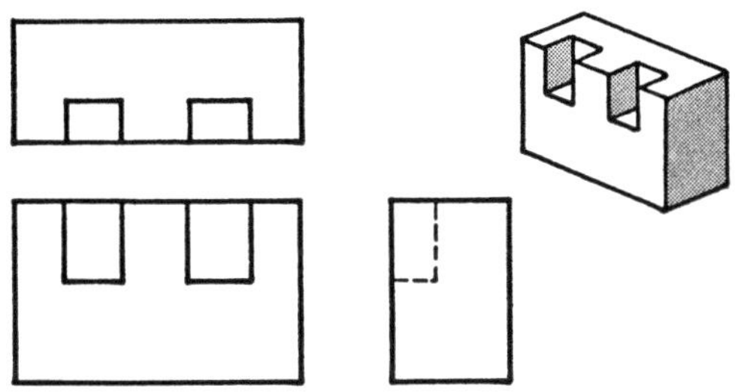

① When two hidden edges are in line with each other, draw only one.

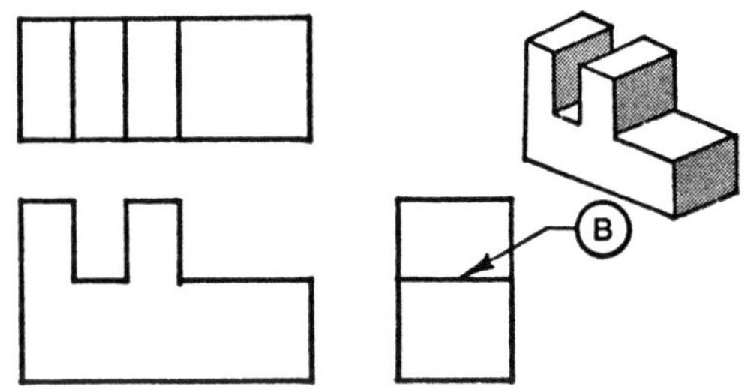

② When a visible edge and a hidden edge are in line, draw the visible edge. (Note line "B.")

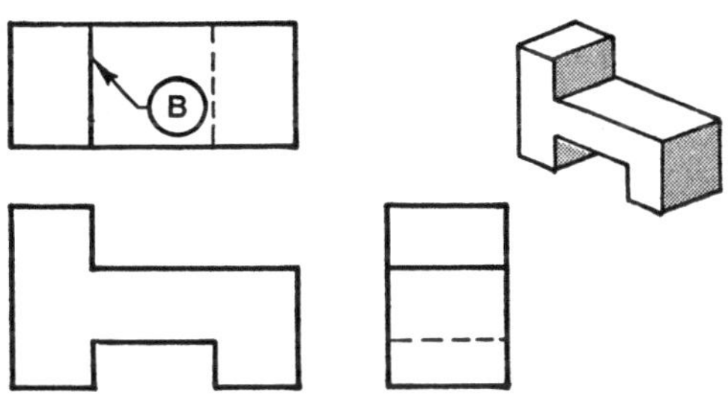

③ The line marked "B" in the top view is another illustration of the problem in ②.

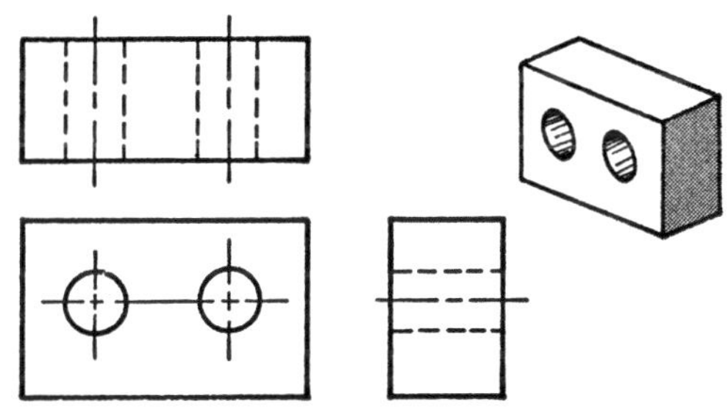

④ When two holes of the same diameter are aligned, only one appears in the end view.

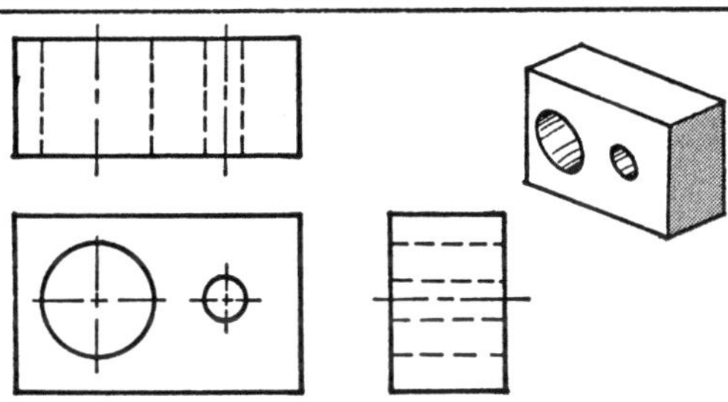

⑤ When two holes of different diameters are aligned, only the center lines coincide in the end view.

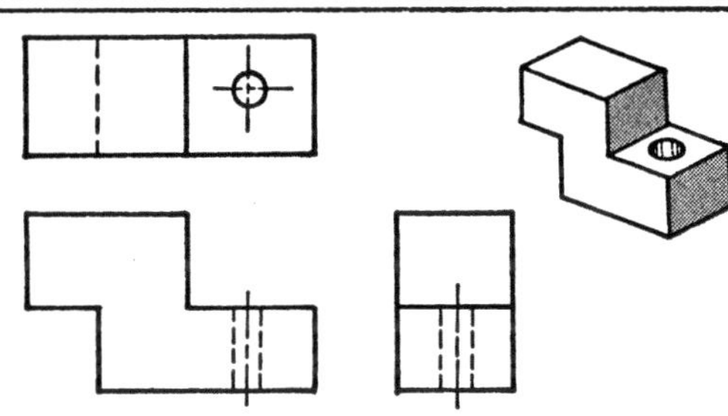

⑥ An interesting situation. Study the top view and the end view.

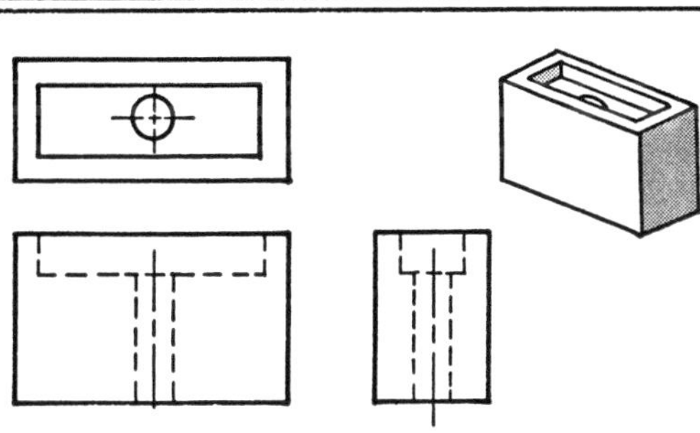

⑦ The rectangular hole goes part way through the block. The circular hole goes through the remainder of the block.

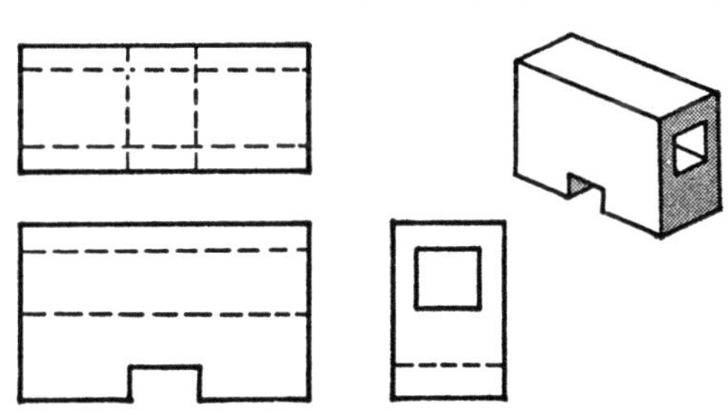

⑧ Hidden edges may cross each other. Study the top view. Hidden lines should not cross at their intersections.

PROJECTING HOLES FROM ONE VIEW TO ANOTHER VIEW

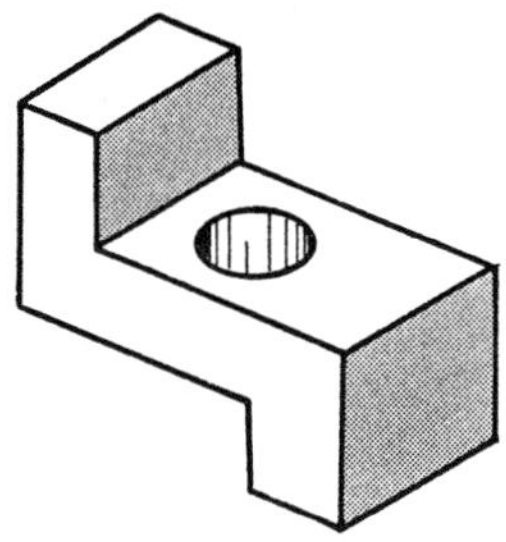

(1) Revolving will not eliminate hidden edges.

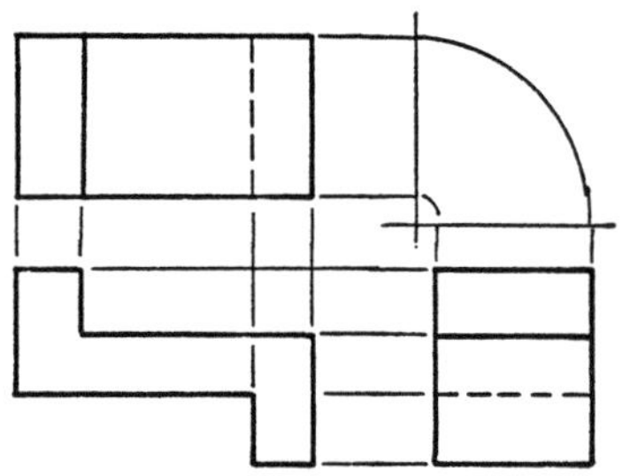

(2) Draw the front view. Project in order to draw the other two views.

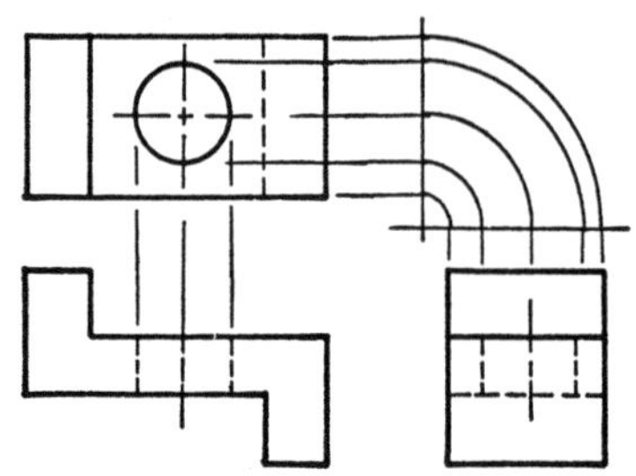

(3) Draw the circular view of the hole. Project to the other two views.

A BASIC METHOD FOR PROJECTING HOLES

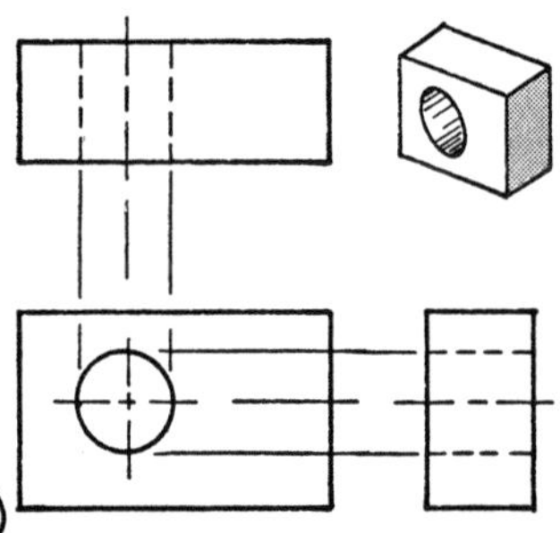

(1)

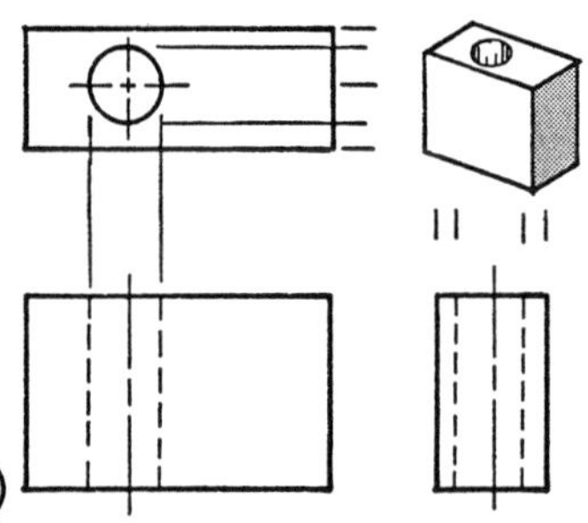

(2)

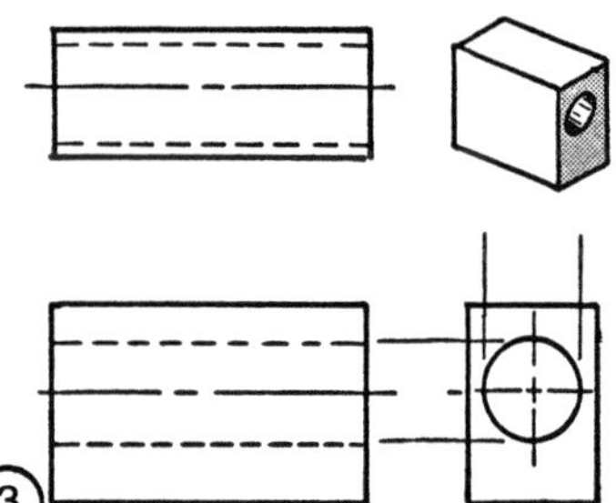

(3)

First: Draw the three basic views without drawing the circular hole.
Second: Draw the circle in whichever view it occurs.
Third: Project the hole to the other views. Project the center lines.

ASSIGNMENT

Draw the orthographic view.

(1) DIE

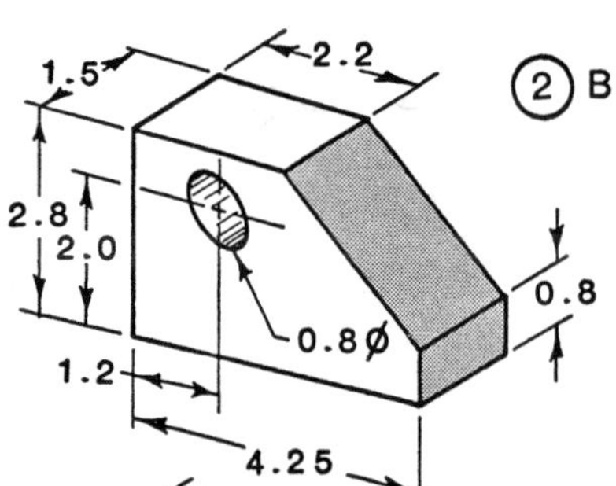

(2) BENCH REST

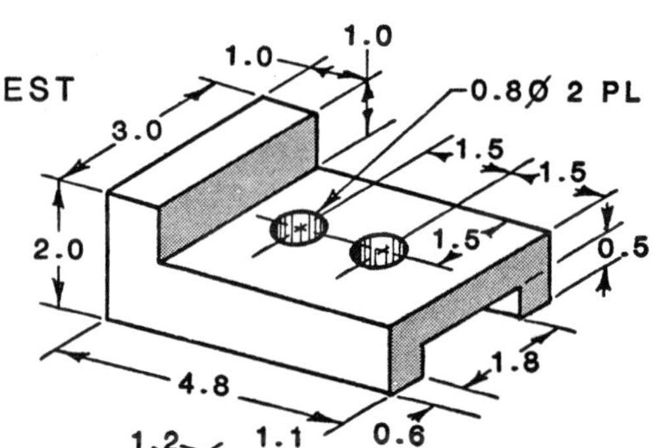

(3) BEARING MOUNT

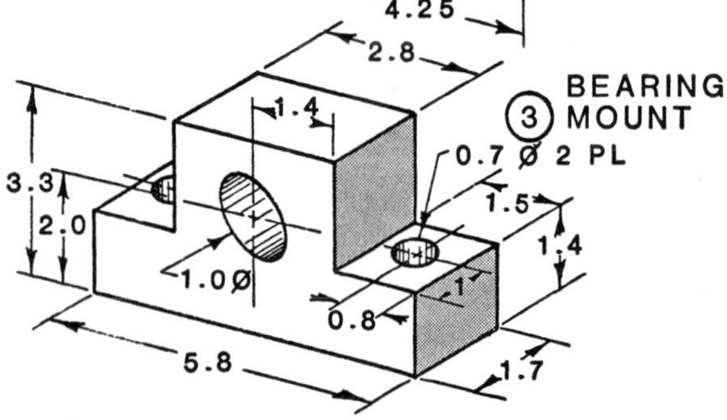

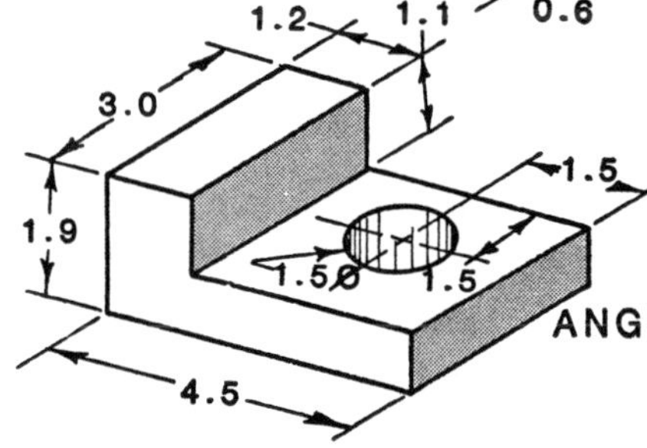

(4) ANGLE BRACE

HOW TO DRAW OBJECTS HAVING ROUNDED ENDS

Whenever a complete circle or a large arc is an important part of the basic shape, it is best to draw first the view showing the circle or arc. Then project to the other views. This method promotes accuracy in drawing.

DRAWING AN OBJECT HAVING A SEMICIRCULAR END

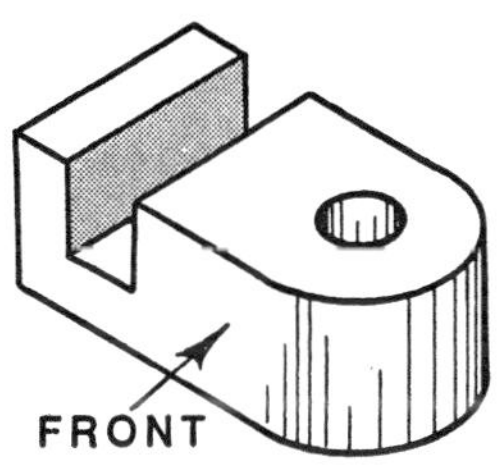

(1) One end of the object is semicircular in shape.

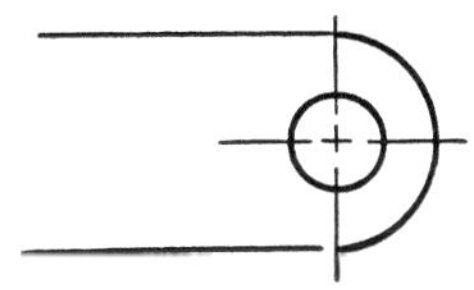

(2) Draw the semicircle in the top view. Complete the top view.

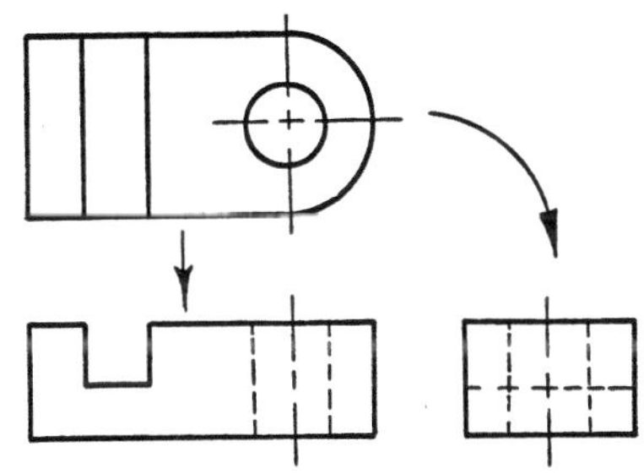

(3) Draw the other two views by projecting.

DRAWING AN OBJECT HAVING A CYLINDRICAL PART

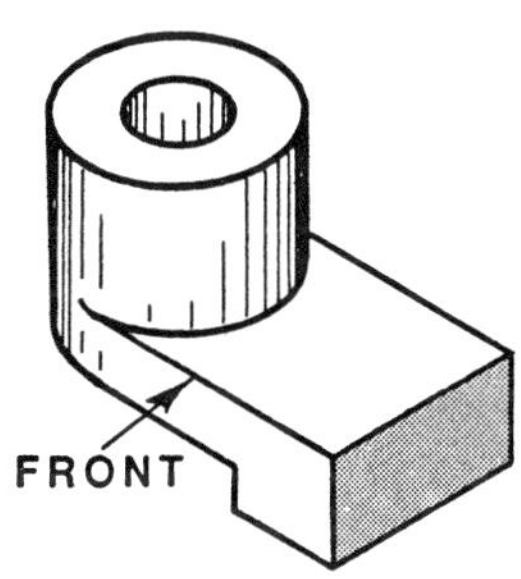

(1) One end of this object is cylindrical.

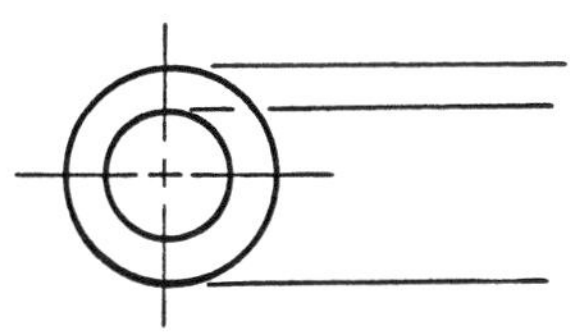

(2) Draw the circles in the top view. Complete the top view.

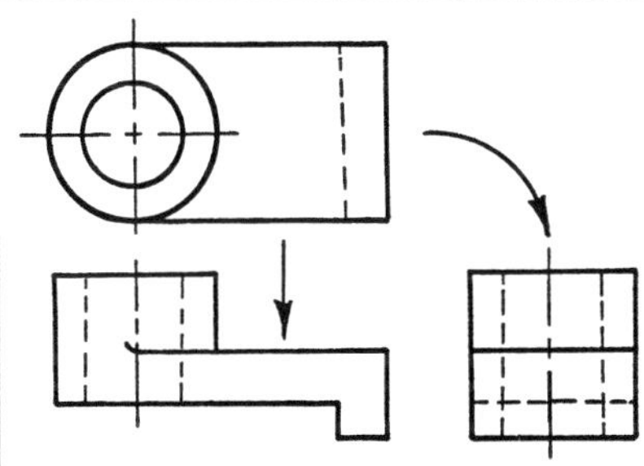

(3) Project to the other two views. Greater accuracy and speed will result from using this method.

DRAWING AN OBJECT HAVING TWO ROUNDED ENDS

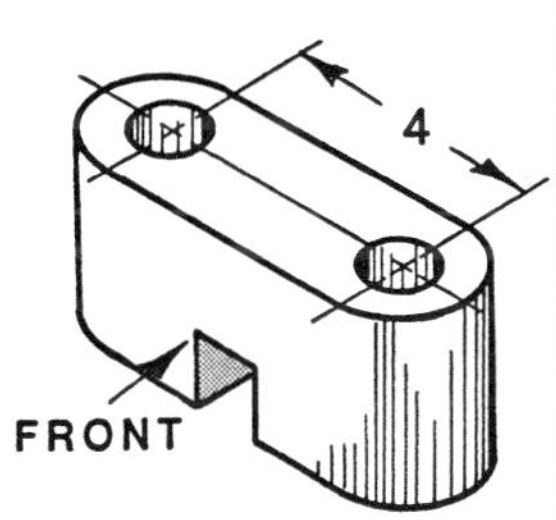

(1) Both ends of the object are rounded. Note that the center lines are 4″ apart.

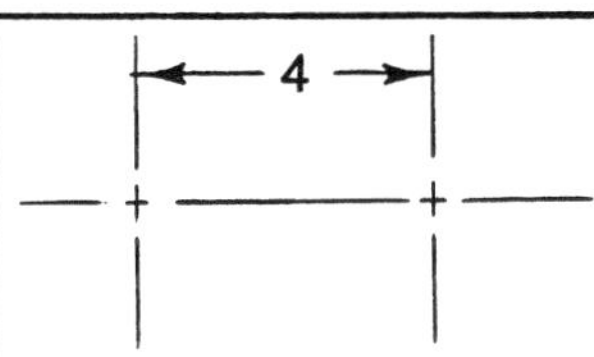

(2) Draw the horizontal center line. Space the vertical center lines 4″ apart.

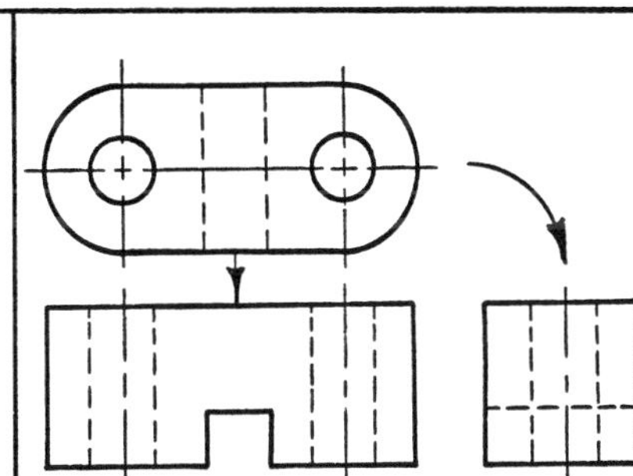

(3) Complete the top view. Project to the other two views.

DRAWING AN OBJECT HAVING SMALL ROUNDED CORNERS

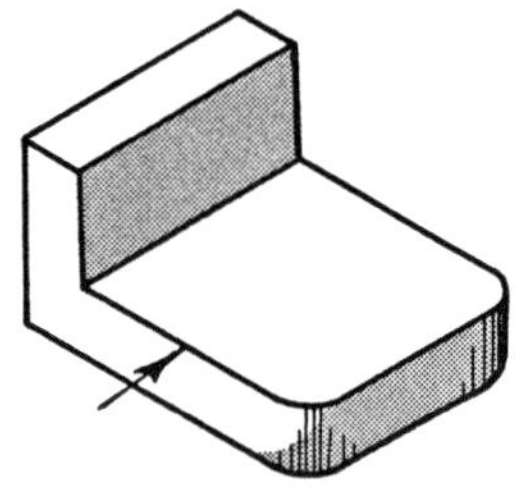

① Study the object.

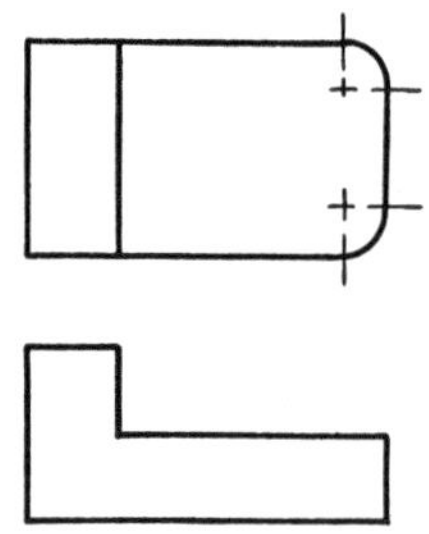

② The small, rounded corners are not important enough to dictate which view to draw first.

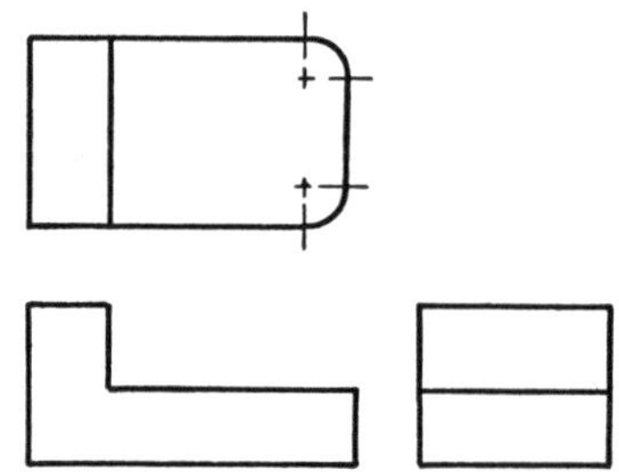

③ The corners will be rounded in the top view only.
Do you know why?

ASSIGNMENT

Make a three-view orthographic projection of these objects.

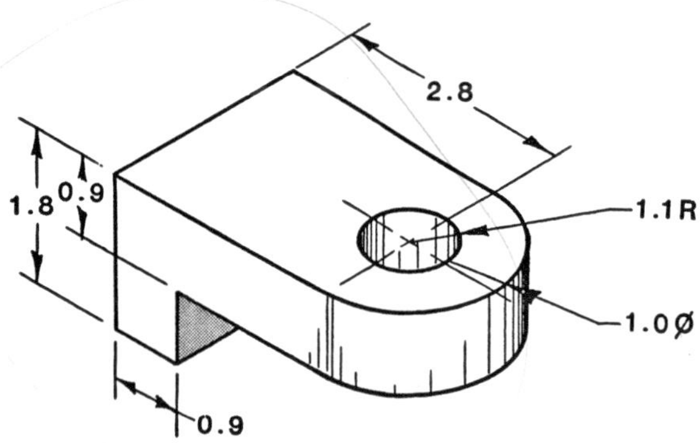

① SUPPORT

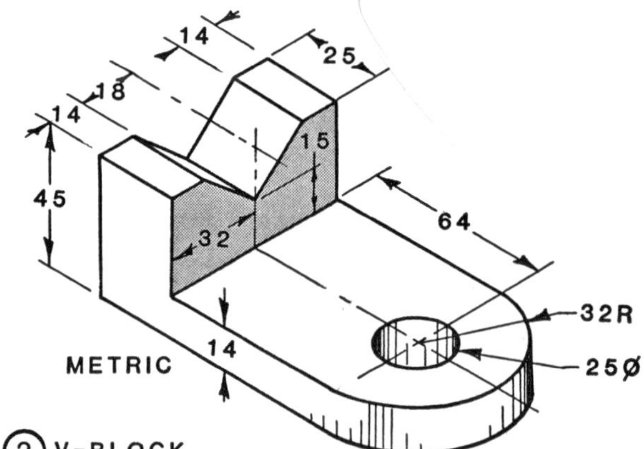

② V-BLOCK

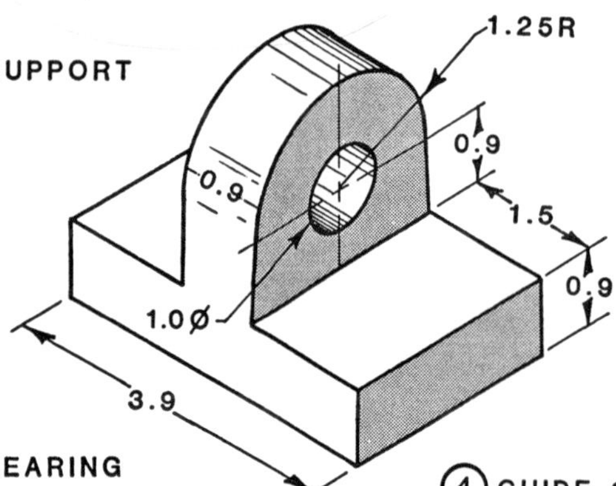

③ BEARING

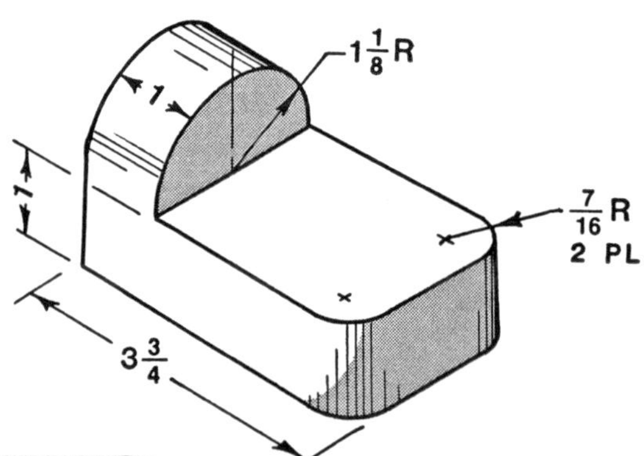

④ GUIDE STOP

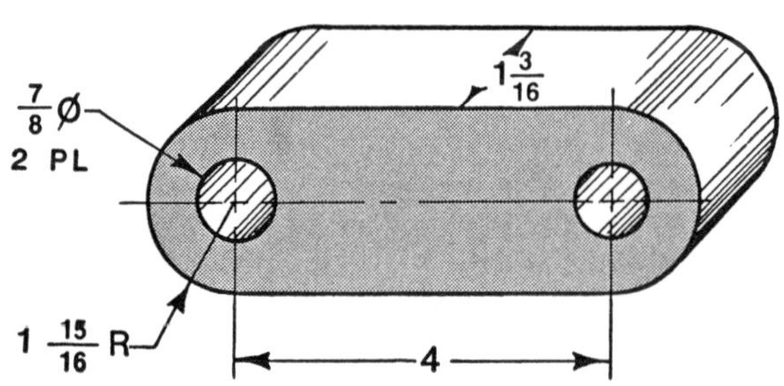

⑤ LINK

SYMMETRICAL OBJECTS AND TWO-VIEW DRAWINGS

Symmetrical means having the same form on each side of a central line or plane. In other words, if the two halves of an object are exactly alike, the object is symmetrical. Usually, only two views of a symmetrical object are needed. It is important to remember that two views may be used only when the object is symmetrical.

THREE WAYS OF DRAWING THE TWO VIEWS

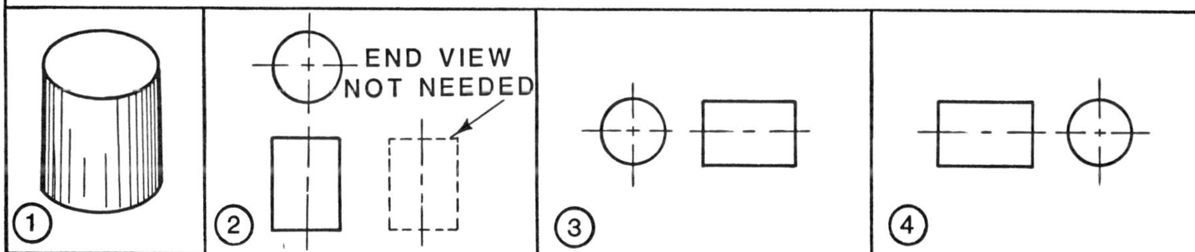

A cylinder is a symmetrical object.	Since only two views of a symmetrical object are needed, you will realize that the object may be drawn in any of the three ways as shown at ②, ③, or ④ above.

ANOTHER EXAMPLE WHERE ONLY TWO VIEWS ARE REQUIRED

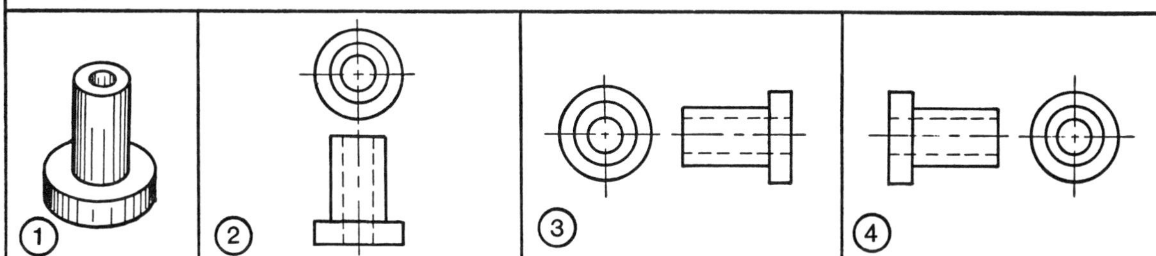

The object is symmetrical.	The two views may be arranged in any of the three ways shown above. (Note how the center lines are used.)

ONLY TWO VIEWS WOULD BE REQUIRED ON ANY OF THESE

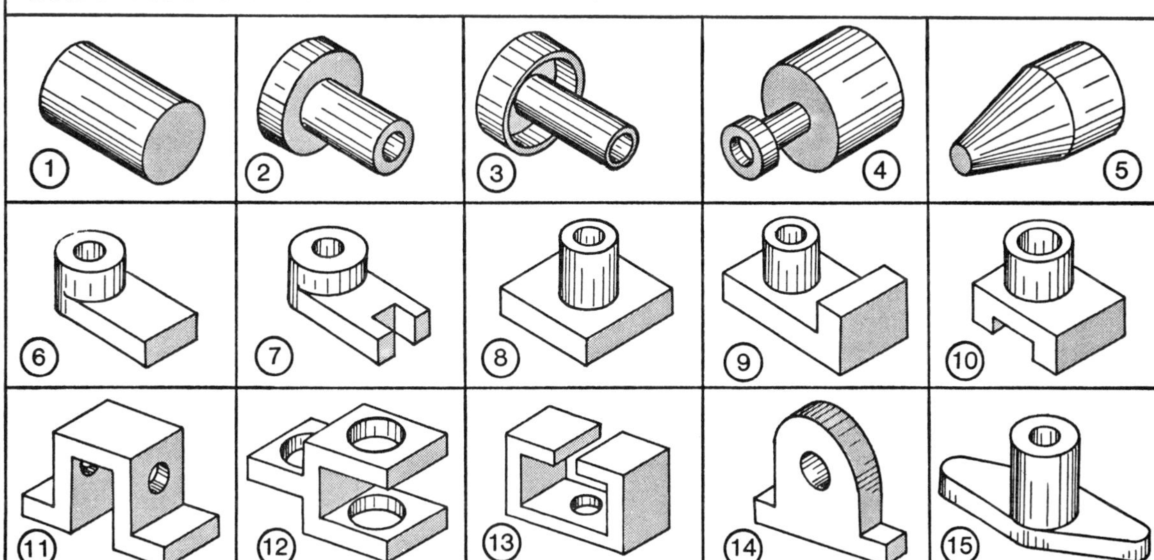

SELECTING THE CORRECT TWO VIEWS OF A SYMMETRICAL OBJECT

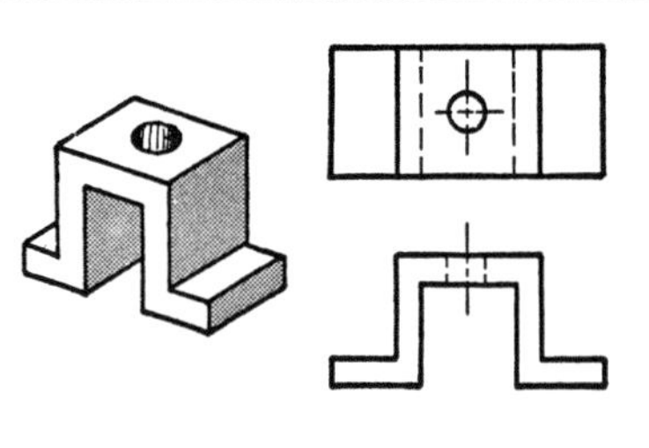

① Be sure that one view shows the hole as a complete circle.

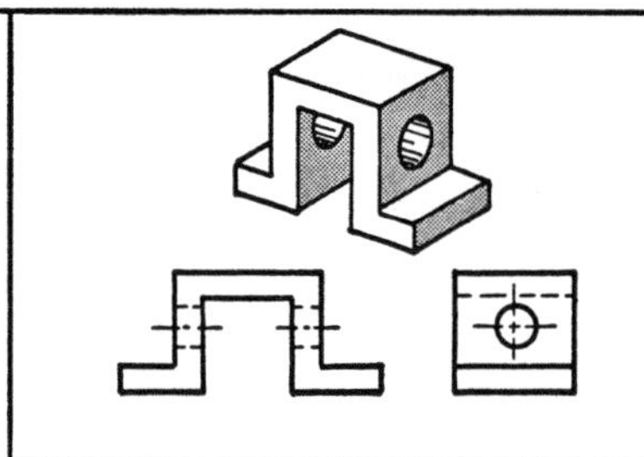

② One view must show the holes as complete circles.

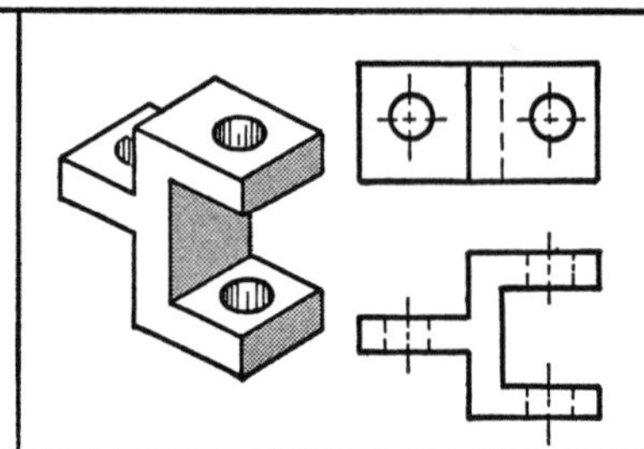

③ This is one of six possible answers. (Revolve the book in order to get three more.)

A REVIEW OF METHODS

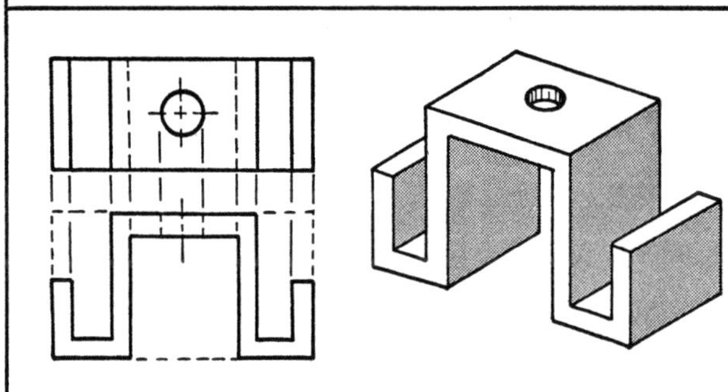

① When the object is rectangular in shape, draw the basic shape of one view and project to the other view.

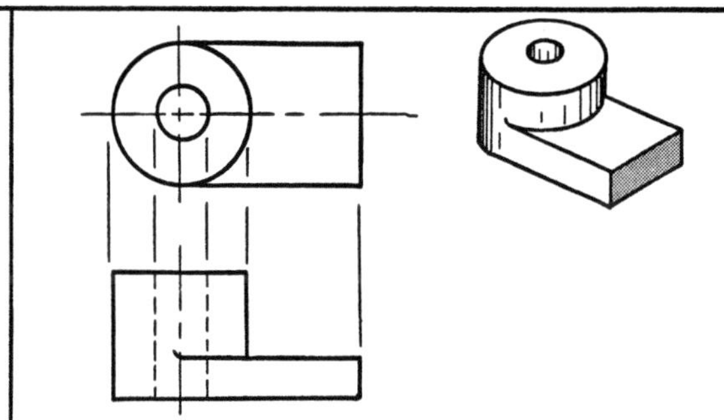

② When the circles are an important part of the basic shape, draw the circles first, and then project to the other view.

ASSIGNMENT

Make a two-view orthographic projection of one of these objects. (Do not dimension.)

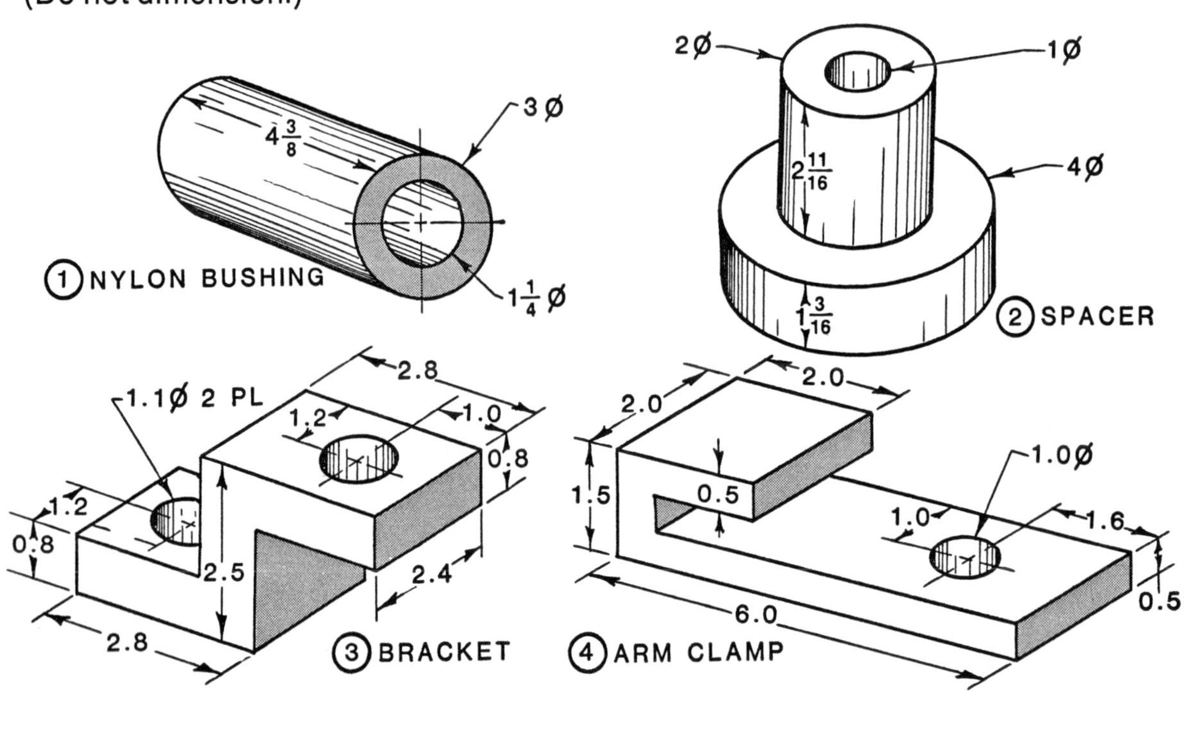

PLAN YOUR DRAWING BEFORE YOU BEGIN TO DRAW

Practice making a mental plan of how you intend to carry a drawing to completion. This is an important skill of drafting that you should develop.

HOW TO PLAN WHEN CYLINDERS ARE INVOLVED

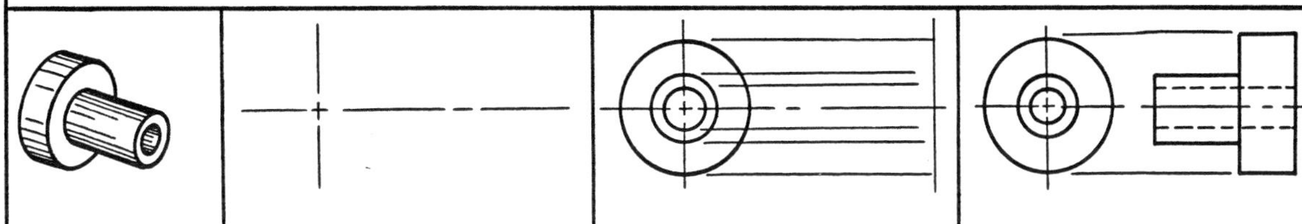

1. Study the shape of the object.
2. Plan a well-centered drawing. Draw the center lines.
3. Draw the circles. Project to the right for the end view.
4. Complete the end view by projecting to the left.

HOW TO PLAN WHEN CIRCLES AND ARCS ARE IMPORTANT

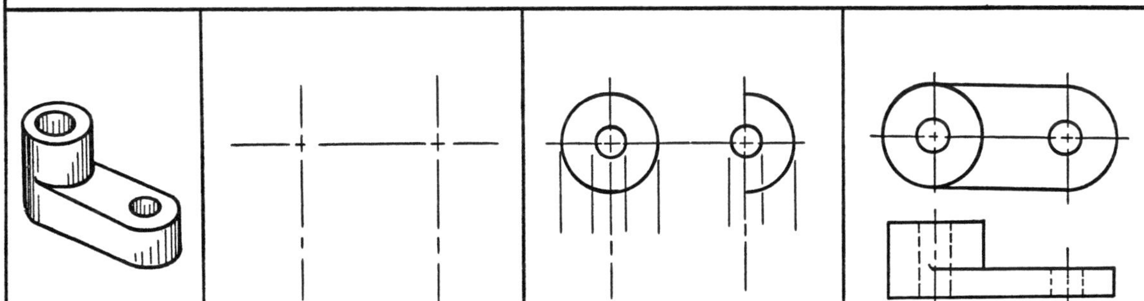

1. Study and plan.
2. Draw the center lines for the top view.
3. Draw the circles and the arc. Project downward.
4. Complete both views. Make an extra effort at having the points of tangency smooth.

The profile view is sometimes called the outline or contour view. It is the view whose outline best describes the shape of the object.

Learn to recognize the profile view in every orthographic projection that you draw. When possible, make the profile view the front view of your drawing. Study every object with this in mind.

A STUDY OF THE ORTHOGRAPHIC VIEW

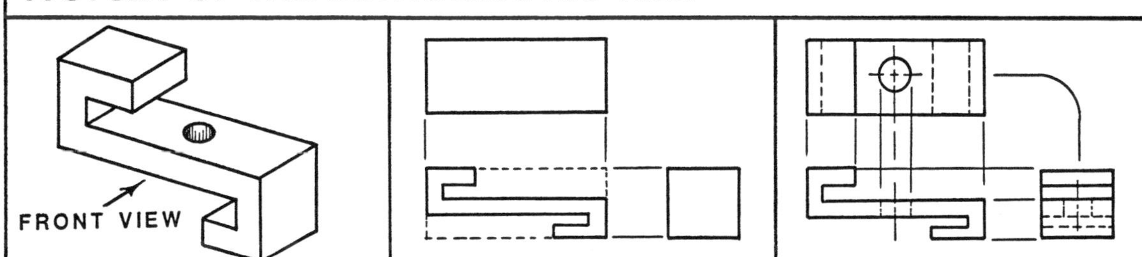

1. Select the profile front view. The right-hand profile view and top view will be projected from the front view.
2. Draw the basic shape of the front view. Project the basic shape to the other views.
3. Project the details from the front view to the other views. Project from the circles.

A LOST EDGE

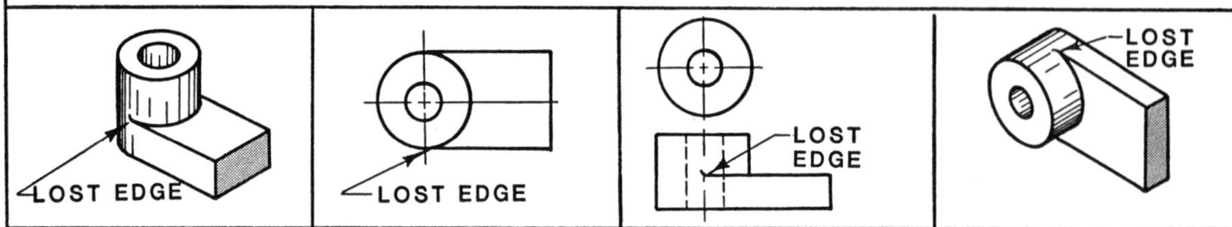

A lost edge occurs when a straight line is drawn tangent to a circle. Learn to recognize a lost edge when it occurs.

PRACTICE IN VISUALIZING RELATED FORMS

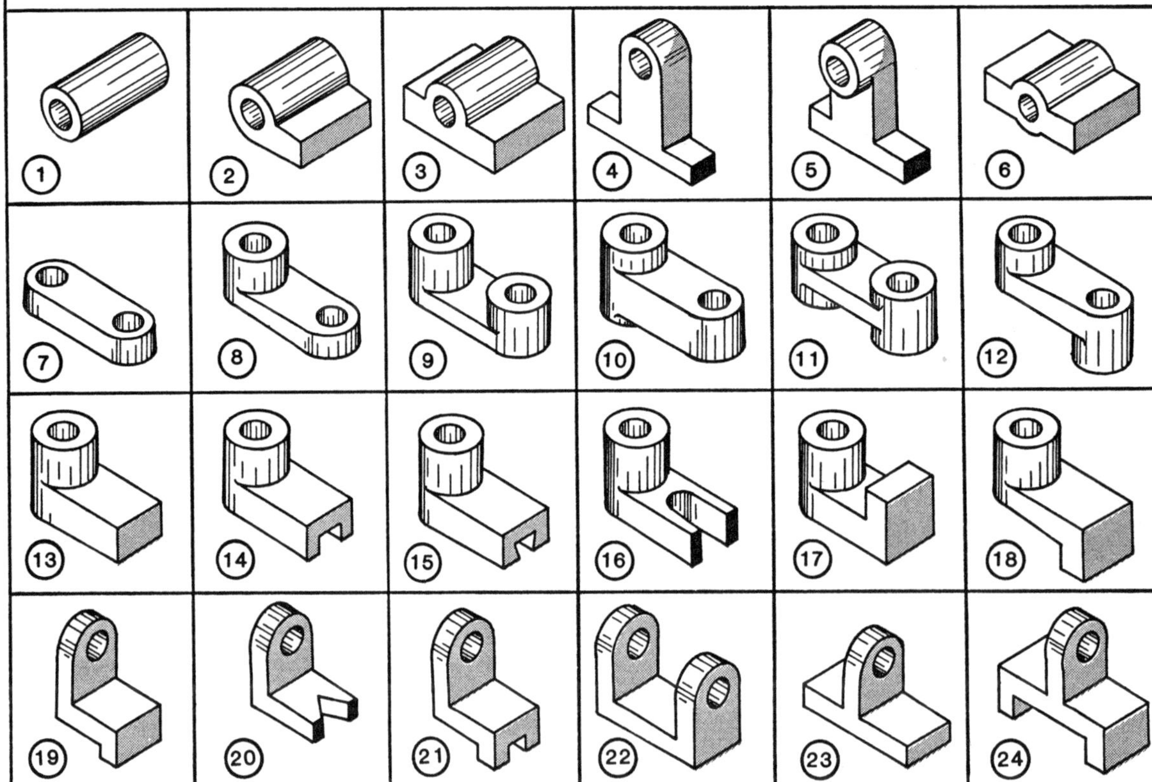

Learn to recognize a form and any variation of that form. There are about twelve basic forms, but thousands of variations of each.

ASSIGNMENT

Make a two-view orthographic projection.

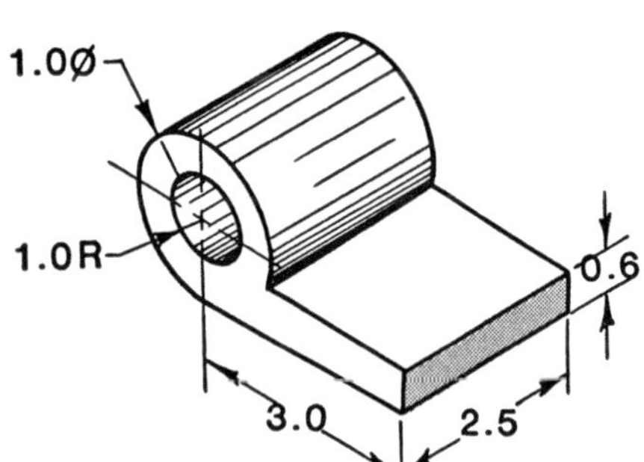

(1) HINGE BLANK

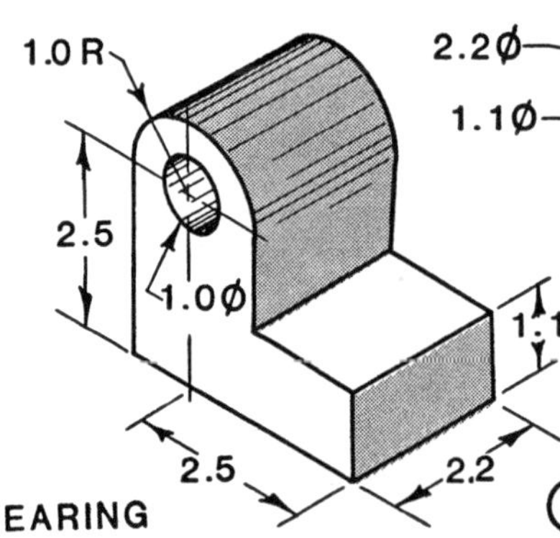

(2) BEARING

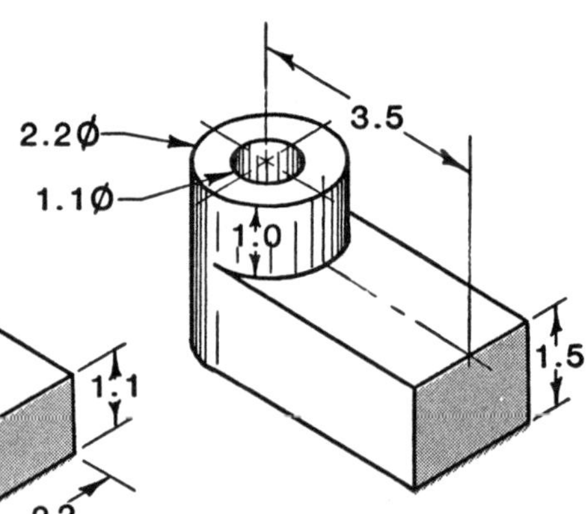

(3) BEARING PLATE

A STUDY OF PRISMS

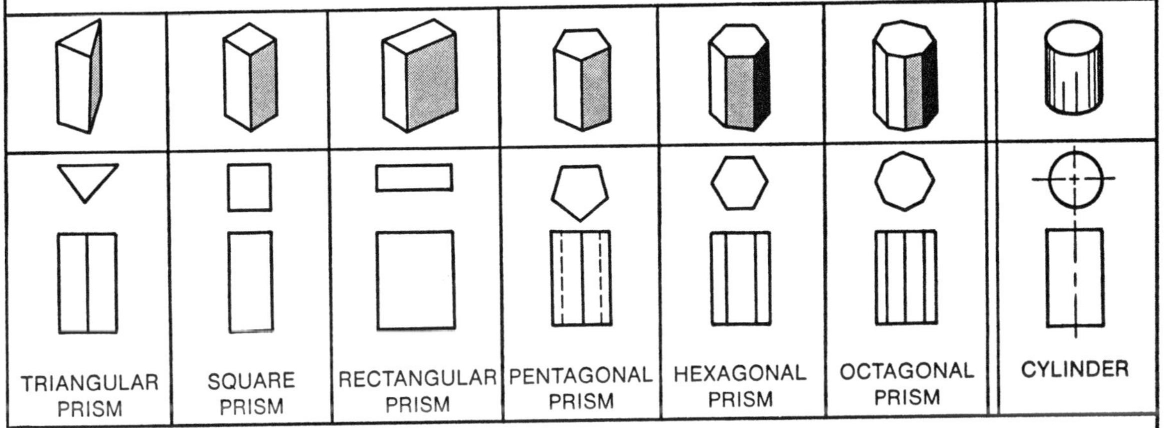

Prisms are classified according to the shape of the base. Only two views are needed.

A STUDY OF PYRAMIDS

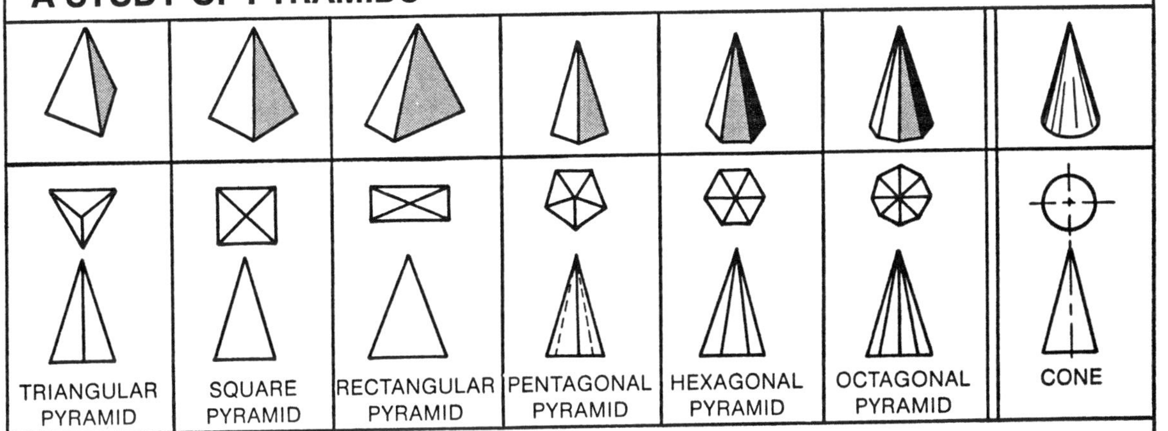

Pyramids are classified according to the shape of the base. Only two views are needed.

ASSIGNMENT

Make a two-view orthographic projection.

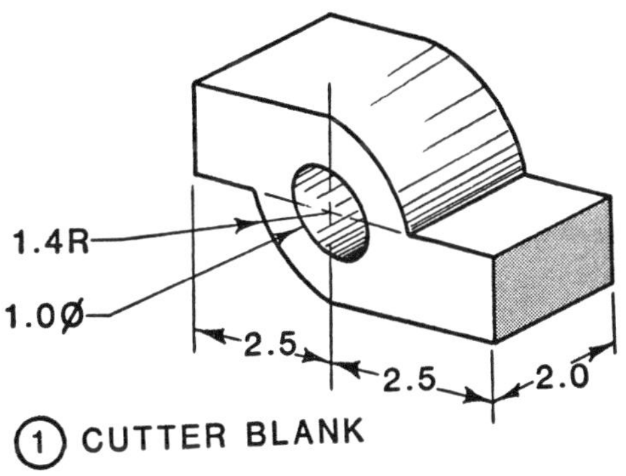

(1) CUTTER BLANK

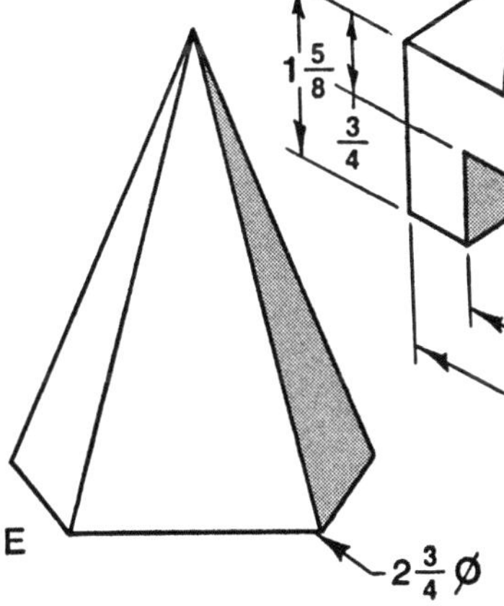

(2) HEXAGON PYRAMID 3" ALTITUDE

(3) CLAMP

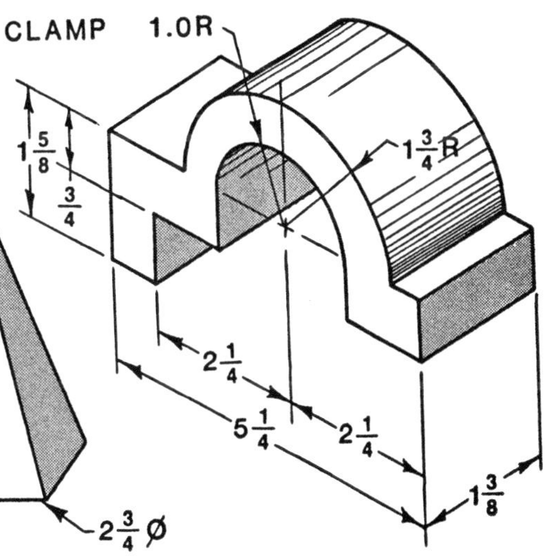

AN IMPORTANT APPROACH TO VISUALIZATION

Drafters have trained themselves to see an object in relation to its most important characteristics.

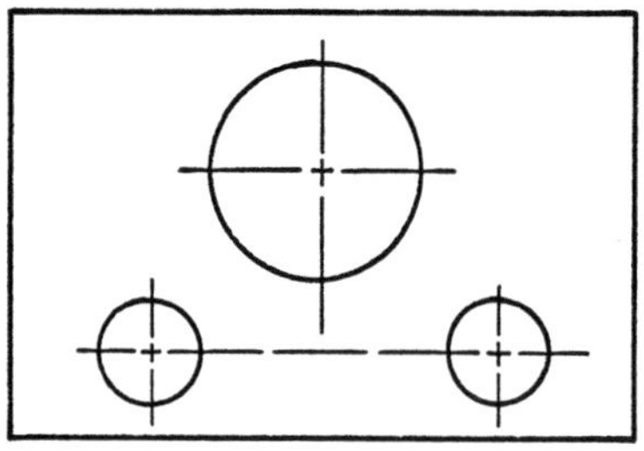

(1) This object may be seen as a rectangular surface with three holes in it. In that case, the rectangle should be drawn first. The holes are then located and drawn. The most important characteristic, then, is the rectangular shape. The second most important is the holes within the rectangle.

(2) A study of this object reveals that the circular shapes are more important than the rectangular ones. The circular shapes are the key to the entire drawing.

When such a situation occurs, always draw the circular shapes first, then attach the rectangular shapes to them. This is an important idea in visualization.

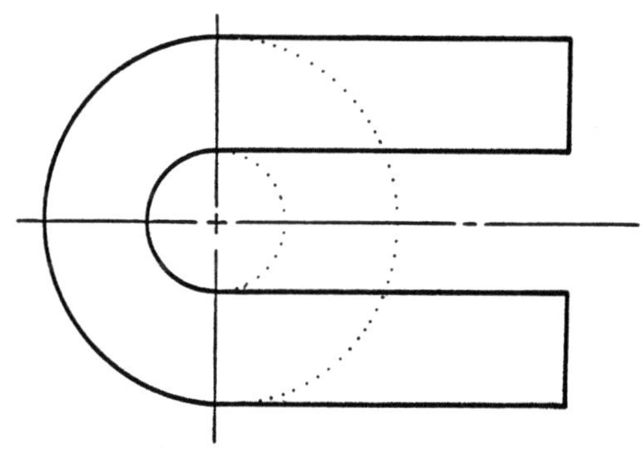

CENTER LINES ARE OFTEN THE FIRST LINES TO BE DRAWN

If the holes and curves are an important part of an object, the center lines for those holes should be the starting point of the drawing.

BUILDING AROUND THE CENTER LINES

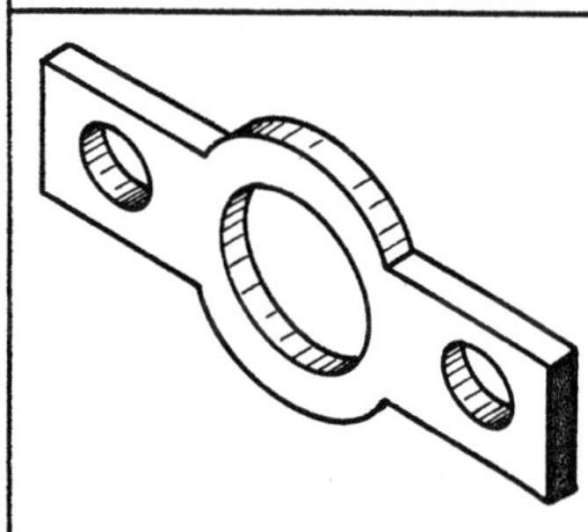

(1) The circular parts are the important parts of this object.

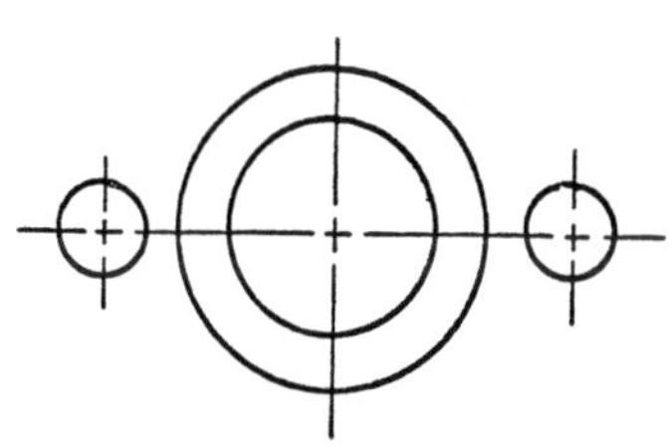

(2) Locate the center lines. Draw the circles.

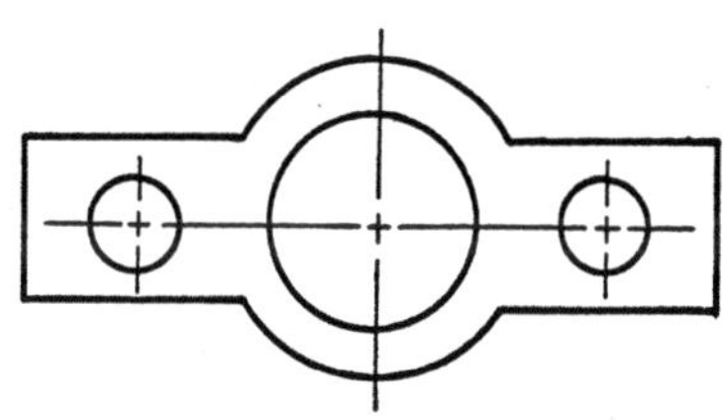

(3) Build the rectangular shape around the center lines.

Remember: When studying an object before drawing, see its most important characteristics, and then select the profile view and the other required views. If important circular parts are involved, locate and draw the circles first.

ONE-VIEW DRAWINGS

There would be no reason for making a two-view drawing of an object that is very thin, for example, 1/64" thick. The end view would hardly be thicker than the width of the pencil lines. A note indicating the thickness of the material is used in place of an end view on such drawings.

GOOD JUDGMENT AND ONE-VIEW DRAWINGS

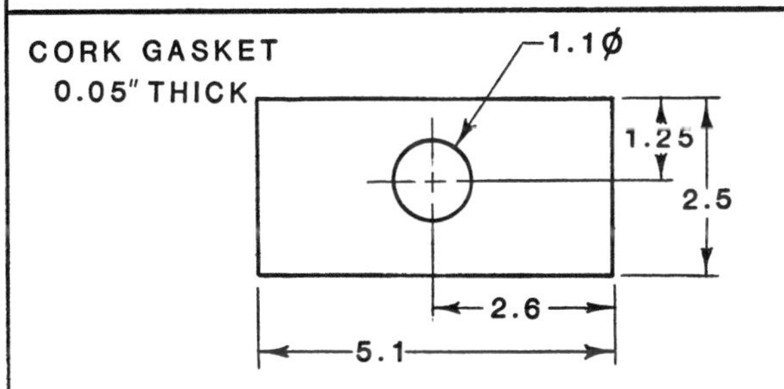

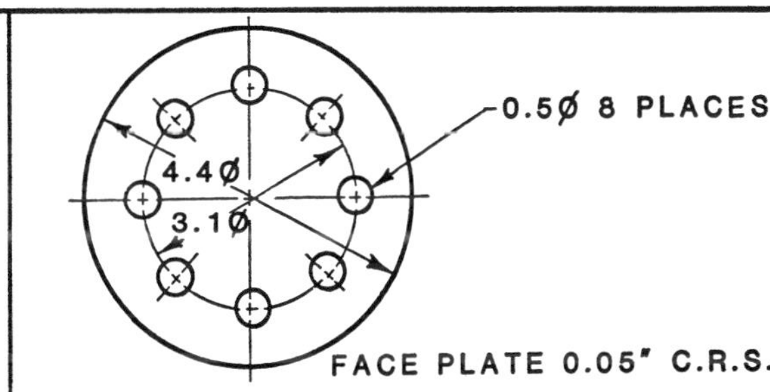

(1) An end view of this object would be of little value since the view itself is only 0.05" thick, as indicated by the note.

(2) The drawing, the dimensions, and the notes serve to give a complete description of this object. (C.R.S. = Cold Rolled Steel)

WORKING FROM THE CENTER LINES

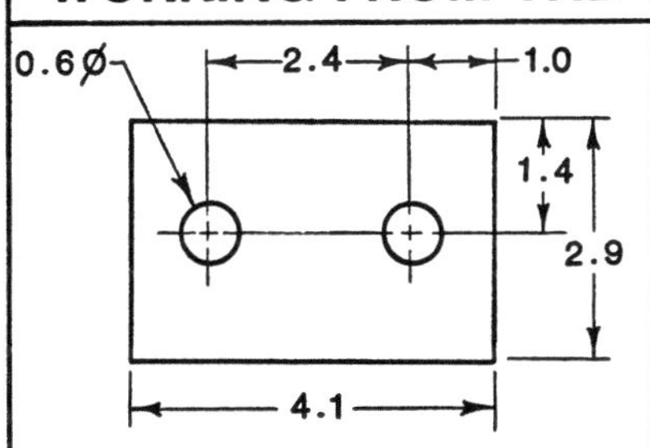

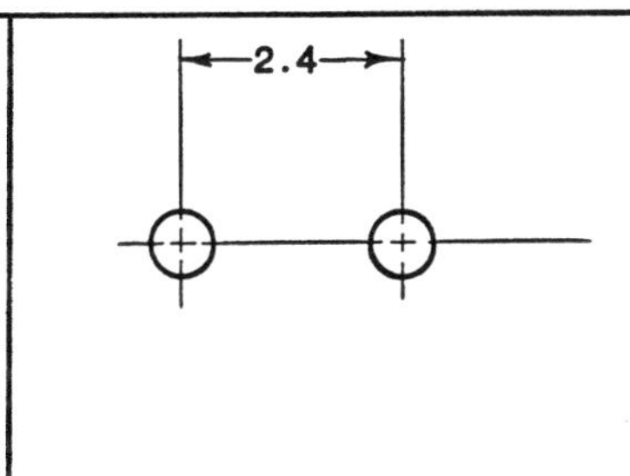

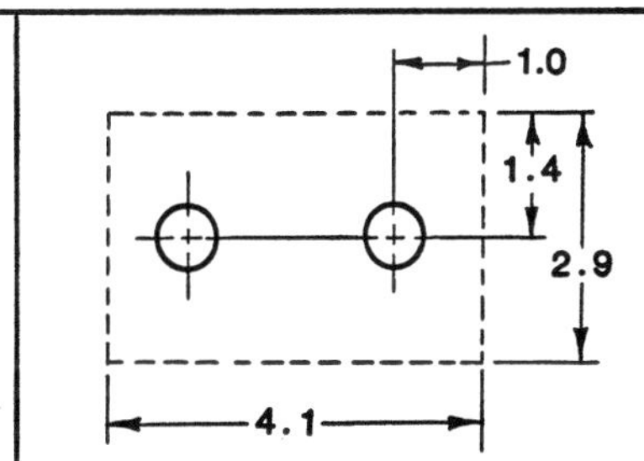

(1) The relationship of the holes is the important feature of this object.

(2) Draw the center lines for the circles.
Draw the circles.

(3) Draw the rectangle around the arrangement of the two holes.

EVENLY SPACED HOLES WITHIN A CIRCLE

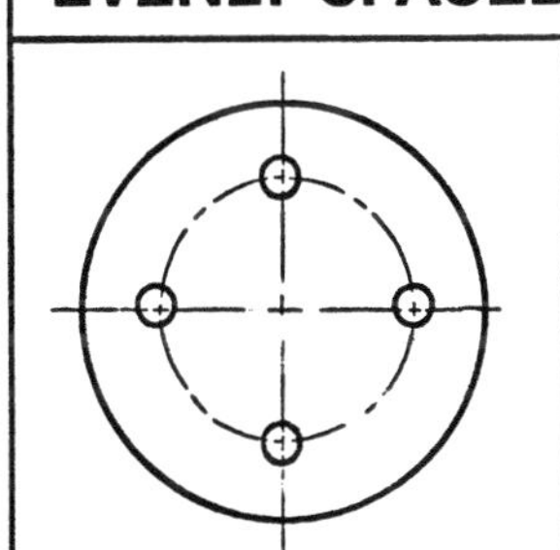

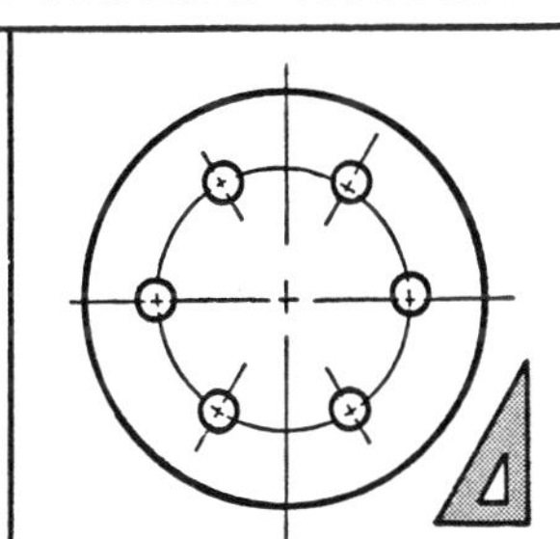

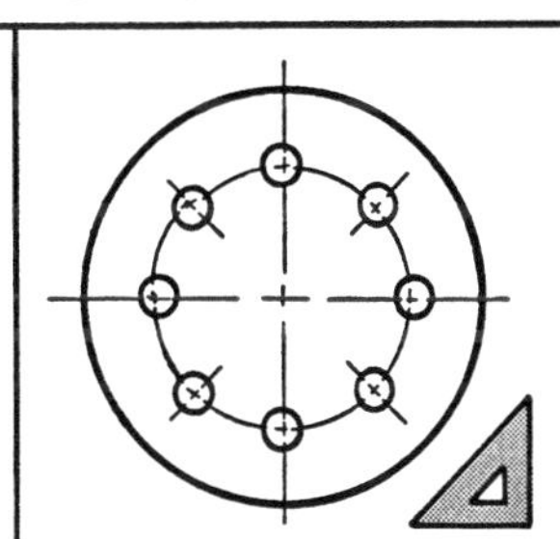

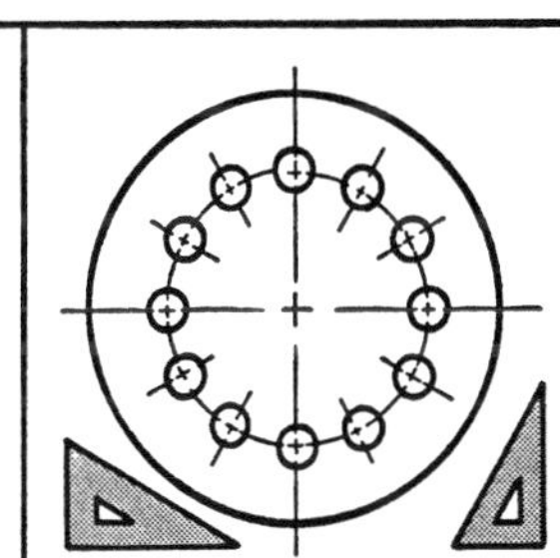

(1) **Four Holes:** Use a triangle and the T-square.

(2) **Six Holes:** Use the 60° angle and the T-square.

(3) **Eight Holes:** Use the 45° triangle and the T-square.

(4) **Twelve Holes:** Use the 30°, 60°, 90° angles and T-square.

THE CURVED CENTER LINE

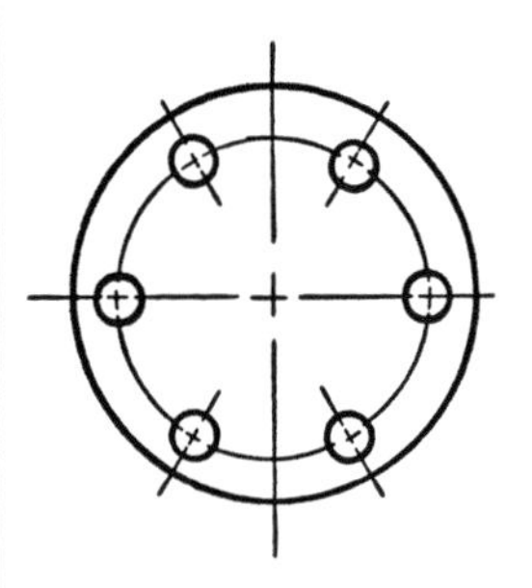	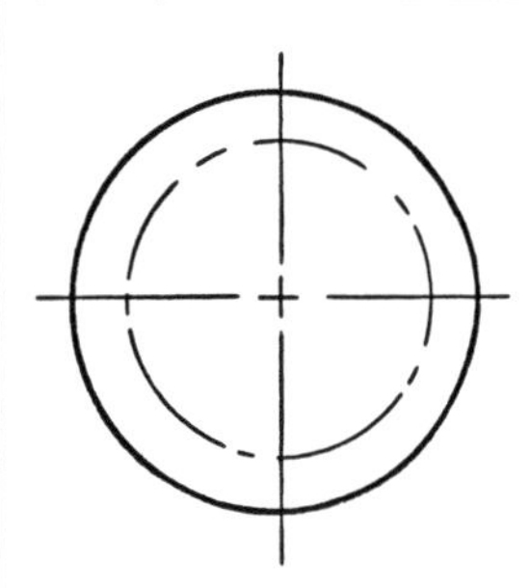	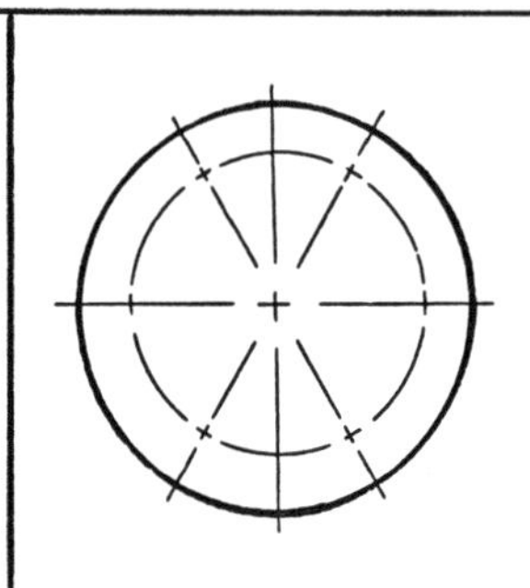	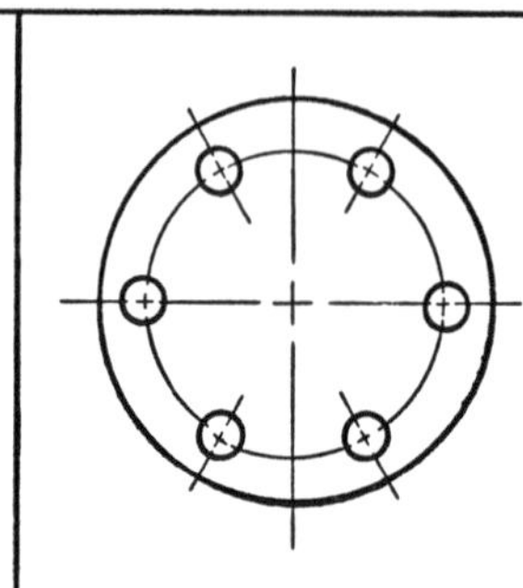
(1) Seven center lines are needed on this drawing for the following reasons:	(2) The curved center line is needed in order to locate the small holes.	(3) The straight center lines are needed for the small holes.	(4) The large center lines are needed for the large circle.

ASSIGNMENT

Make a one-view orthographic projection.

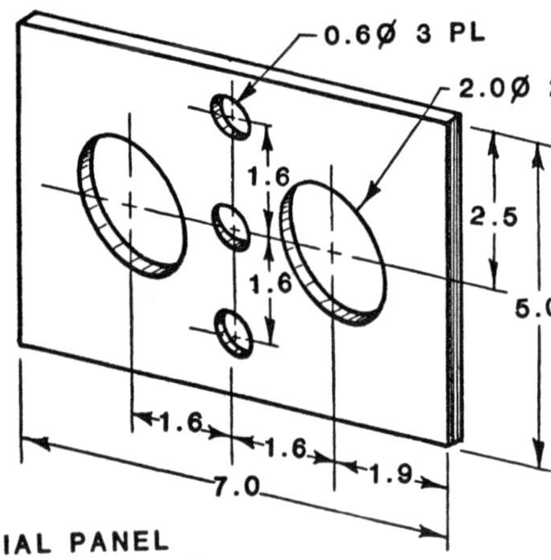

(1) DIAL PANEL
1/8" BAKELITE

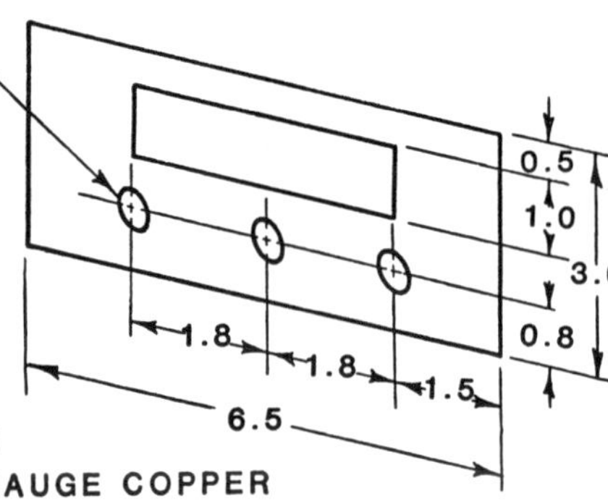

(2) SHIM
20-GAUGE COPPER

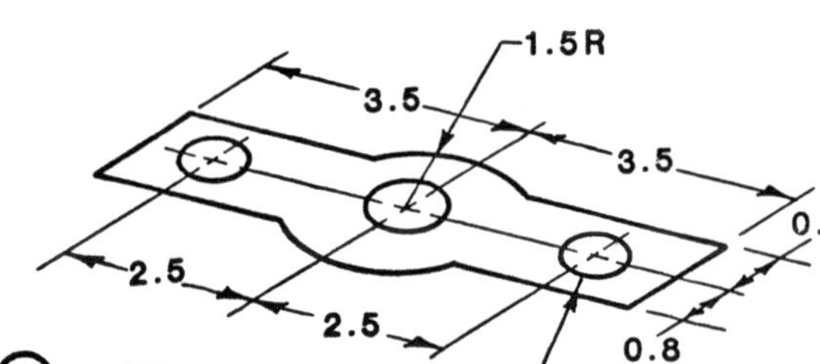

(3) SHIM
0.04Ø BRASS

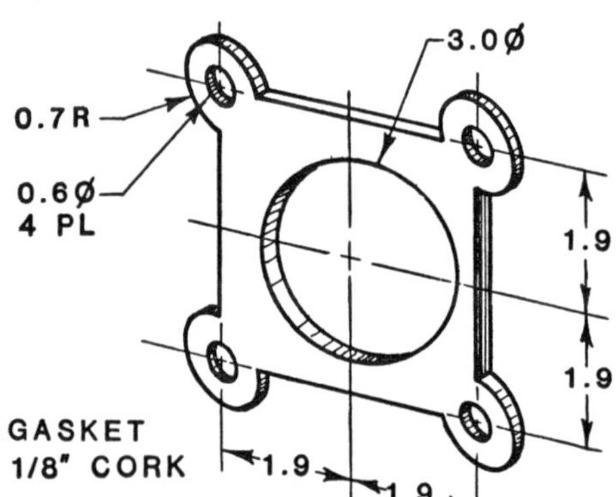

(4) GASKET
1/8" CORK

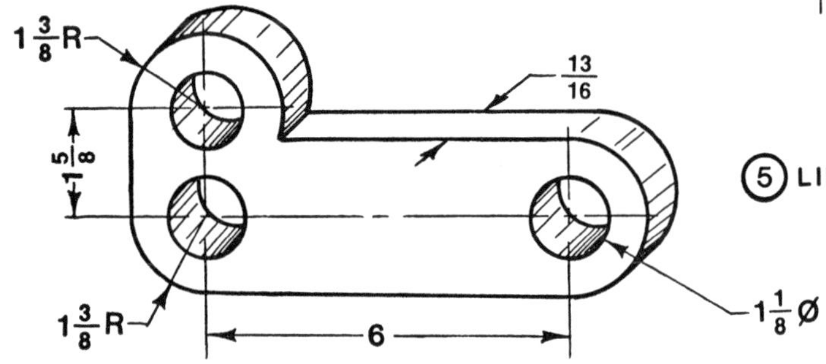

(5) LINK $\frac{1}{2}$ C.R.S.

STUDY EACH OF THESE DRAWINGS

Plan the method of drawing each. Make a one-view orthographic projection.

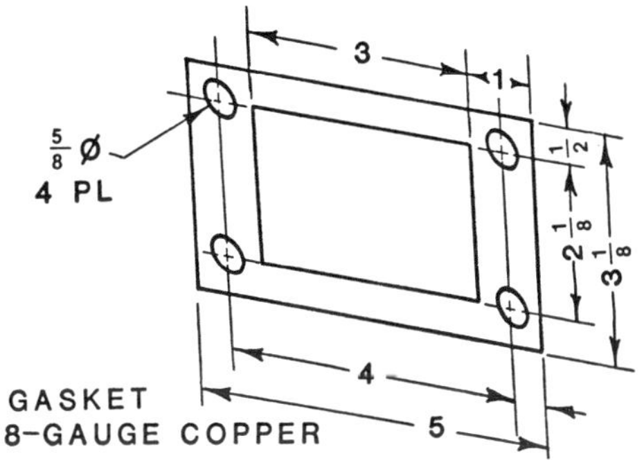

(1) GASKET
18-GAUGE COPPER

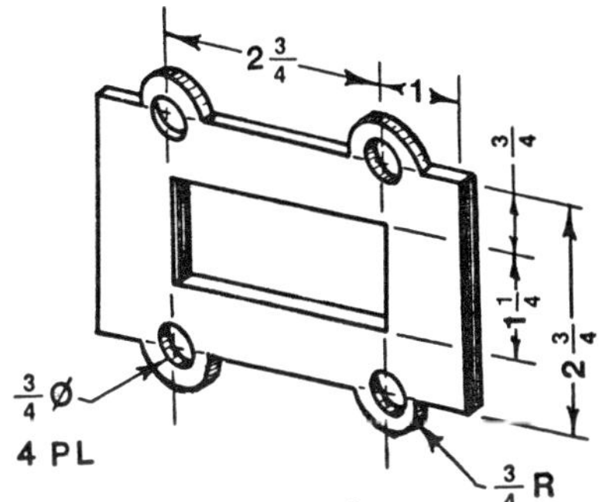

(2) RUBBER GASKET 1/32" THK

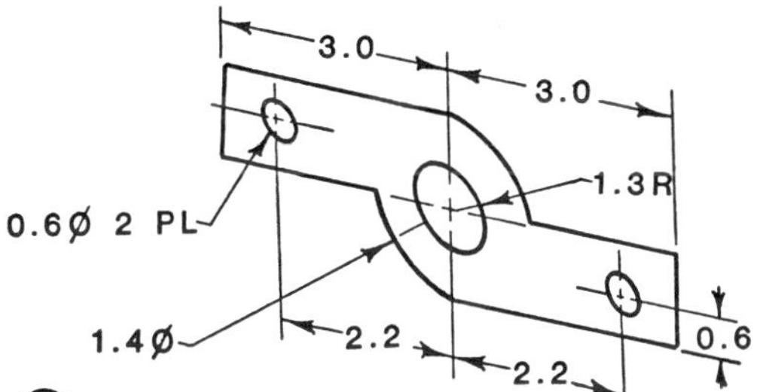

(3) BRASS SHIM 0.05 THK

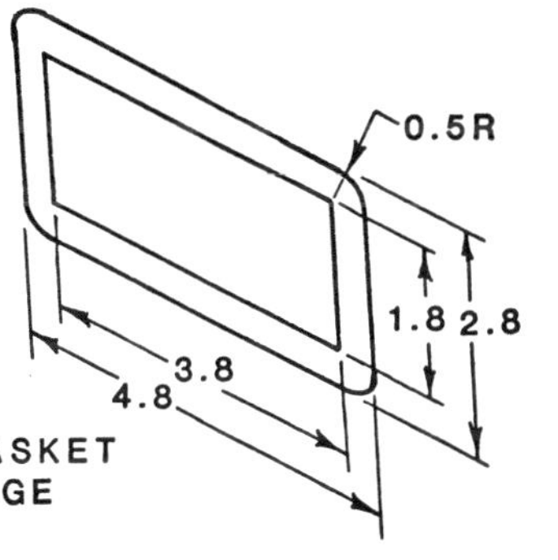

(4) BRASS GASKET
20-GAUGE

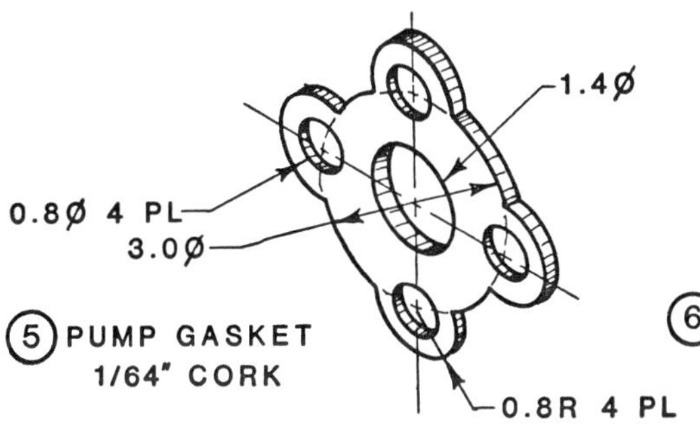

(5) PUMP GASKET
1/64" CORK

(6) MICA SHIM
0.04" THK

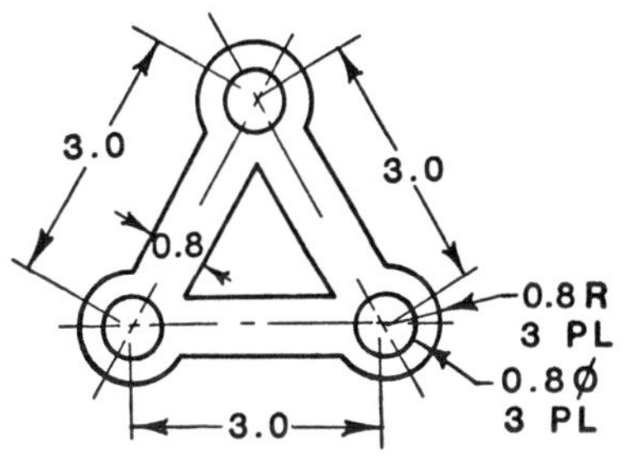

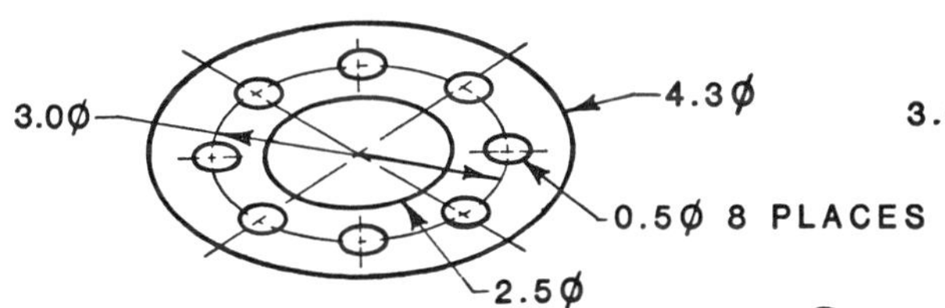

(7) CORK GASKET 0.07" THK

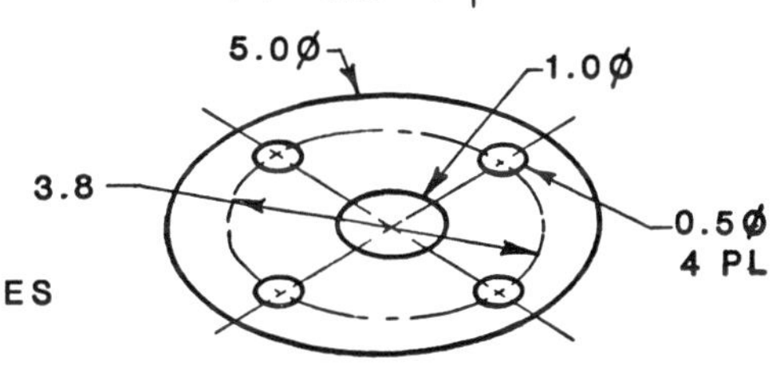

(8) PLASTIC GASKET 0.05" THK

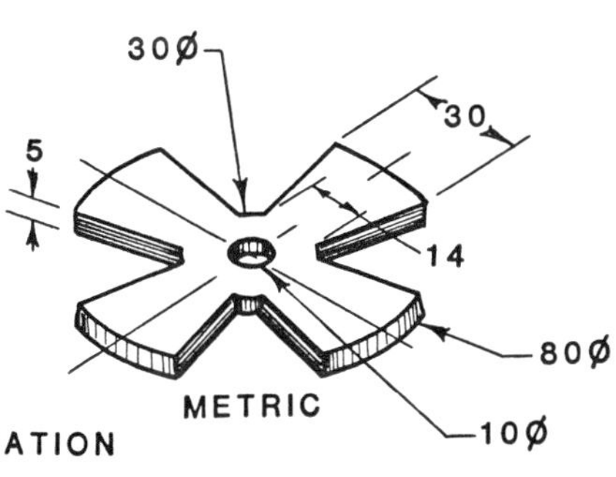

(9) LAMINATION

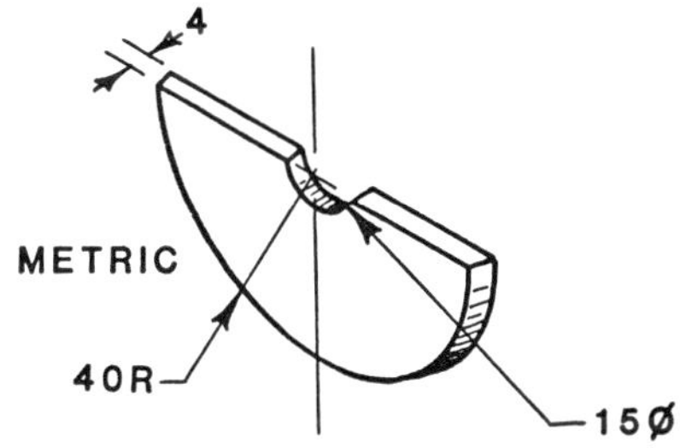

(10) CORK GASKET

A REVIEW OF ORTHOGRAPHIC PROJECTION

You cannot say that you have learned very much about orthographic projection until you are able to observe an object, to plan mentally the steps necessary in making the drawing, and, finally, to make the drawing.

The final goal in drafting is to visualize an object that does not exist (except in your mind), and then to make the drawing.

BEFORE DRAWING, DETERMINE THE NUMBER OF VIEWS NEEDED

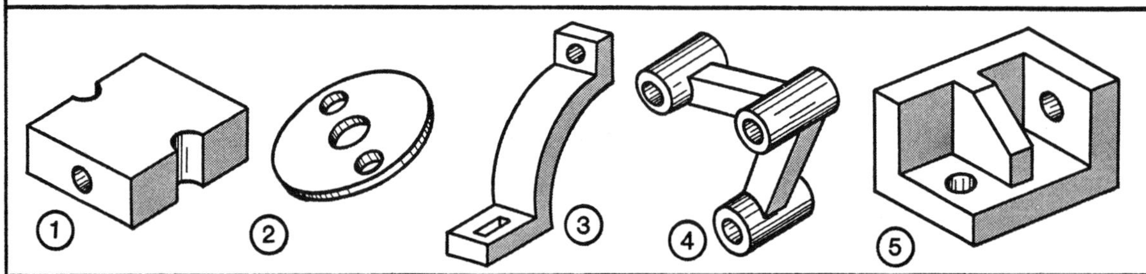

Knowing the basic method of drawing a front view, a top view, and an end view, the question remains, How many views of the object should be drawn in order to describe it accurately?

BEFORE DRAWING, DECIDE ON THE FRONT VIEW

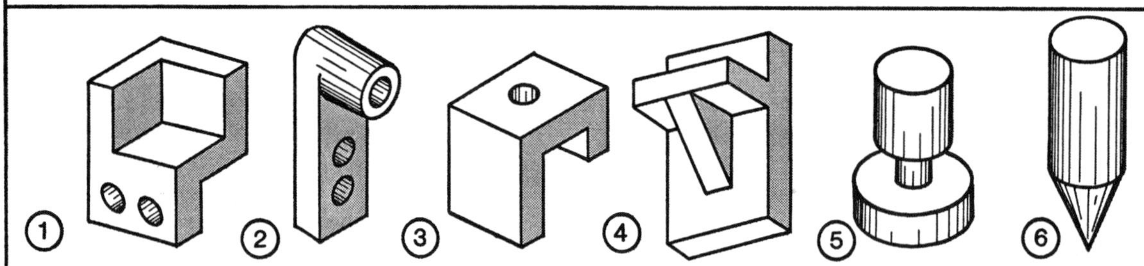

Several questions must be answered before the front view can be selected. Should the object be revolved? Will the circular parts appear as true circles in at least one view? Will there be a good contour view? Will the number of hidden edges be reduced to a minimum?

BEFORE DRAWING, PLAN THE METHOD OF MAKING THE DRAWING

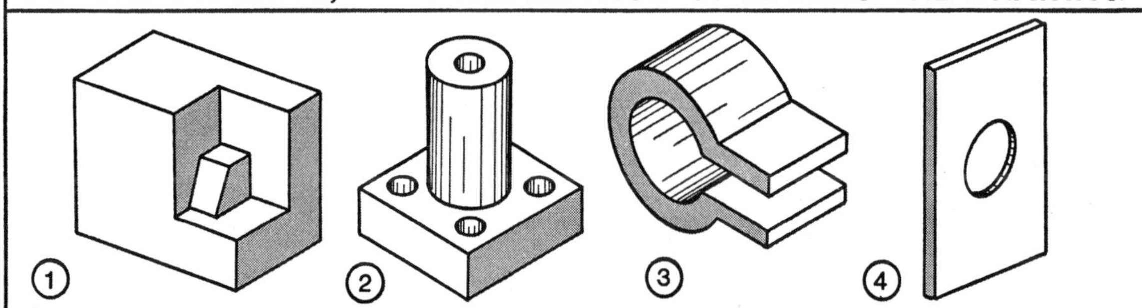

What is the basic shape of the object? Are there any center lines? Are the center lines the key to the problem? Which is the contour view? How many views are needed?

BEFORE DRAWING, PLAN FOR A WELL-CENTERED DRAWING

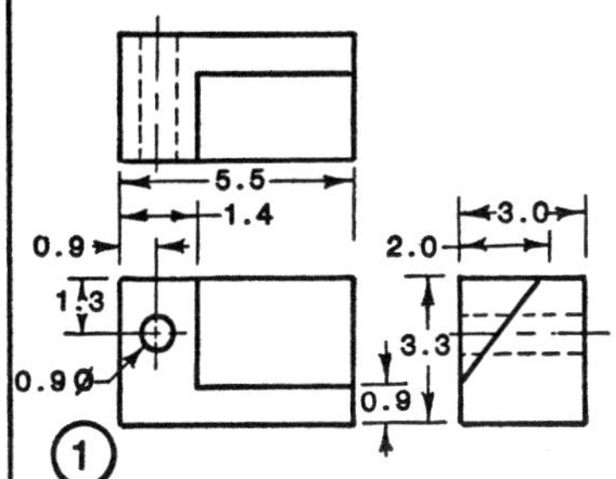

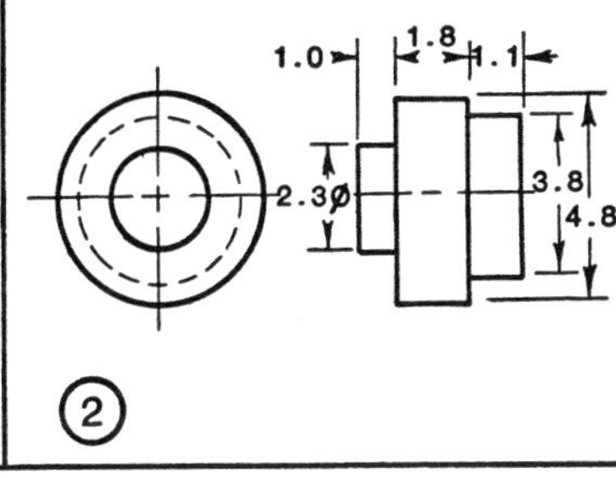

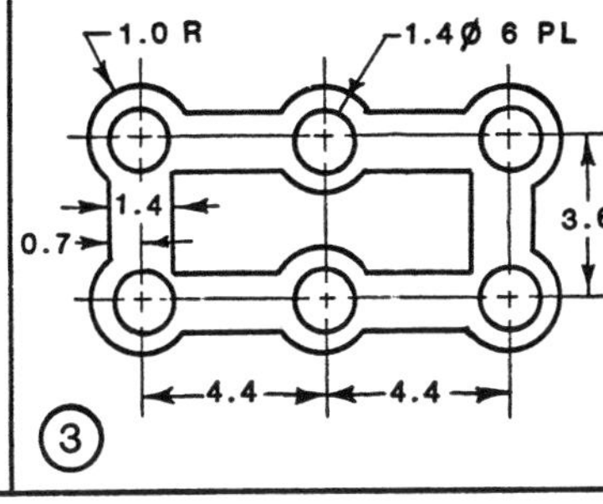

Determine the location of the starting point of the drawing.
Enough room must be left for dimensions and notes.
Every drawing must be centered.

MAKE THE DRAWING FOLLOW A SYSTEMATIC PLAN OF WORK

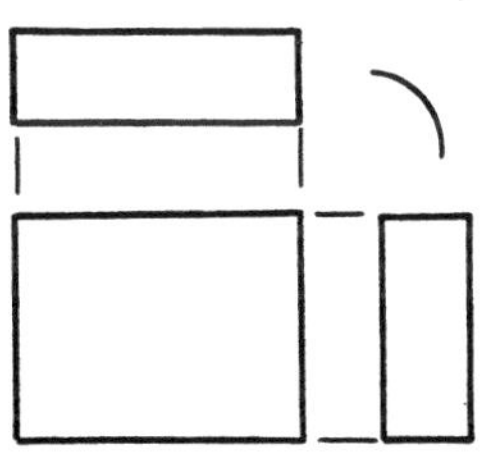

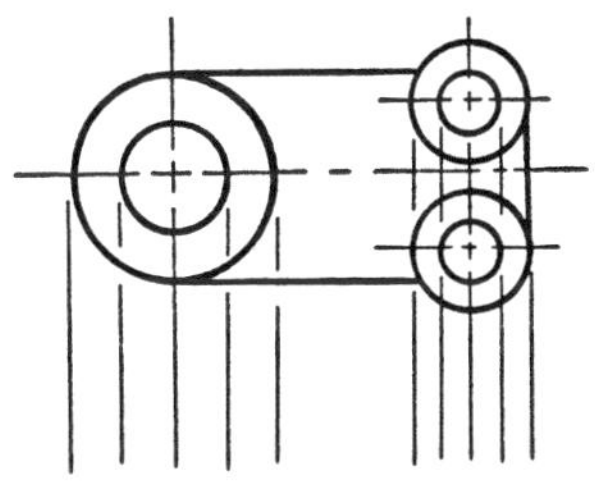

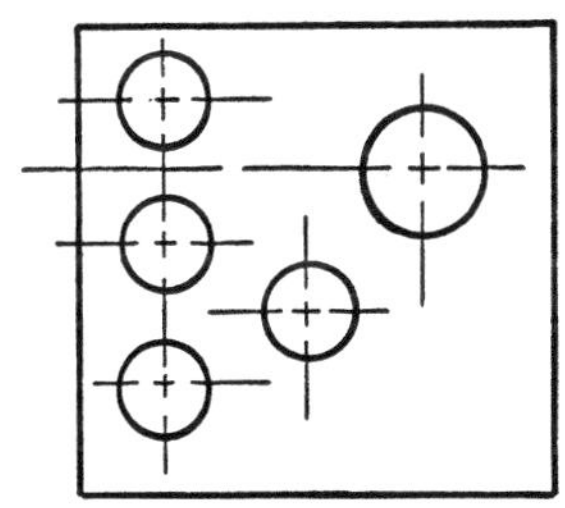

(1) You may start by drawing the three basic shapes.

(2) You may start by drawing the circular elements.

(3) You may build the drawing around the center lines.

ASSIGNMENT

Make an orthographic projection of one of these. Use the necessary number of views.

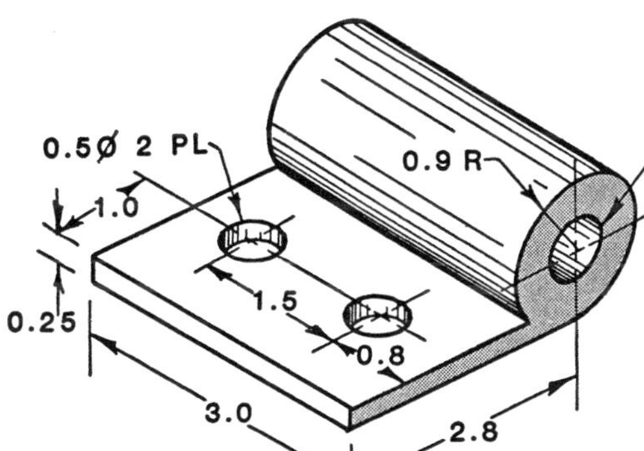

(1) BEARING

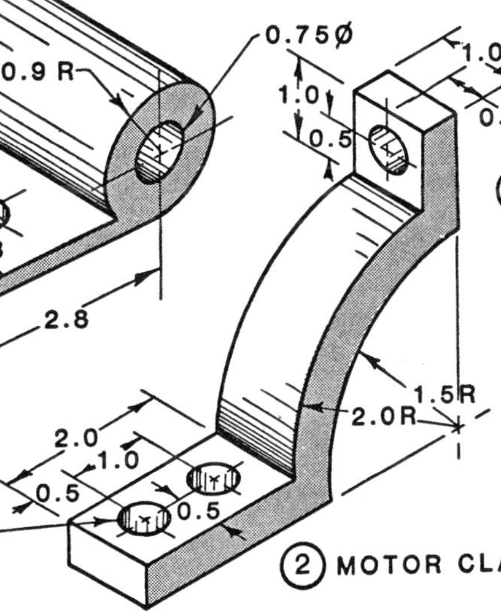

(2) MOTOR CLAMP

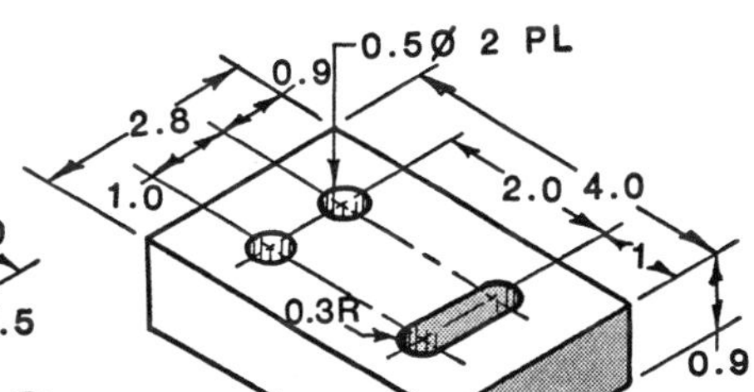

(3) GUIDE

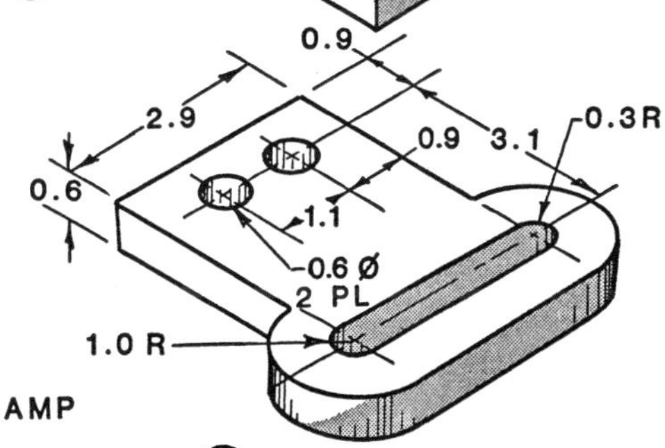

(4) SLOTTED GUIDE

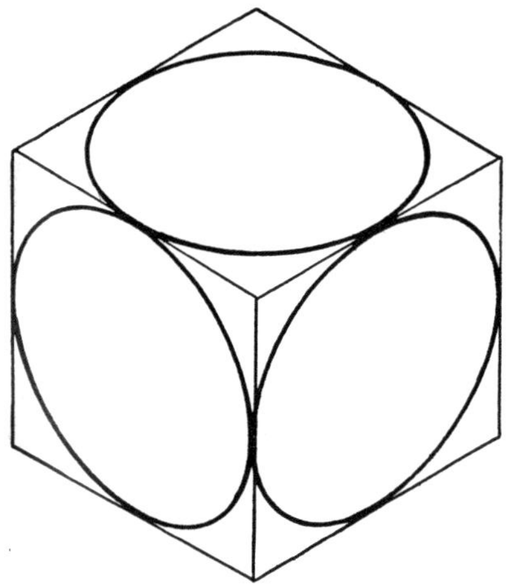

UNIT 11

PICTORIALS

An artist can draw a picture of an object with such skill that the drawing will look like the actual object. In doing this, the artist uses knowledge of drawing in perspective. Such a drawing has the appearance of looking like the object itself. This kind of drawing may be classified as a pictorial drawing. Pictorial drawings are an excellent aid to visualization and are helpful to those persons who have not learned to visualize an object from an orthographic projection.

The drafter has devised three basic ways of drawing a likeness of an object. However, all of these mechanical methods of pictorial drawing have one shortcoming: The drawing produced lacks the qualities of perspective that are to be found in a fine arts drawing. Although mechanical pictorial drawings appear to be distorted, this distortion does not affect their usefulness. The advantage is that the mechanical pictorial is faster and easier to draw.

The ability to make mechanical pictorial drawings will be a useful addition to your basic knowledge of drawing. This unit will cover oblique and isometric drawings.

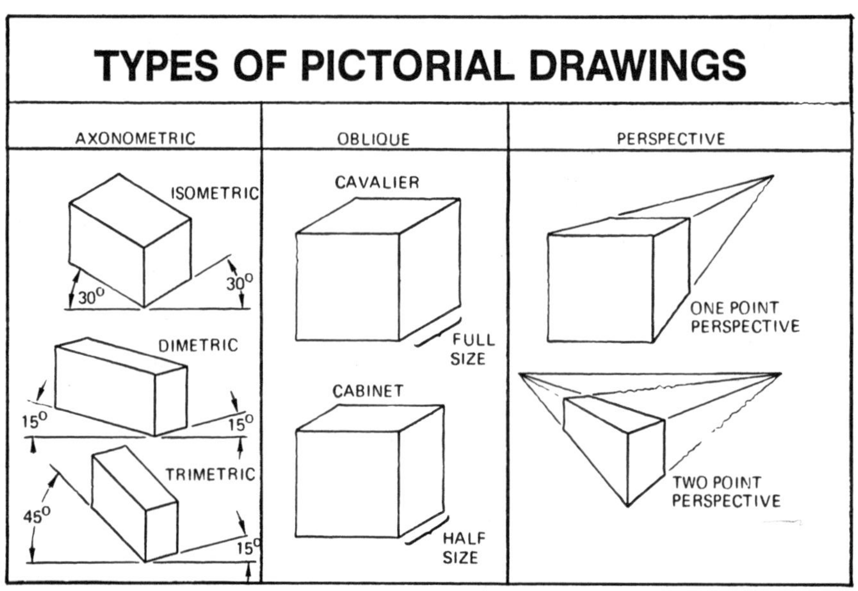

OBLIQUE DRAWINGS

There are two kinds of oblique drawings: cavalier and cabinet. A cavalier drawing and a cabinet drawing are drawn in the same manner, except that a cavalier drawing is always made to the exact measurements of the object. In a cabinet drawing, receding surfaces are drawn one-half size.

HOW TO MAKE AN OBLIQUE DRAWING (Step by Step)

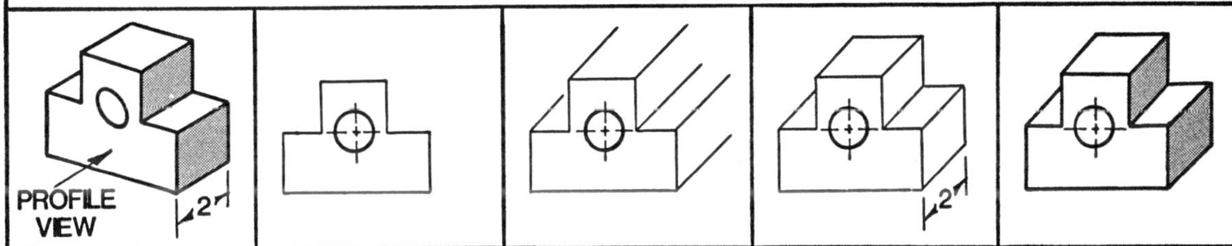

① To make an oblique drawing of this object, select the **profile view**.	② Draw the **profile view** as you would draw the **front view**.	③ Project back and to the right at 45° (30° may be used).	④ Measure the thickness and draw the back edge.	⑤ Darken the finished oblique drawing (cavalier).

REVOLVE THE OBJECT TO OBTAIN THE FRONT VIEW

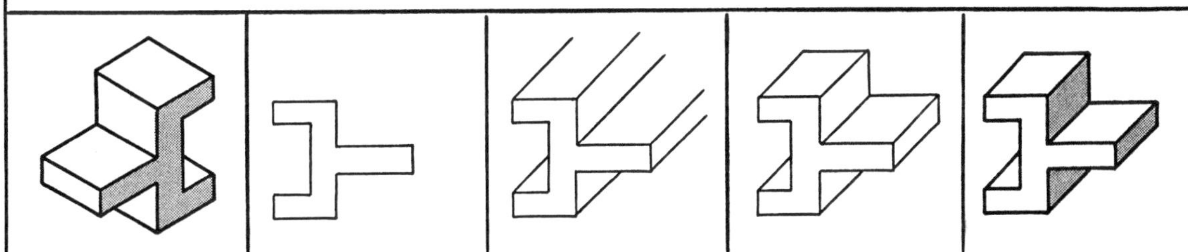

① Before the oblique drawing is made, this object should be revolved.	② Draw the **front view.**	③ Project back at 45° (or 30°).	④ Measure the thickness and draw the back edge.	⑤ Darken the oblique drawing (cavalier).

NOTICE THE DETAILS OF THESE CAVALIER DRAWINGS

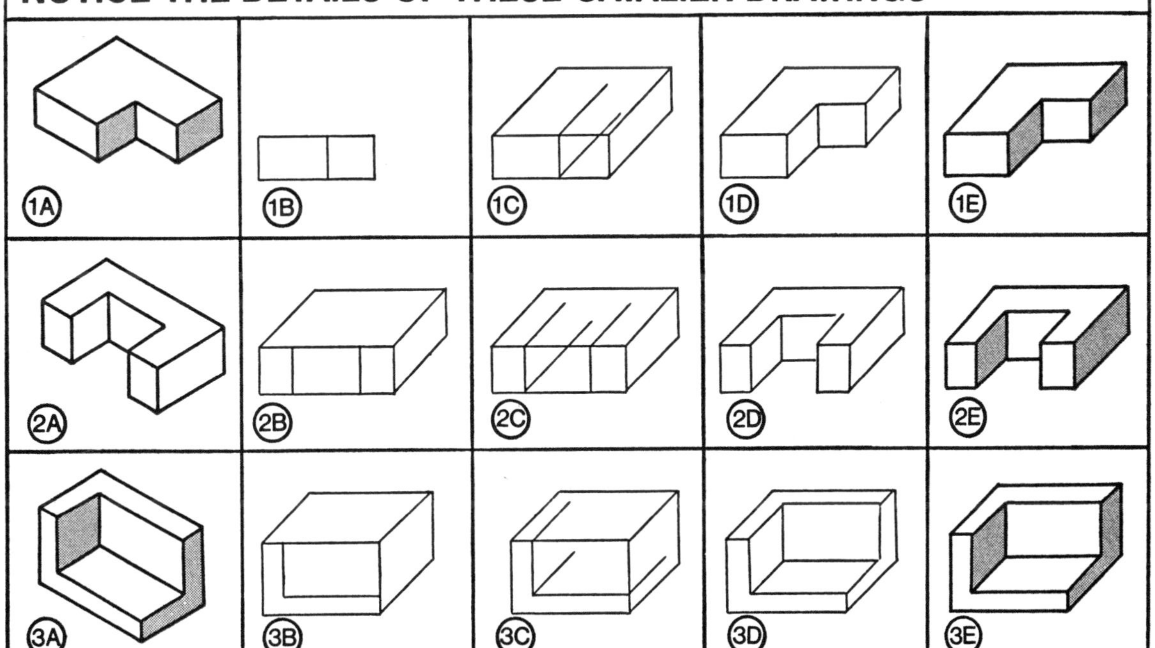

PROJECTING FORWARD AND BACKWARD

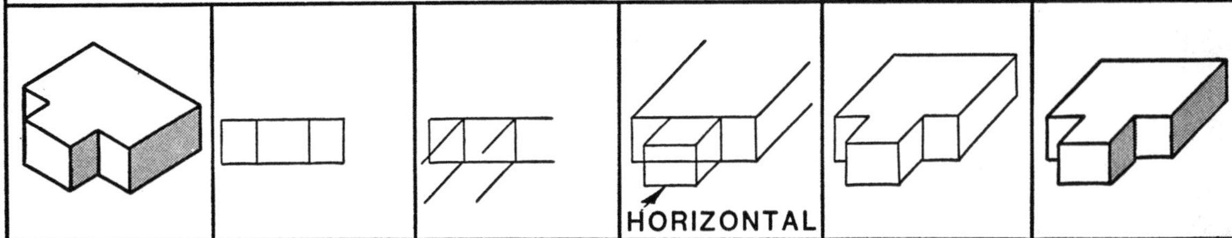

(1) Study the basic shape.	(2) Draw the **front view.**	(3) Project forward (45° or 30°).	(4) Keep the front face horizontal.	(5) Complete the back edges.	(6) Darken the oblique (cavalier) drawing.

STUDY THE STEPS USED IN MAKING THESE CAVALIER DRAWINGS

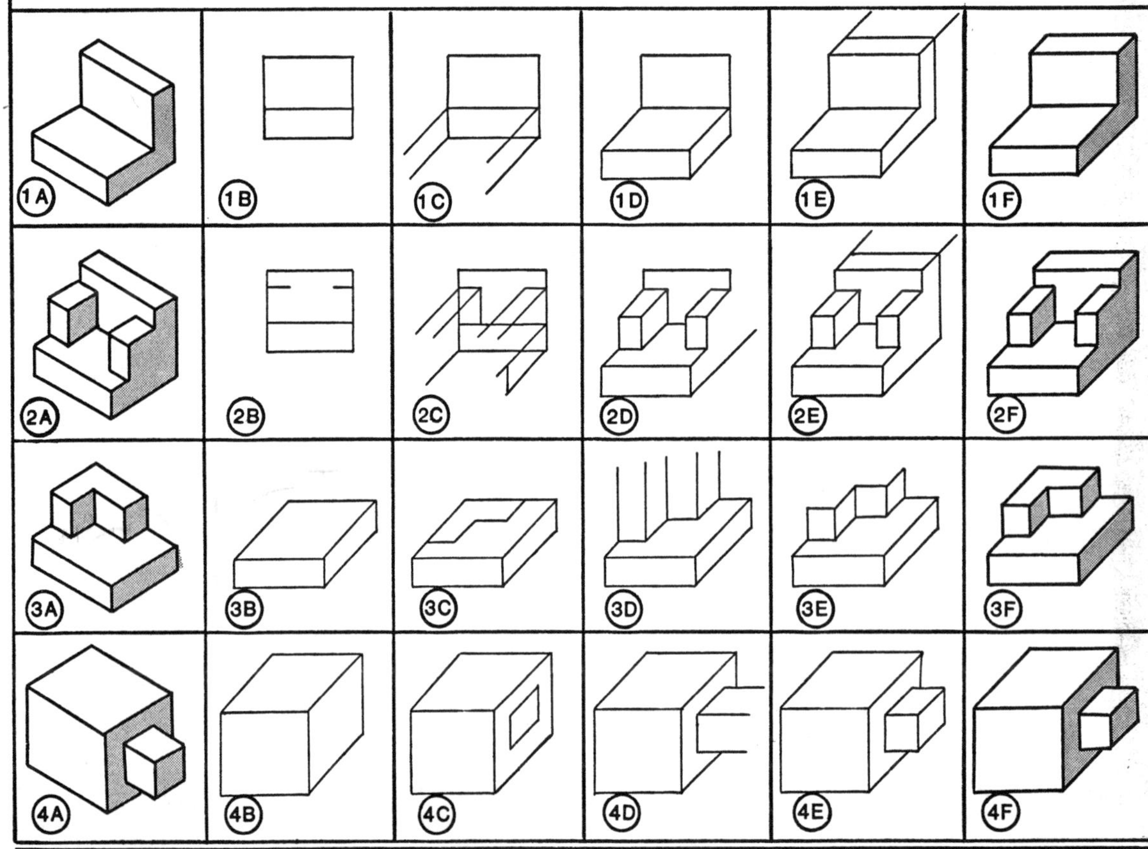

ASSIGNMENT

Make a cavalier drawing.

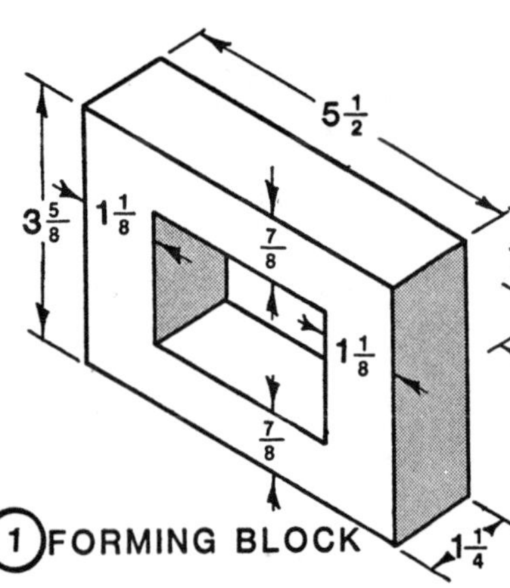

(1) FORMING BLOCK

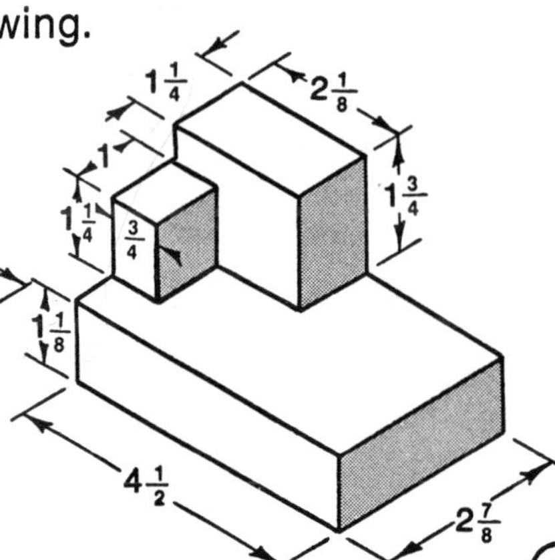

(2) STEP BLOCK

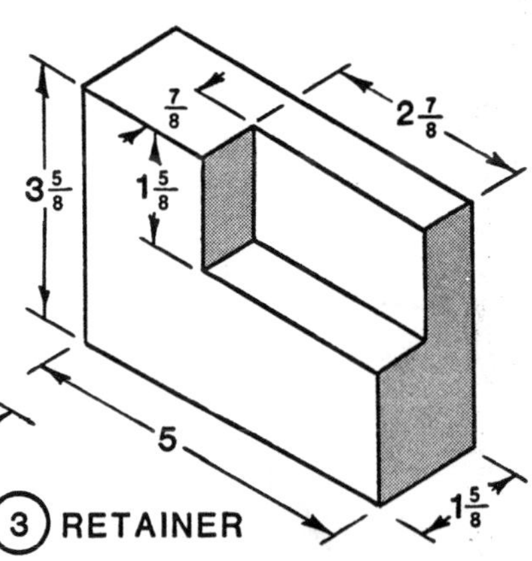

(3) RETAINER

OBLIQUE DRAWINGS OF CYLINDERS

There is a physical limit to the range of material that can be taught in one book on basic drafting. Because of this, only one position for cylinders and cylindrical holes will be shown. Other positions are advanced problems.

STEP-BY-STEP METHOD FOR CYLINDRICAL OBJECTS

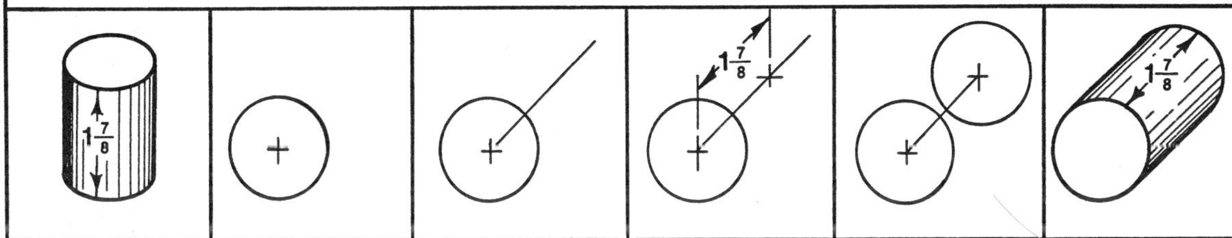

① Revolve the cylinder.	② Draw a complete circle as the front view.	③ Project the center back at 45°.	④ Measure the thickness on the center line.	⑤ Draw the second circle.	⑥ Darken the outline of the object.

STUDY THE STEP-BY-STEP DRAWINGS OF THESE FIVE OBJECTS

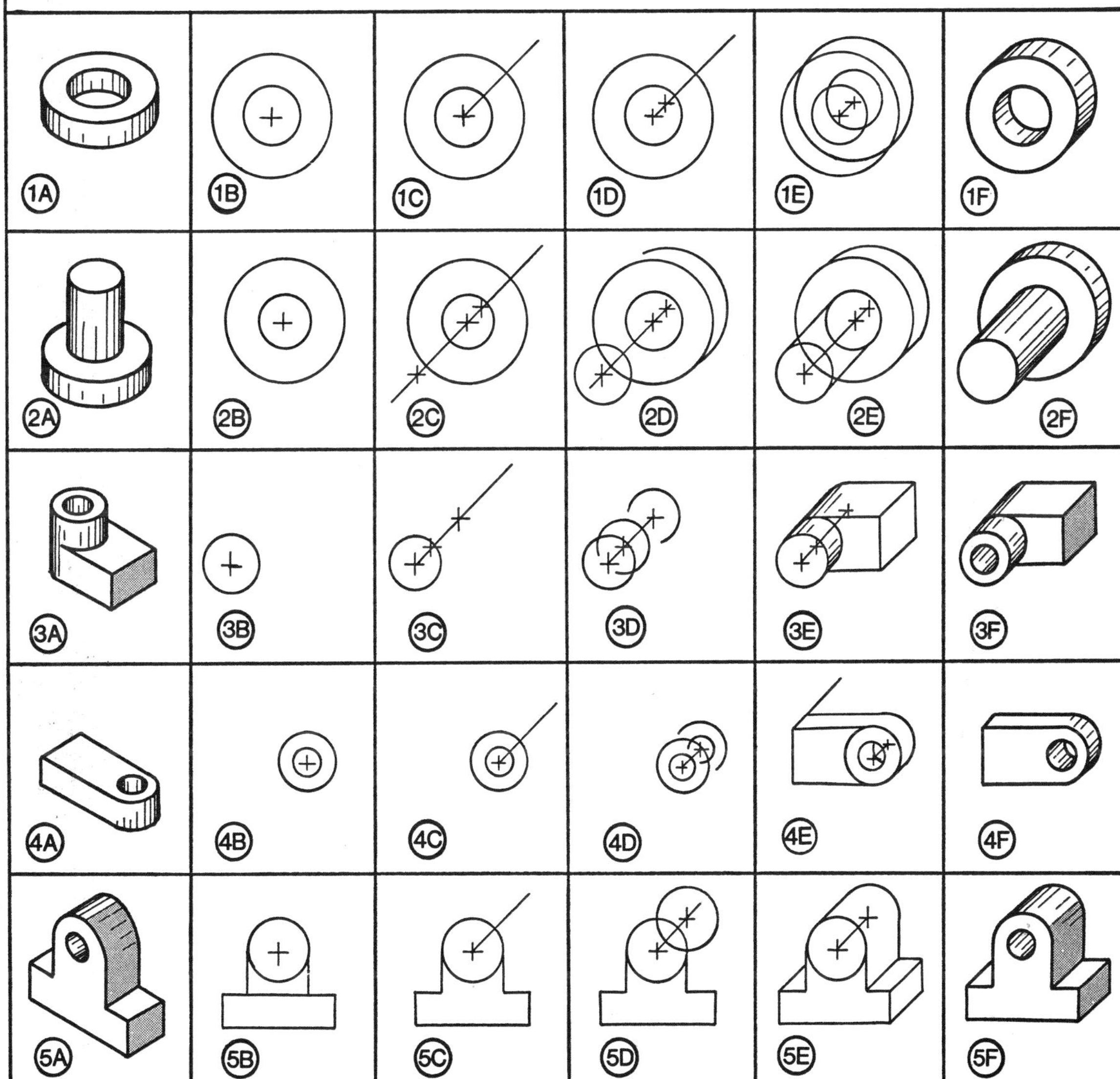

SLOPING LINES IN OBLIQUE DRAWING

A sloping line in an oblique drawing presents a slight problem. The drawings below are presented in order to help you with this problem.

SLOPING LINES IN OBLIQUE DRAWING

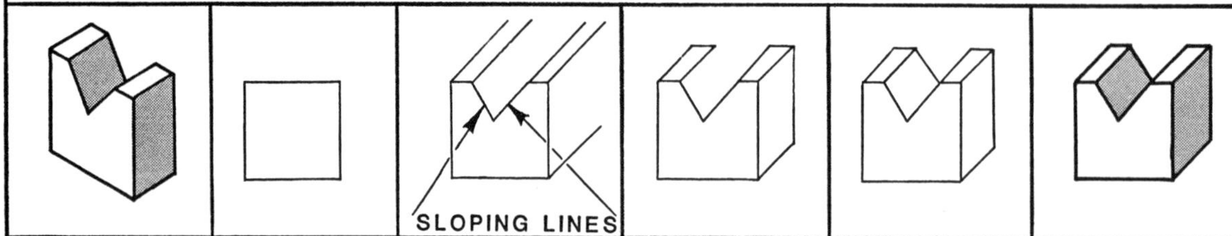

① Study the basic shape.	② Draw the basic shape.	③ Draw the notch and project corner backs.	④ Draw the thickness.	⑤ Draw the rear sloping edges parallel to the front ones.

Be certain that the back edge of the sloping surface is parallel to the front edge. Note the optical illusion of the depth between the isometric and oblique drawings. They are the same dimension.

STUDY THE SLOPING LINES ON THESE OBLIQUE DRAWINGS

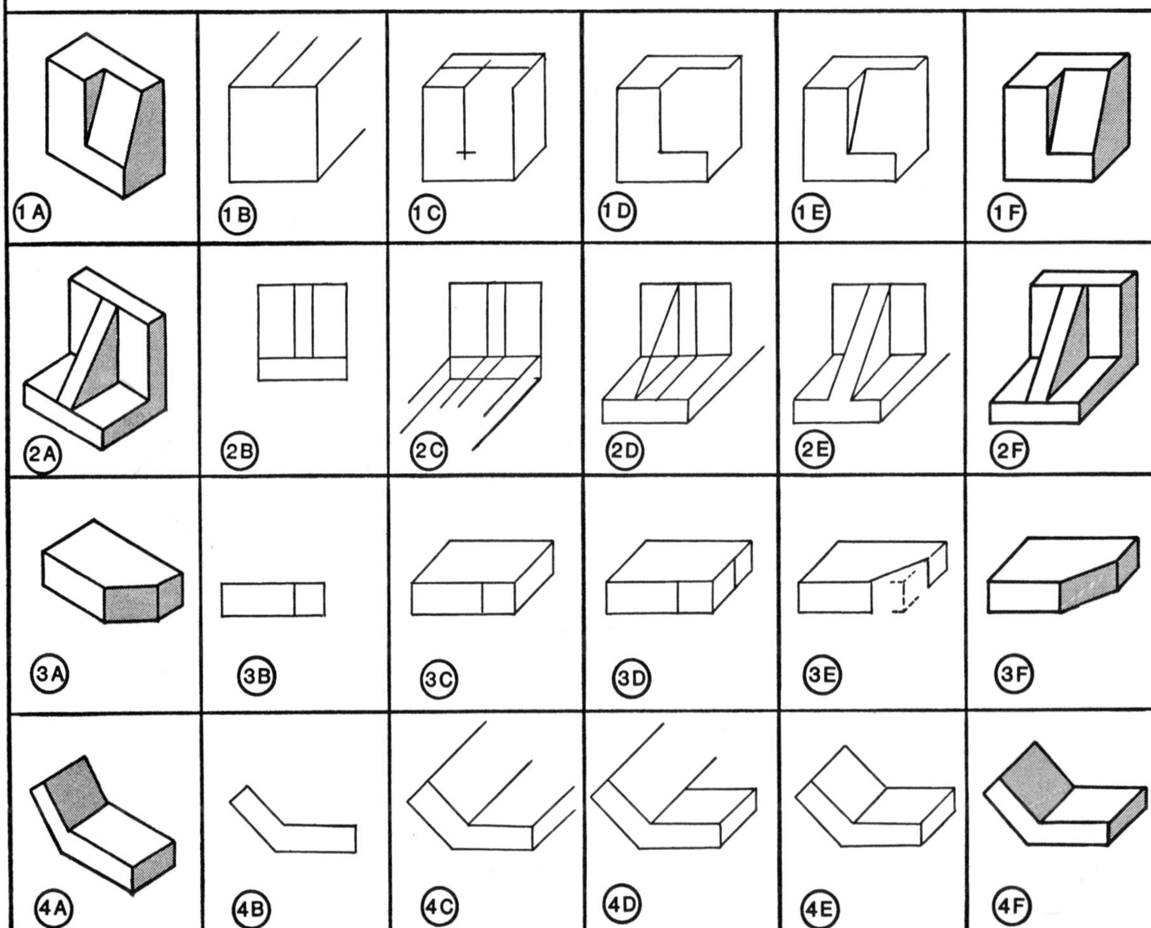

The back edge of an object often repeats the outline of the front edge. Always study the shape of the object. Plan your method of drawing before starting to draw.

CABINET DRAWINGS

Cabinet drawings are made to reduce the effect of distortion found in cavalier drawings.

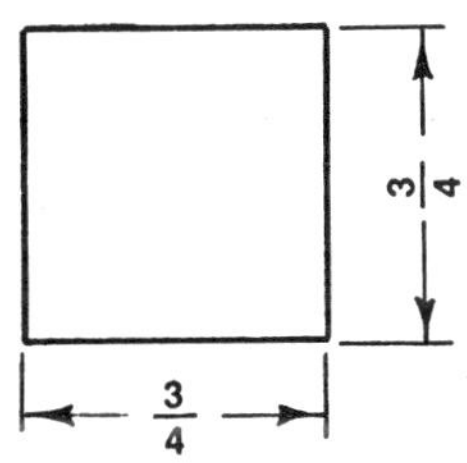

(1) This is a 3/4" square.

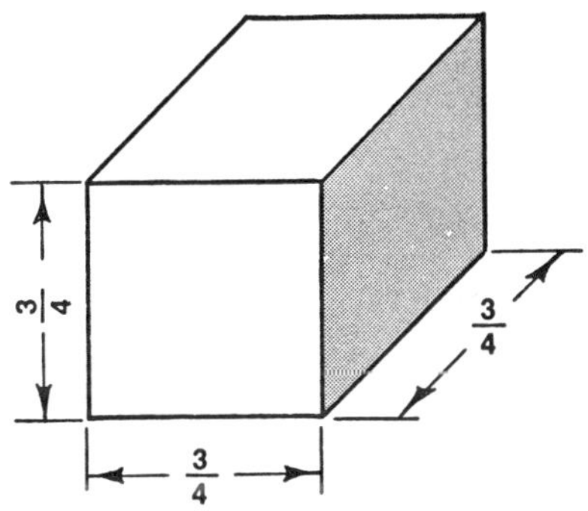

(2) This is a cavalier drawing of a 3/4" cube.

It is 3/4" high, 3/4" wide, and 3/4" thick. Note the optical illusion in its depth.

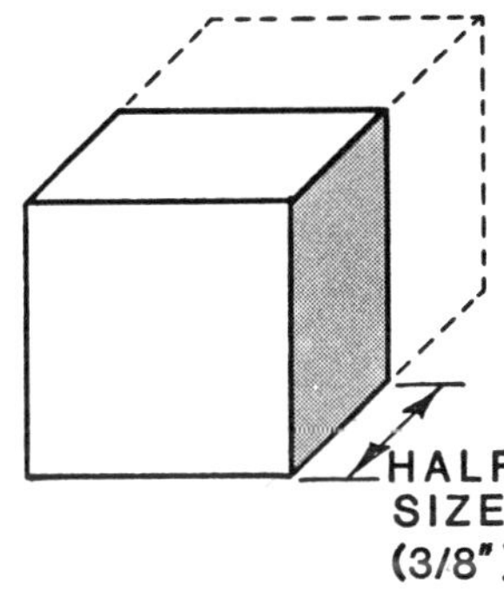

(3) A cabinet drawing of a 3/4" cube is drawn only one half as thick as the dimension indicates. Does it look like a cube now?

CABINET DRAWINGS ARE ONLY HALF AS THICK AS DIMENSIONED

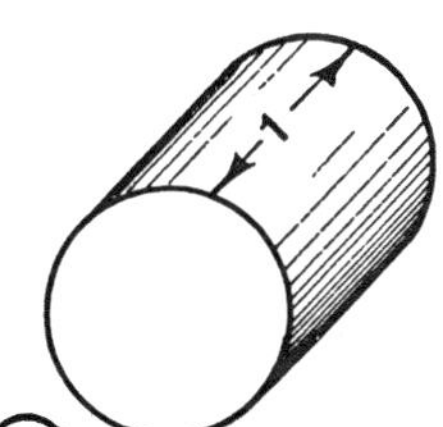

(1)

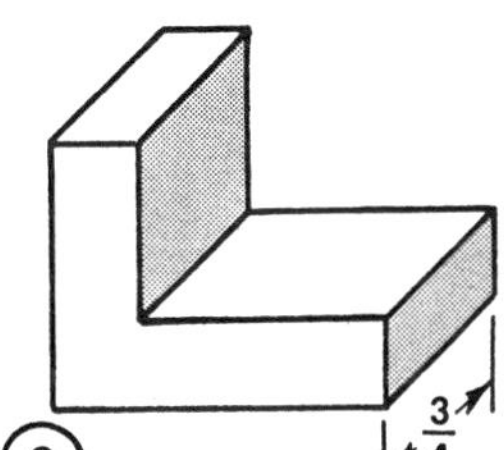

(2)

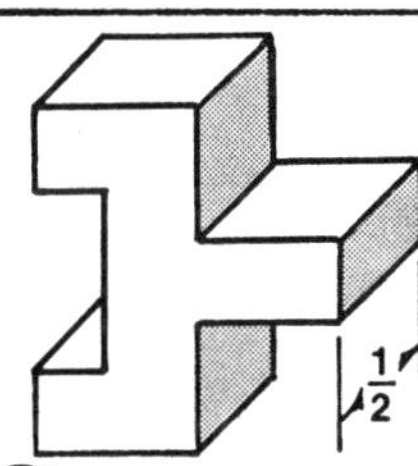

(3)

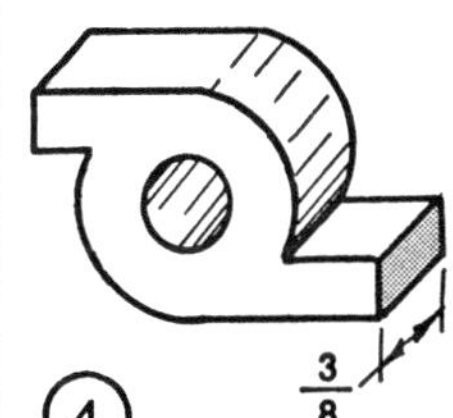

(4)

These are cabinet drawings. The horizontal and vertical lines are true size. The receding lines are only one half as thick as the dimension on the original drawings.

COMPARISONS BETWEEN CAVALIER AND CABINET DRAWING

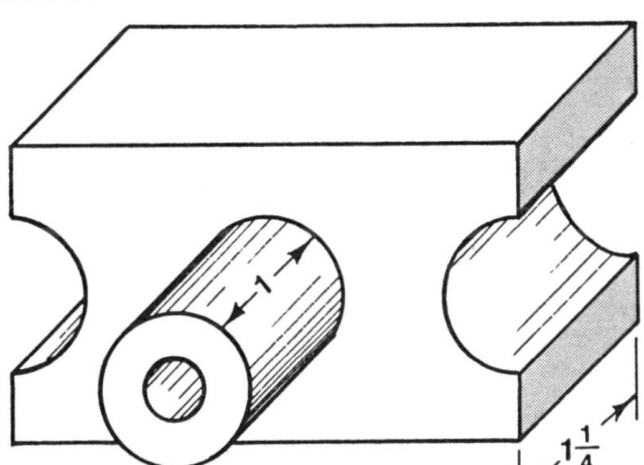

(1) CAVALIER DRAWING

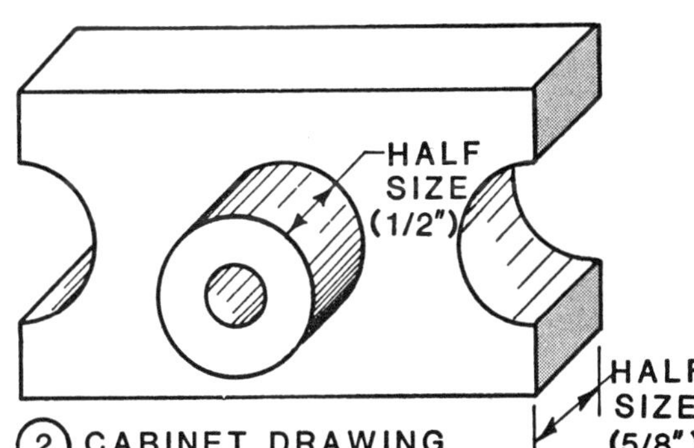

(2) CABINET DRAWING

A cavalier drawing and a cabinet drawing are drawn in the same manner with the exception that in the cabinet drawing, receding surfaces are drawn one-half size. The true size dimensions are placed on the cabinet drawing.

REMINDERS AND SUGGESTIONS

1. Always start by drawing what would normally appear to be the front view of an orthographic projection.
2. Always revolve a cylinder so that the circle faces you.
3. Revolve a long object so that the long dimension is either horizontal or vertical. Never let the long dimension recede into the picture.
4. Make all cabinet drawings one-half of the thickness indicated.

HOW WOULD YOU PLAN AND DRAW EACH OF THESE?

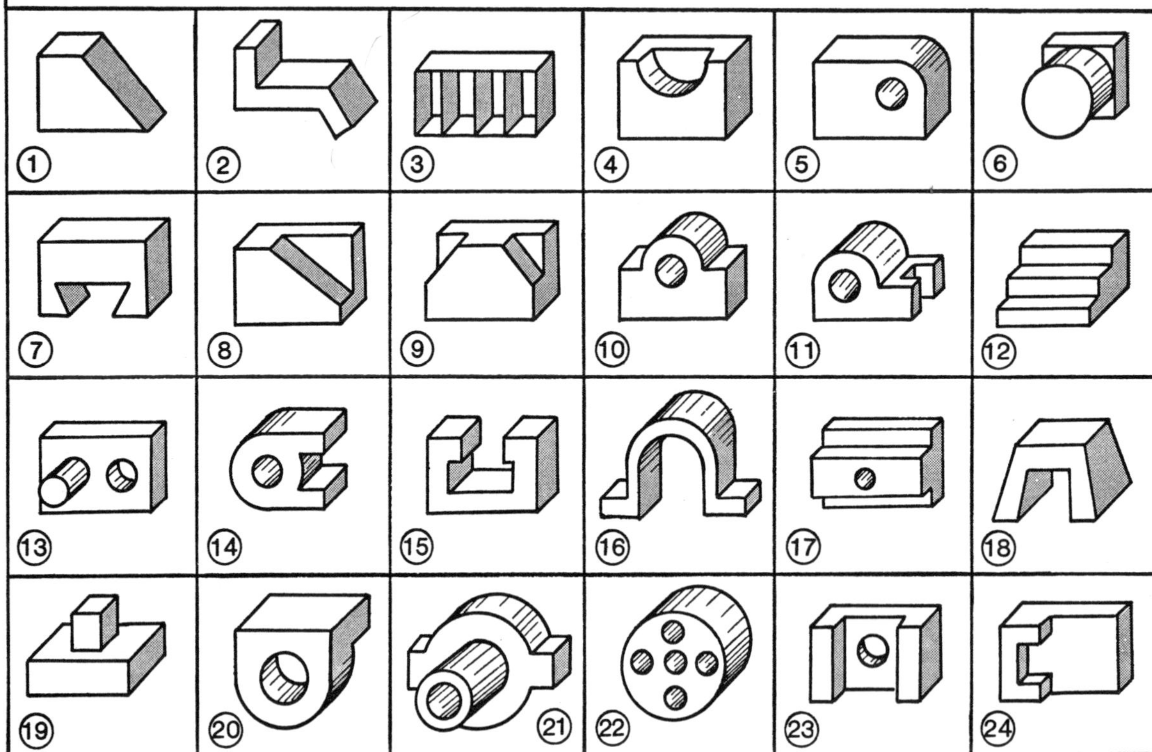

ASSIGNMENT

Make a full-scale cabinet drawing or cabinet drawing of one of these objects.

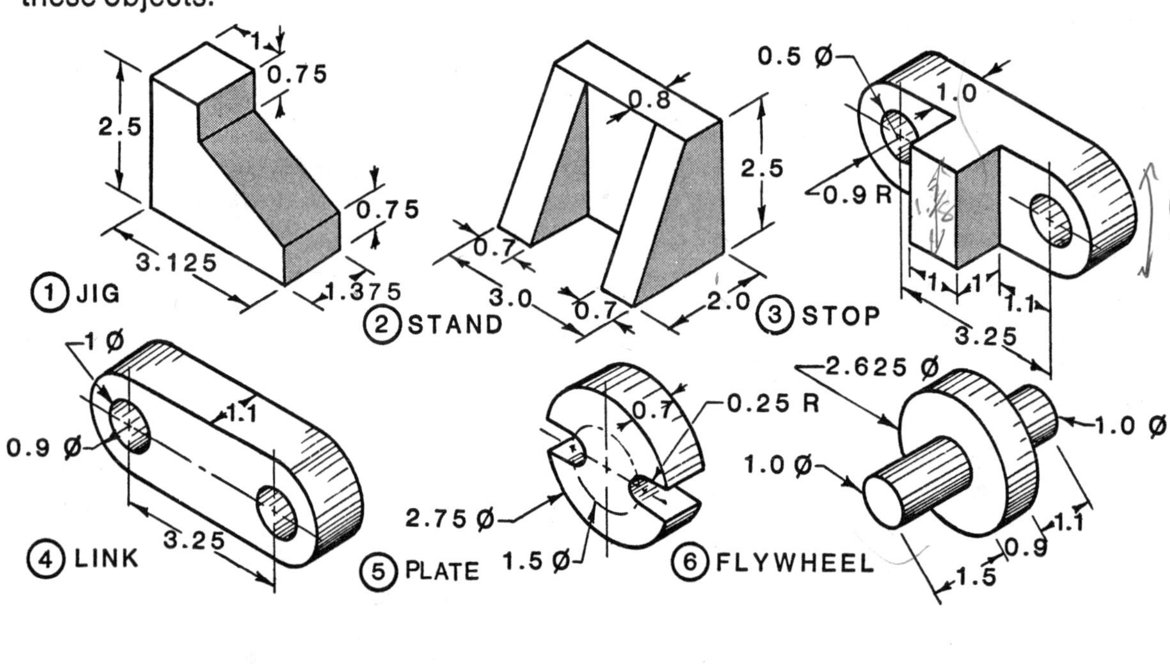

ISOMETRIC DRAWINGS

Isometric drawings are the most common type of pictorial drawings. Although there is a small amount of distortion present in isometric drawings, it is not as noticeable as that found in oblique drawings or cabinet drawings. All vertical and isometric (30° lines) are true size.

ISOMETRIC DRAWING STEP BY STEP

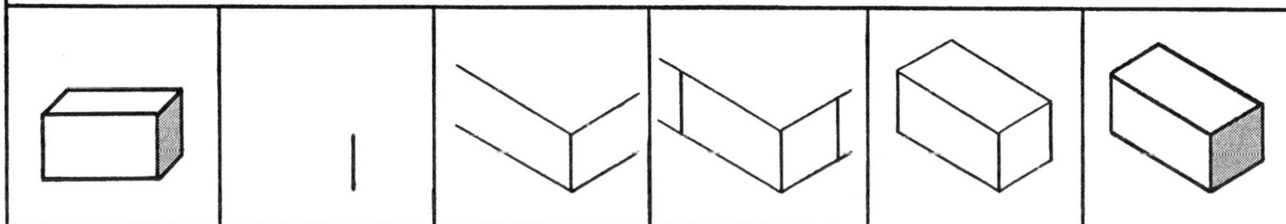

(1) This is a cabinet drawing of a block.	(2) Draw the nearest vertical edge. All dimensions are full size.	(3) Project to the right and the left at 30°.	(4) Measure and draw the vertical back edges.	(5) Draw each remaining edge with a 30° line.	(6) This is a complete isometric drawing.

DRAWING NOTCHES

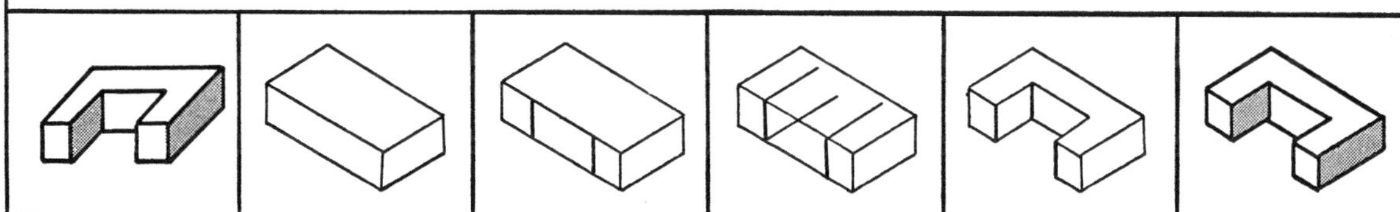

(1) Study the basic shape.	(2) Draw the basic shape in isometric.	(3) Locate the notch.	(4) Project at 30°.	(5) Measure and draw the back of the notch.	(6) This is an isometric drawing.

DRAWING RIDGES

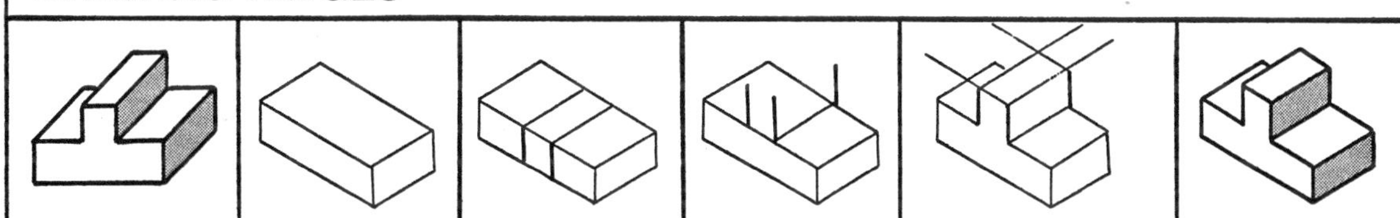

(1) Study the object.	(2) Draw the basic shape of the base.	(3) Locate the ridge.	(4) Raise the ridge, using vertical lines.	(5) Finish the ridge as a separate object.	(6) This is an isometric drawing.

STUDY THESE ISOMETRIC DRAWINGS

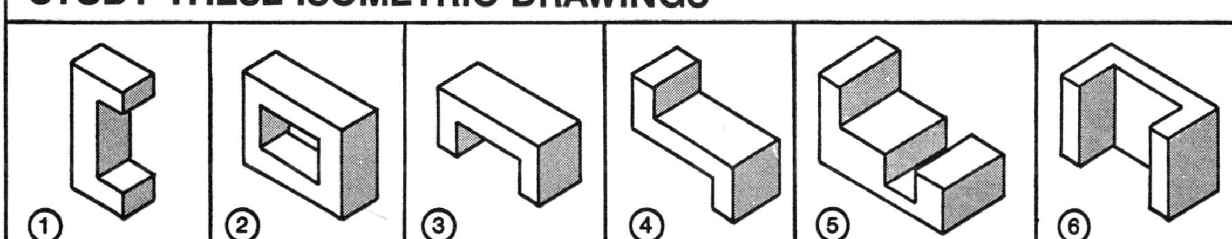

THE BASIC IDEA IN ISOMETRIC DRAWING

1. All vertical lines remain vertical.
2. The right side of an object slants to the right at 30°.
3. The left side of an object slants to the left at 30°.
4. All measurements are full-length measurements.

NONISOMETRIC LINES (SLOPING LINES) ARE NOT TRUE SIZE					
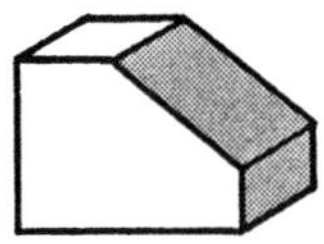	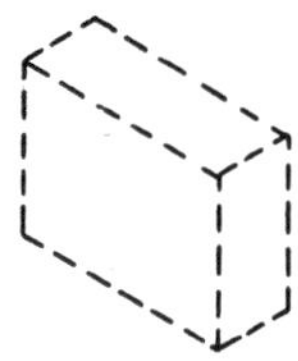	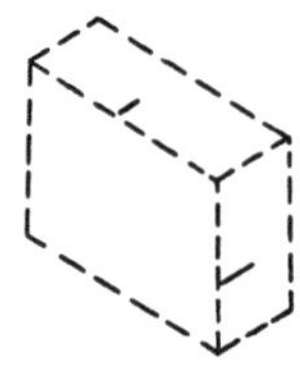	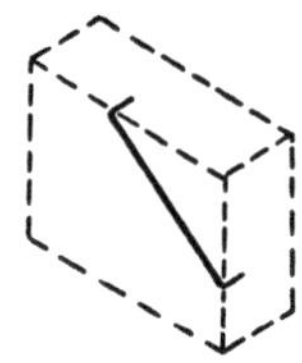	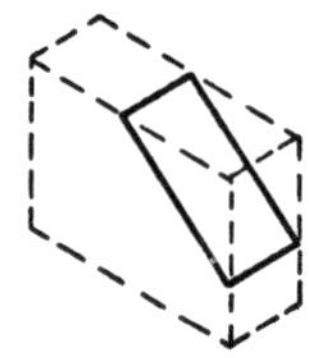	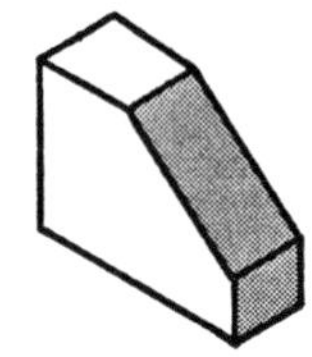
(1) Study the basic shape.	(2) Draw the basic shape.	(3) Mark off the position of the nonisometric lines on the isometric lines (true size).	(4) Draw the first line with a "free" triangle.	(5) Draw the second line parallel to the first.	(6) This is an isometric drawing.

HOW TO CENTER AN ISOMETRIC DRAWING (A ROUGH ESTIMATE)		
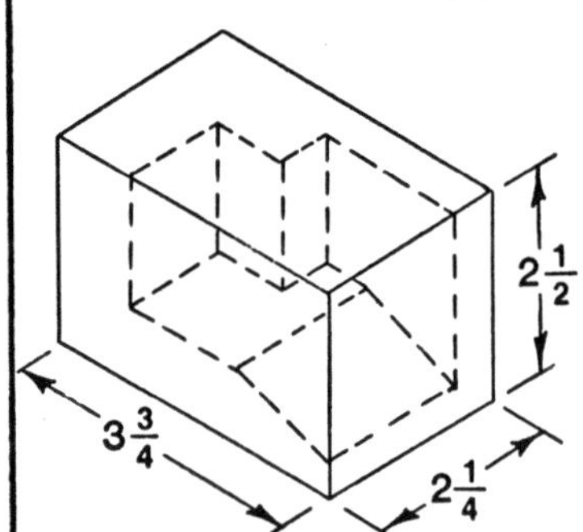	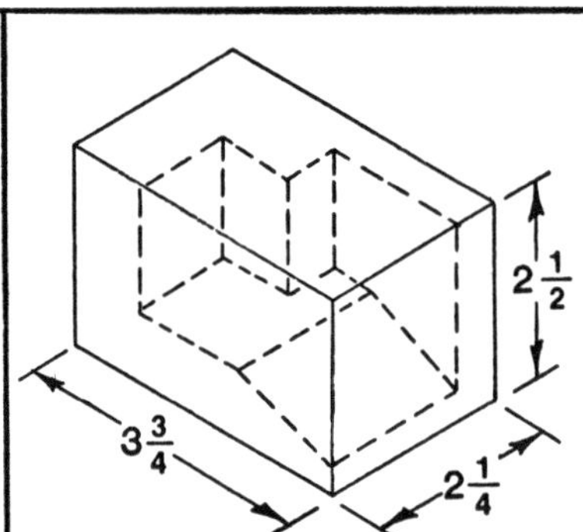	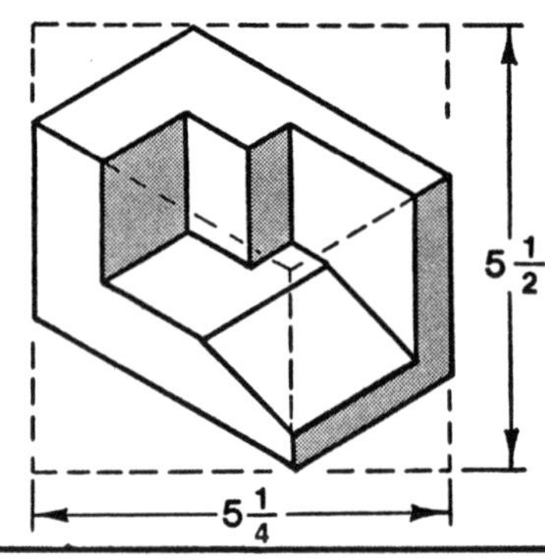
(1) **To obtain the full length:** Add the overall length ($3\frac{3}{4}$) to the overall width ($2\frac{1}{4}$) ($3\frac{3}{4} + 2\frac{1}{4} + 6$). Take seven-eighths of this sum ($\frac{7}{8} \times \frac{6}{1} + 5\frac{1}{4}$). See panel (3).	(2) **To obtain the full height:** Add one-half of the overall width ($1\frac{1}{8} + 1\frac{7}{8} = 3$). Add the full height to this ($3 + 2\frac{1}{2} = 5\frac{1}{2}$). See panel (3).	(3) **The layout shape:** The drawing shown in (1), then, will fit into a rectangle approximately $5\frac{1}{4}$ long by $5\frac{1}{2}$ high.

ASSIGNMENT

Draw the isometric drawing.

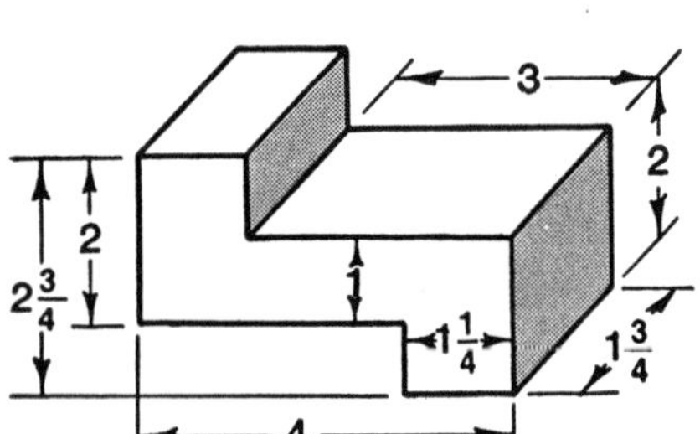

(1) BENCH HOOK

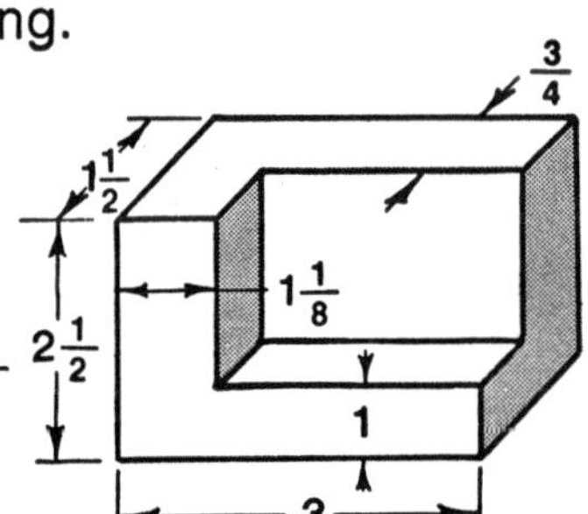

(2) CORNER BLOCK

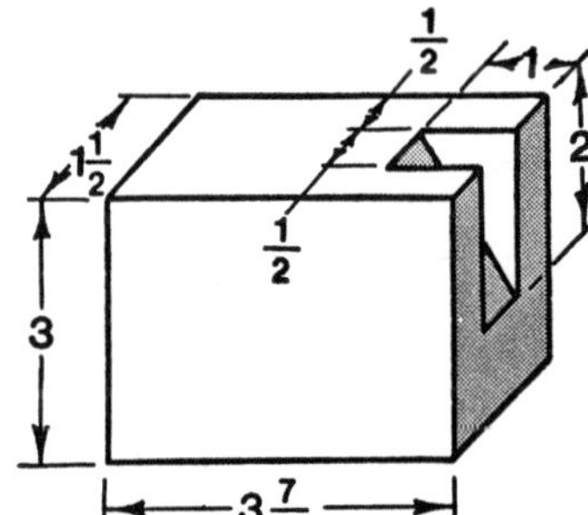

(3) GUIDE BLOCK

ISOMETRIC CIRCLES

It would be impossible to make these isometric drawings without a knowledge of how to construct isometric circles.

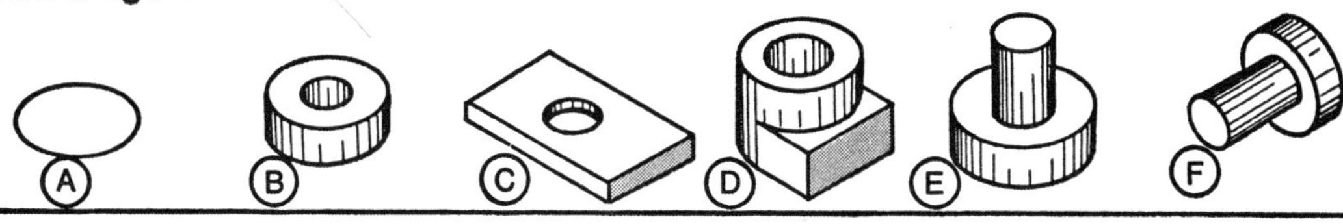

BASIC SHAPES

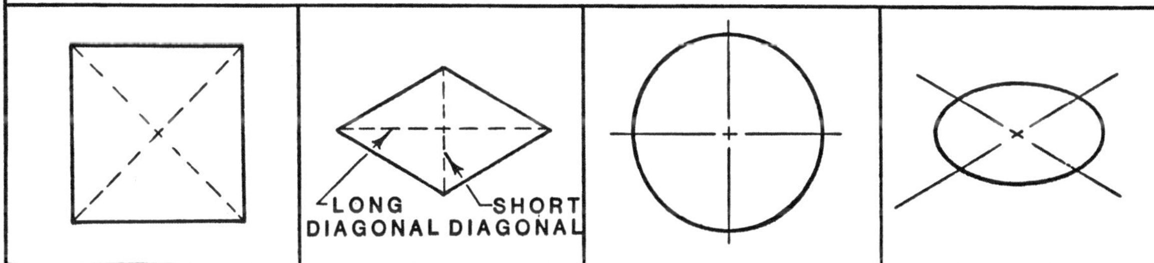

① This is a true square (with diagonals).	② This is an isometric square (with diagonals).	③ This is a true circle.	④ This is an isometric circle.

CONSTRUCTING THE FRAME FOR AN ISOMETRIC CIRCLE

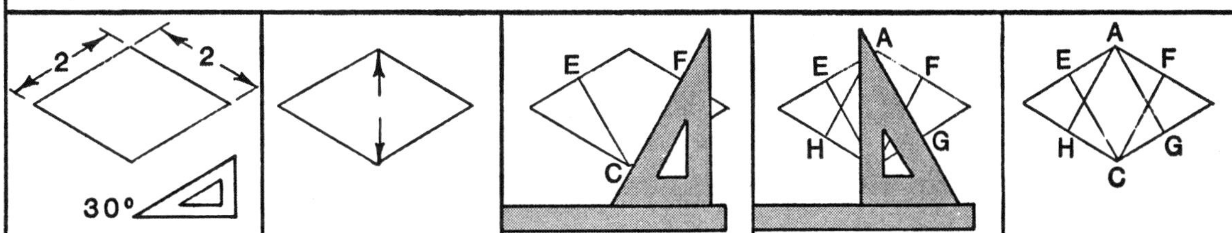

① Draw a 2″ isometric square for a 2″ diameter isometric circle.	② You must use both ends of the short diagonal (obtuse angles).	③ Draw lines "EC" and "CF" each at an angle of 60° with the horizontal.	④ Draw lines "HA" and "AG" each at an angle of 60° with the horizontal.	⑤ The completed framework.

CONSTRUCTING THE ISOMETRIC CIRCLE (The Square Method)

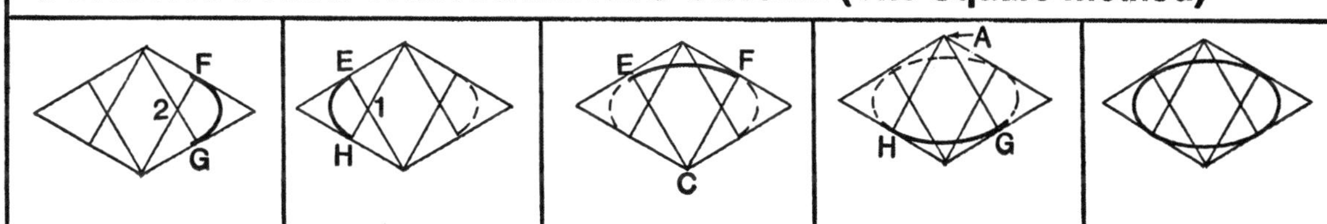

① With radius "2F," swing arc "FG."	② With radius "1E," swing arc "EH."	③ With radius "CE," swing arc "EF."	④ With radius "AG," swing arc "HG."	⑤ The completed isometric circle.

REMINDERS AND SUGGESTIONS

1. The isometric square must be drawn exactly the same size as the diameter of the isometric circle that you wish to produce. For example: A 2½″ isometric square must be drawn in order to produce a 2½″ diameter isometric circle.
2. Measure accurately.

THE CROSS METHOD FOR ISOMETRIC CIRCLES

Students often have difficulty in centering an isometric circle in the exact spot where they wish to center it. Here is a method that will locate an isometric circle exactly where you want it.

THE CROSS METHOD FOR LOCATING AN ISOMETRIC CIRCLE

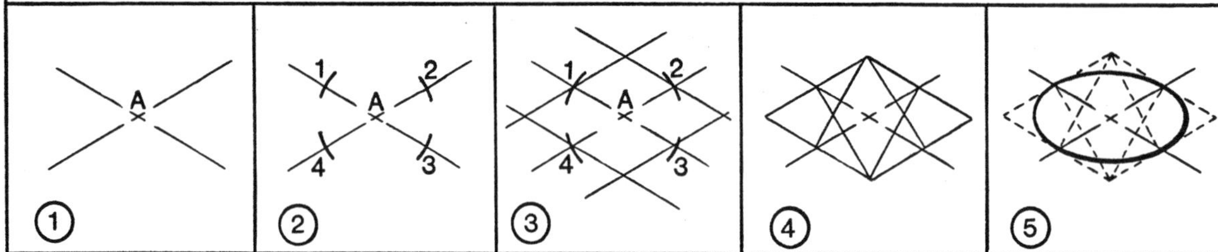

1. In order to draw a 3″ diameter circle at "A," draw two 30° lines through "A."
2. With "A" as center, and with a 1½″ radius, swing four arcs through points 1, 2, 3, and 4.
3. Draw 30° lines through points 1, 2, 3, 4 (you have an isometric square).
4. Draw the frame for an isometric circle in the usual manner.
5. Complete the isometric circle.

WHEN TO USE THE CROSS METHOD FOR ISOMETRIC CIRCLES

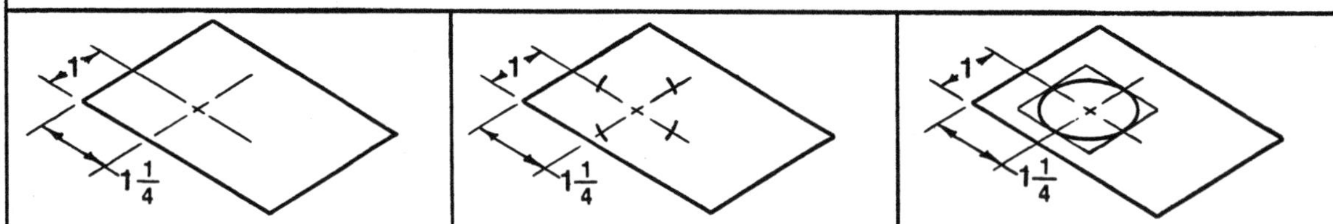

① The exact center of an isometric circle may be accurately located.	② The radius is used to help establish the isometric frame.	③ Use the cross method for locating isometric circles.

WHEN TO USE THE SQUARE METHOD FOR ISOMETRIC CIRCLES

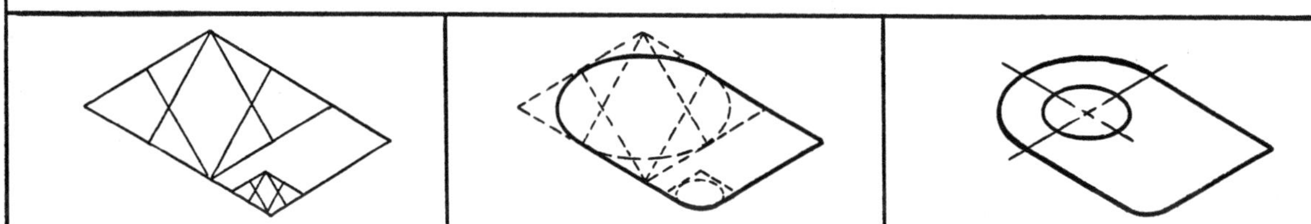

① Use the square method for rounded ends and corners.	② The square method was used here.	③ Both methods may be used on the same drawing.

ASSIGNMENT

Make an isometric drawing of one of these drawings.

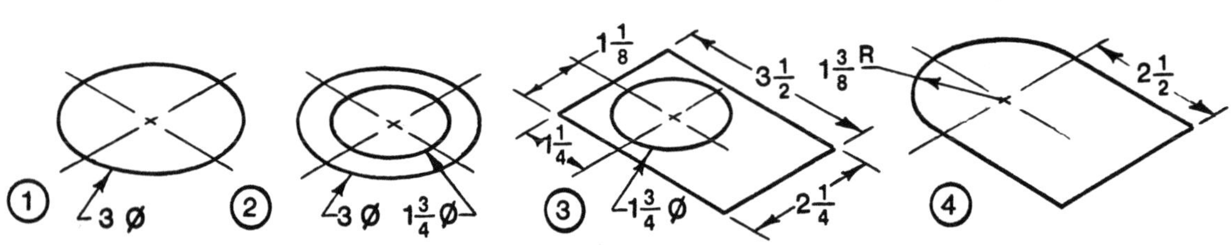

VERTICAL ISOMETRIC CIRCLES

CONSTRUCTING THE FRAME FOR A VERTICAL ISOMETRIC CIRCLE

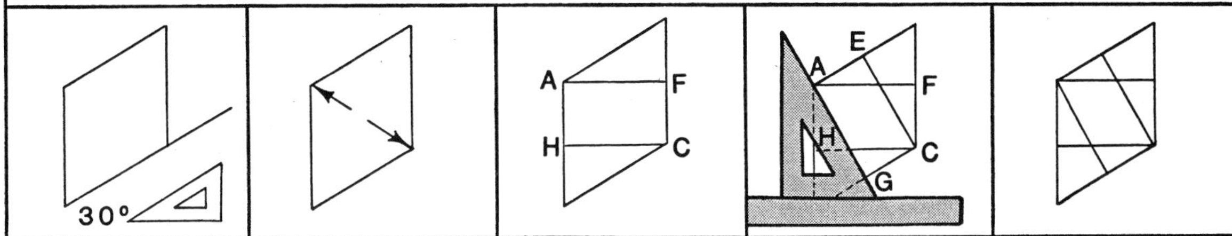

① Construct an isometric square in this position.

② Use the short diagonal only.

③ Draw horizontal lines "AF" and "HC."

④ Draw lines "CE" and "AG," each at an angle of 60° with the horizontal.

⑤ The completed frame.

CONSTRUCTING THE ISOMETRIC CIRCLE

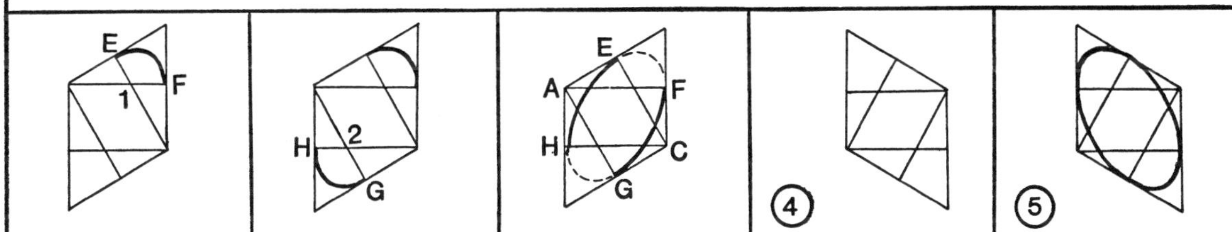

① Swing arc "EF" from center, 1.

② Swing arc "HG" from center, 2.

③ Swing arc "HE" from "C," and "FG" from "A."

Reverse the frame as at ④ in order to reverse the isometric circle as at ⑤.

THE ISOMETRIC CUBE

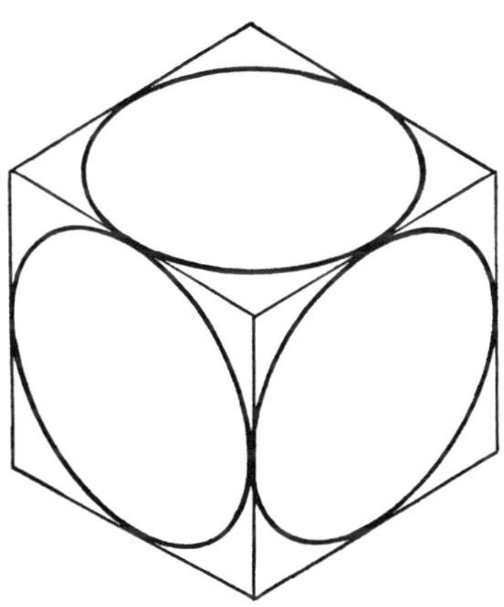

ASSIGNMENT

Draw a 3" diameter isometric cube and inscribe an isometric circle in each of its three surfaces.

This is an extremely important exercise. It is the key to the drawing and to application of the isometric circle in many practical problems.

HELPFUL SUGGESTIONS

1. Always draw the framework for an isometric circle as lightly as possible.
2. Do not punch holes in the paper with your compass.
3. Measure accurately.
4. Keep your pencil sharp.

CROSS METHOD USED FOR CENTERING VERTICAL ISOMETRIC CIRCLES

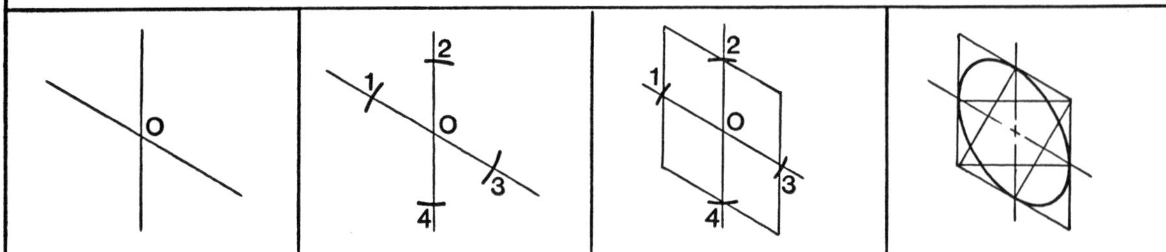

(1) To center a vertical isometric circle at "0": Draw a vertical line and a 30° line crossing at "0."	(2) With "0" as center, and a radius equal to the radius of the required circle, locate points 1, 2, 3, and 4.	(3) Draw vertical lines through 1 and 3 and 30° lines through 2 and 4.	(4) Complete the isometric circle in the usual way.

WHEN TO USE SQUARE METHOD FOR VERTICAL ISOMETRIC CIRCLES

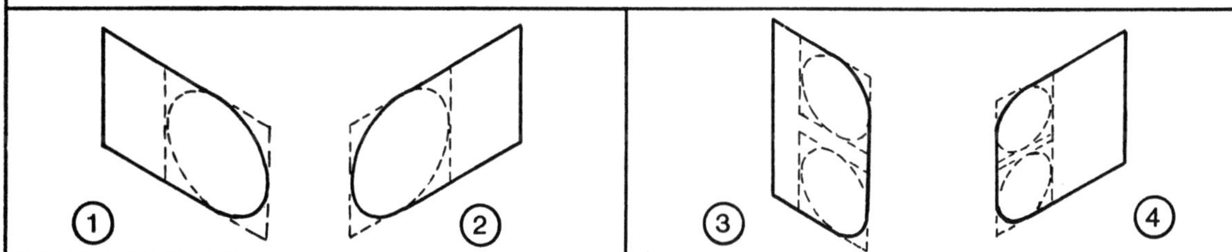

Use the square method for rounded ends and rounded corners.

WHEN TO USE CROSS METHOD FOR VERTICAL ISOMETRIC CIRCLES

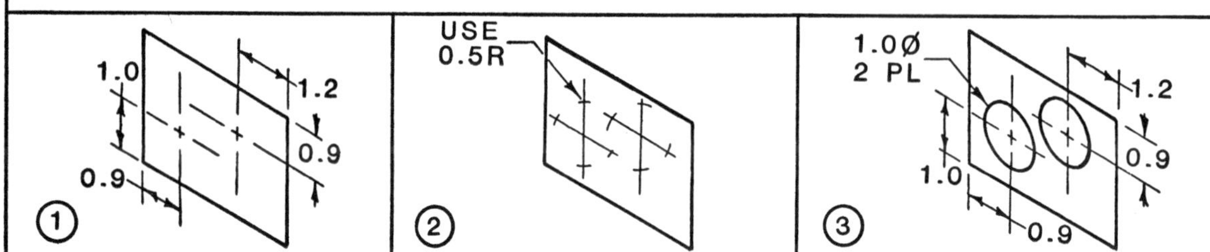

Use the cross method to locate a circle in a definite place.

ASSIGNMENT

Make an isometric drawing of one of these drawings.

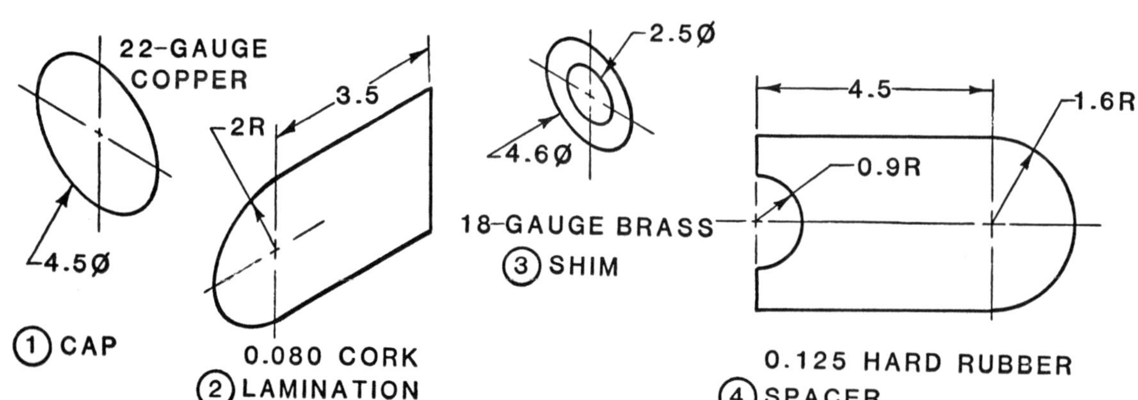

THE APPLICATION OF ISOMETRIC CIRCLES

Many objects have shapes that include only part of an isometric circle. In order to draw these shapes, it is necessary to construct the complete isometric circle. You may use as much of the circle as is needed.

If you study this object with your eyes partly closed, the construction lines will blur. Consider how you would construct the circular element that you are looking at. Open your eyes and the construction lines will become more evident.

(C = Use cross method, S = Use square method.)

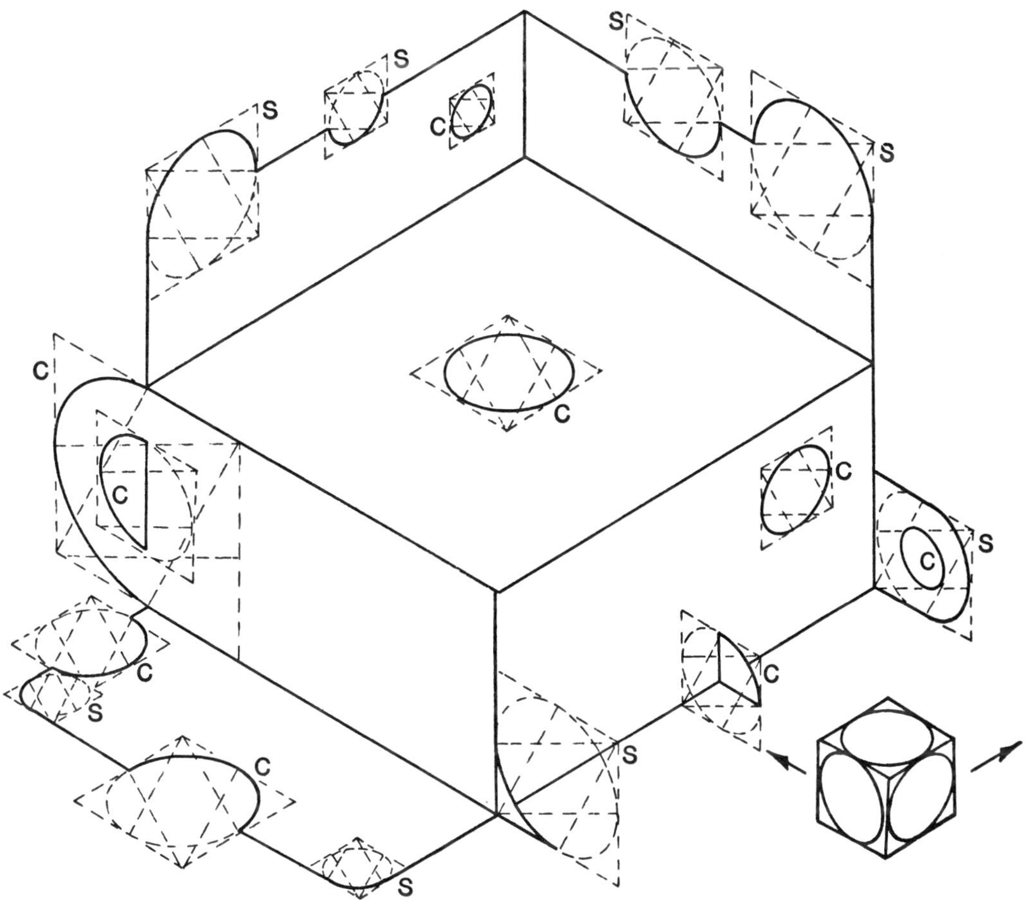

STUDY THESE DRAWINGS (Find the best method above.)

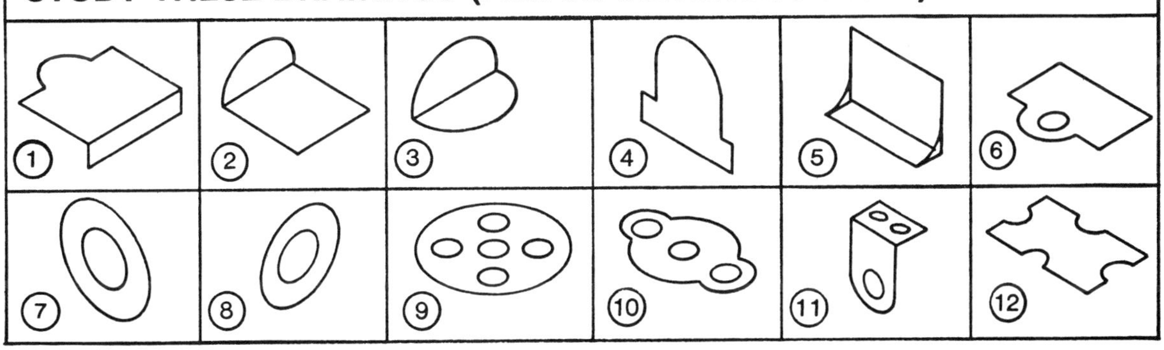

THE THREE POSITIONS OF ISOMETRIC CIRCLES

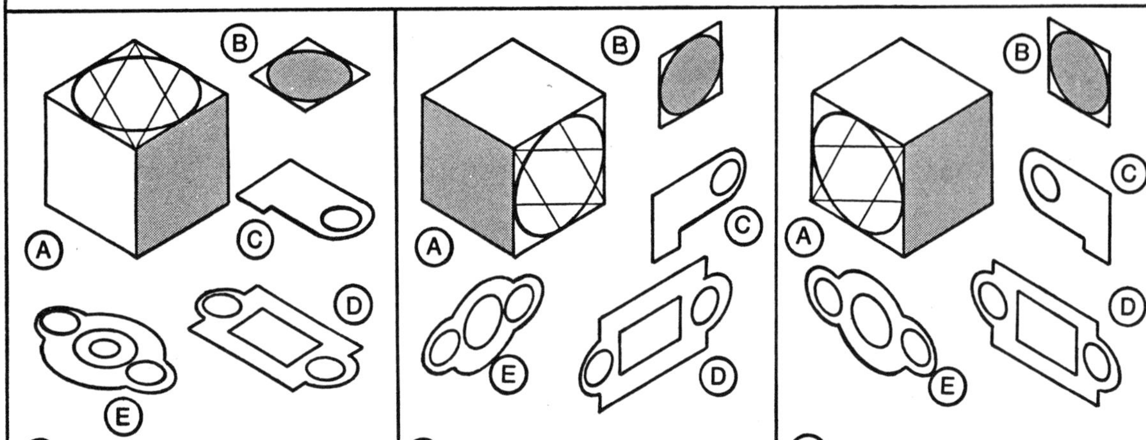

① HORIZONTAL SURFACE | ② RIGHT VERTICAL SURFACE | ③ LEFT VERTICAL SURFACE

① If the surface is a horizontal one, the isometric circle drawn on it must be in a horizontal position.

② If the vertical surface recedes to the right, the isometric circle on that surface must recede to the right.

③ If the vertical surface recedes to the left, the isometric circle on that surface must recede to the left.

(The three surfaces of an isometric cube are the keys to the problem.)

ASSIGNMENT

Make an isometric drawing of one of these drawings. (Use metric dimensions.)

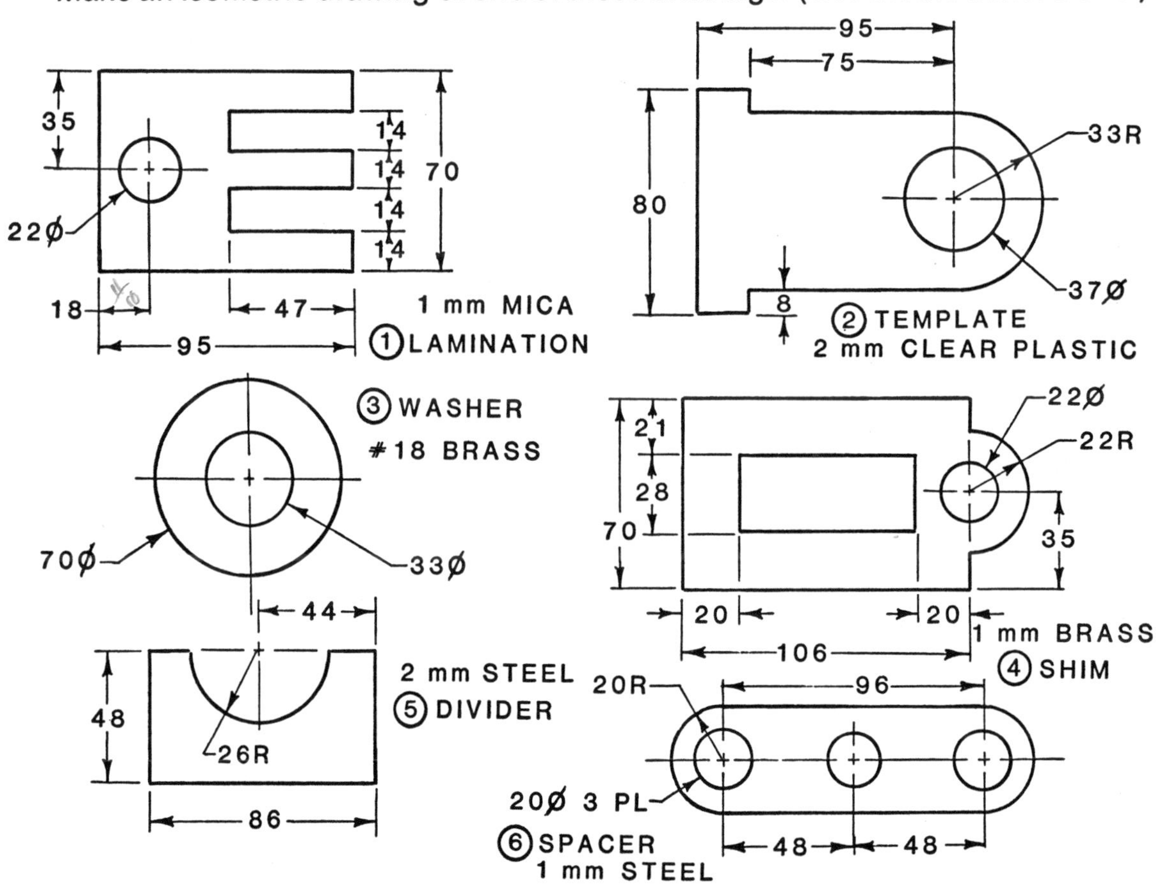

ISOMETRIC CYLINDERS

A cylinder may be considered to be a pile of isometric circles similar to a stack of coins.

CONSTRUCTING AN ISOMETRIC CYLINDER BY DROPPING CENTERS

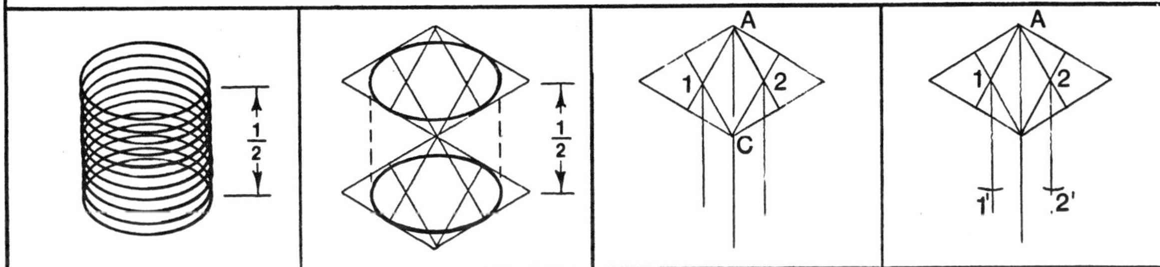

(1) A stack of isometric circles will make a cylinder 1/2" high.

(2) Two isometric circles connected by two vertical lines will make a cylinder.

(3) Drop the four centers of the isometric circle as shown.

(4) Open the compass to a 1/2" spread. From 1 swing 1'. From 2 swing 2'.

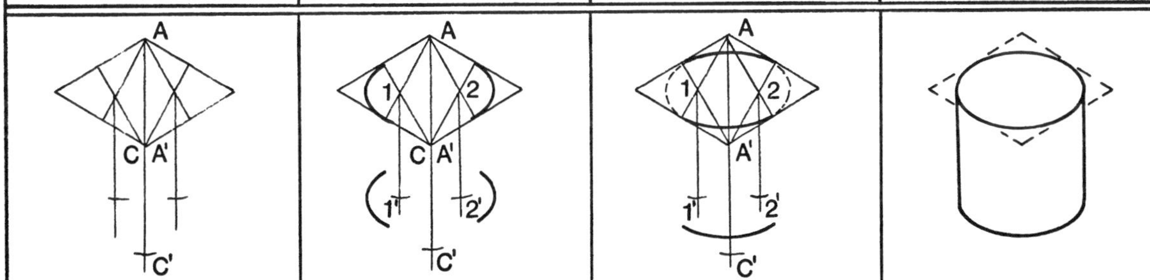

(5) From "A" swing "A'." From "C" swing "C'."

(6) Swing the small arcs from 1, 1', 2, and 2'.

(7) Swing the large arcs from "A," "A'," and "C."

(8) Drop the two vertical sides as shown.

PROJECTING UPWARD TO CONSTRUCT AN ISOMETRIC CYLINDER

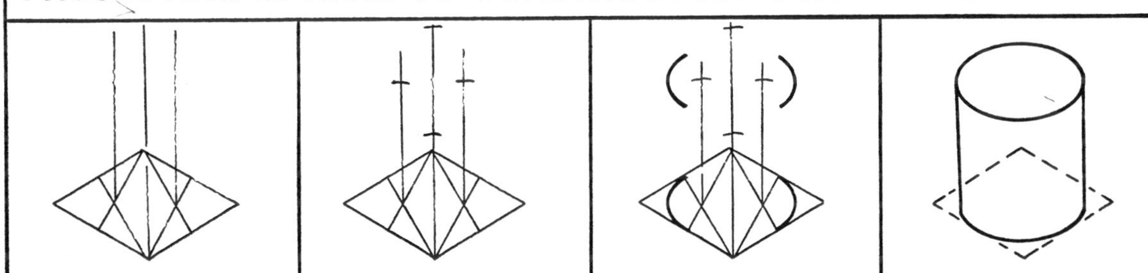

ASSIGNMENT

Make an isometric drawing of one of these objects.

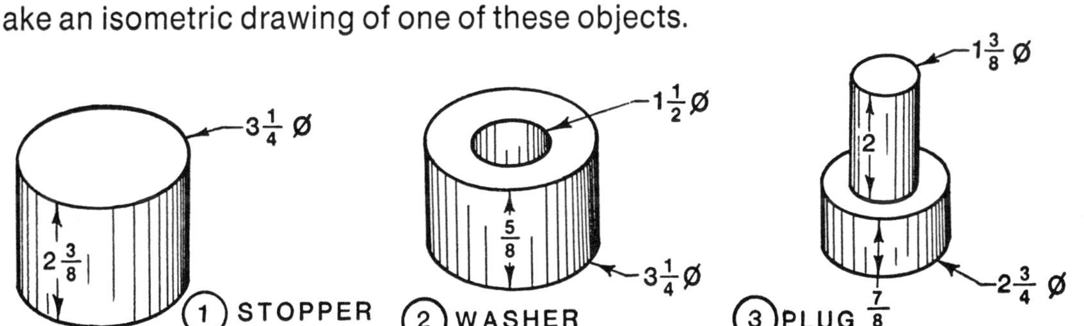

PROJECTING TO THE RIGHT OR LEFT

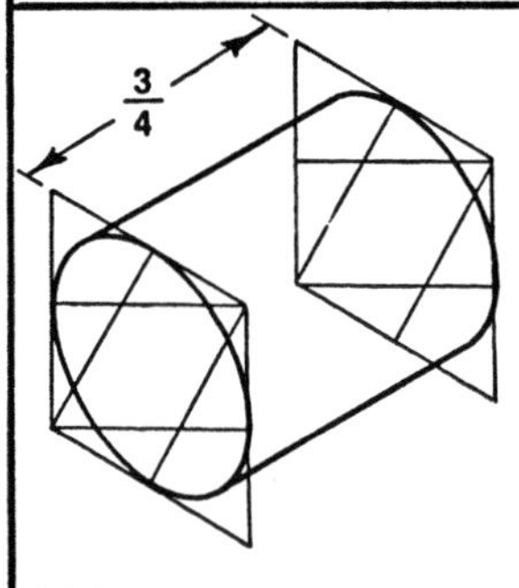

① This cylinder is 3/4″ long, therefore the circles are 3/4″ apart.

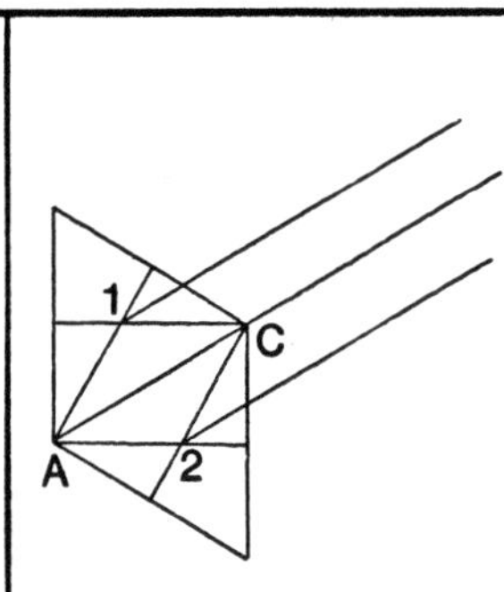

② Project centers 1, 2, "A," and "C" to the right at 30°.

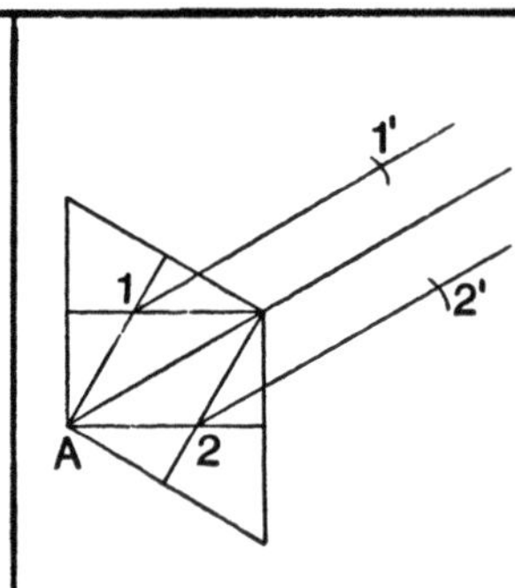

③ Open compass to a 3/4″ span. From 1, swing 1′. From 2, swing 2′.

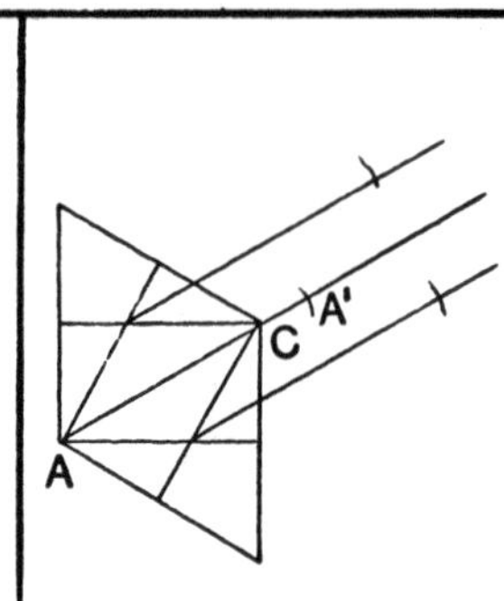

④ From "A" swing "A′." Only three centers will be needed.

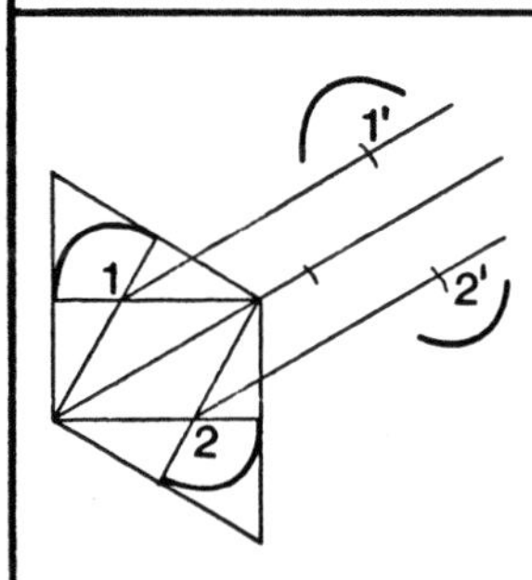

⑤ Swing the four small arcs from 1, 1′, 2, and 2′.

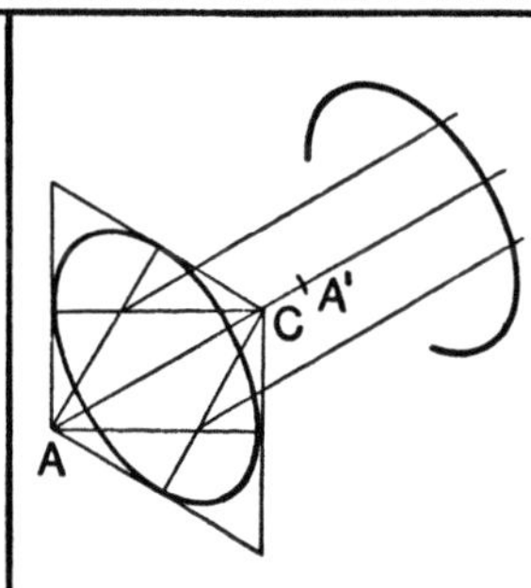

⑥ Swing the large arcs from "A," "A′," and "C."

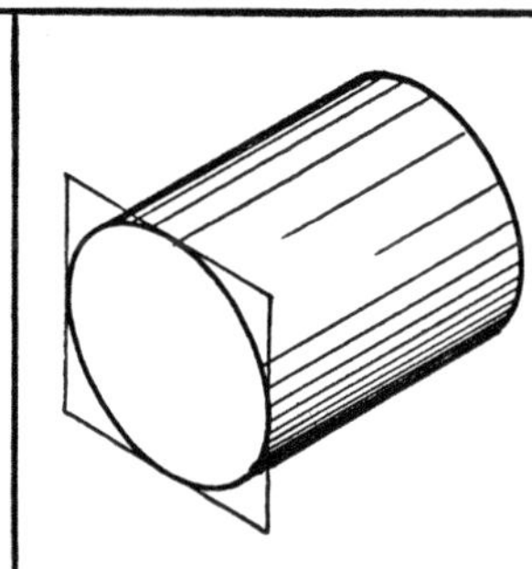

⑦ Draw the sides with 30° lines.

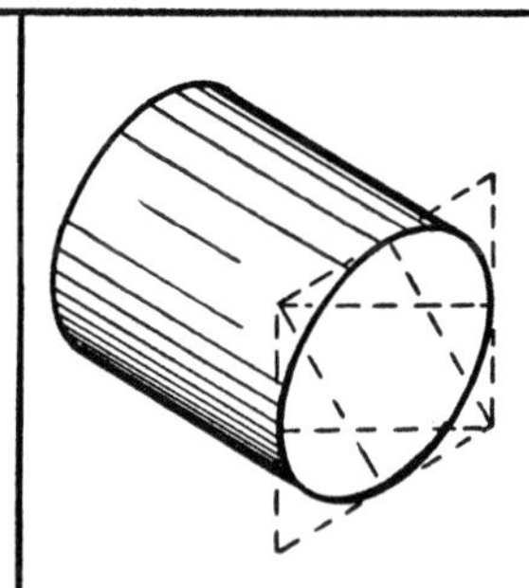

⑧ The same cylinder facing toward the right.

PROJECTING FORWARD

①

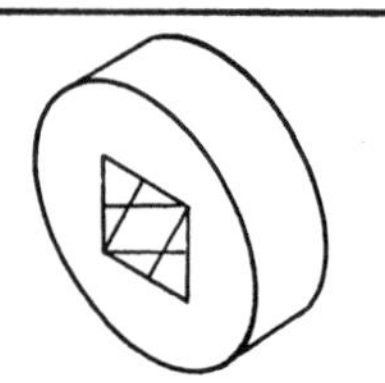

②

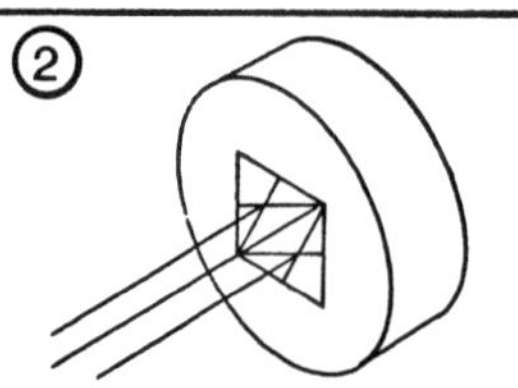

③

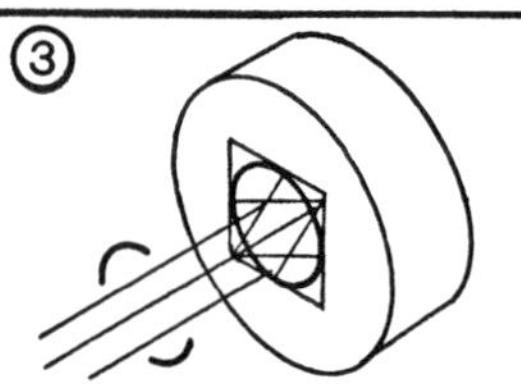

④

ASSIGNMENT

Make an isometric drawing of one of these objects.

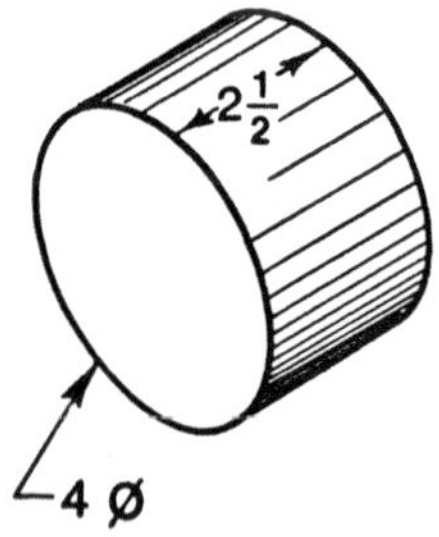

① PLUG

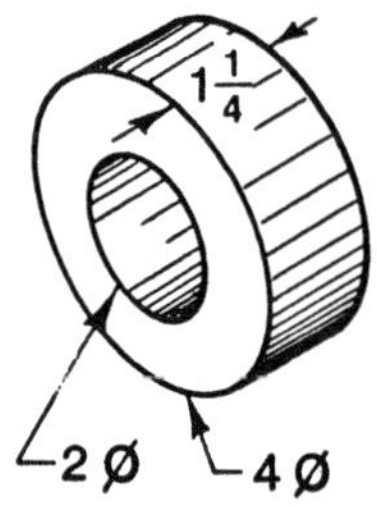

② WASHER

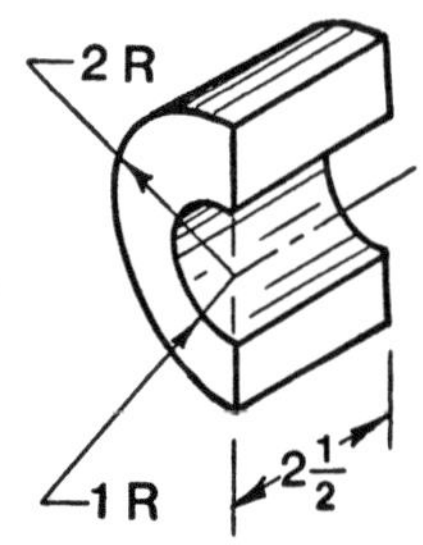

③ SPACER

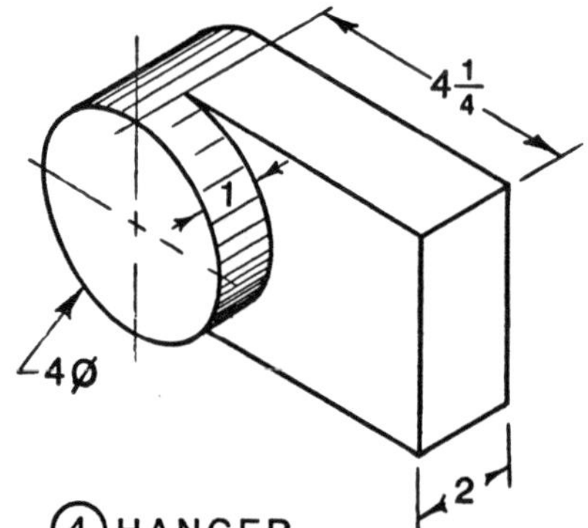

④ HANGER

PROBLEMS IN PICTORIAL DRAWING

You now have enough knowledge to make an oblique, cabinet, or isometric drawing of any of these objects.

ASSIGNMENT

Draw the orthographic, isometric, and cabinet drawings for each problem.

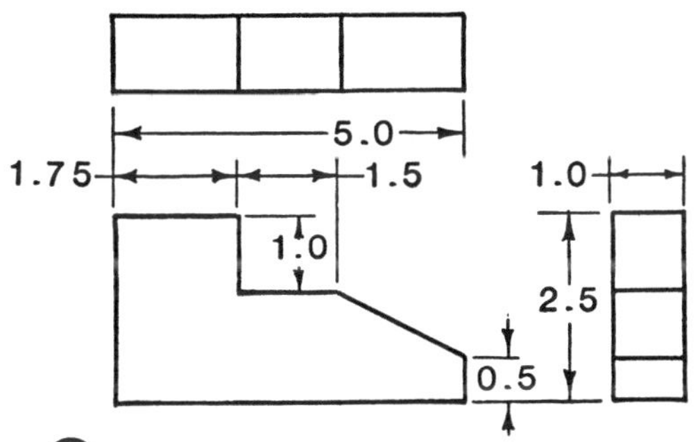

(1) ANGLE BLOCK

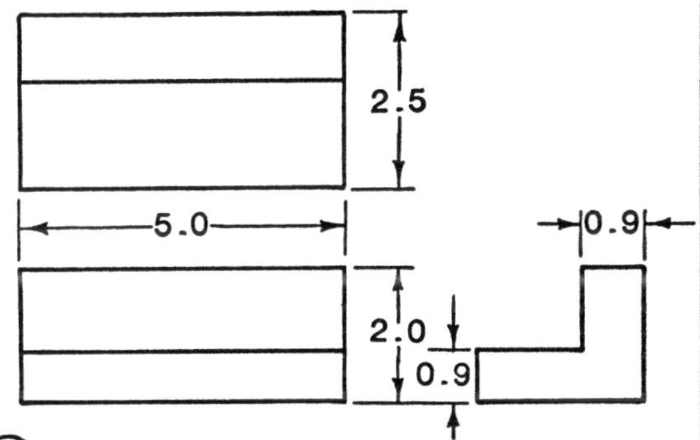

(2) RAIL

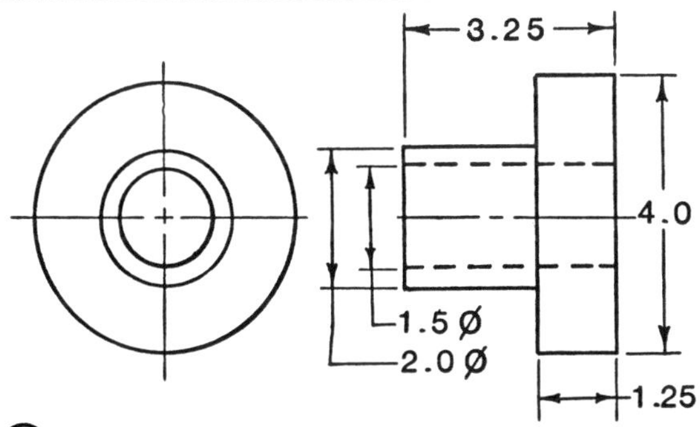

(3) BUSHING

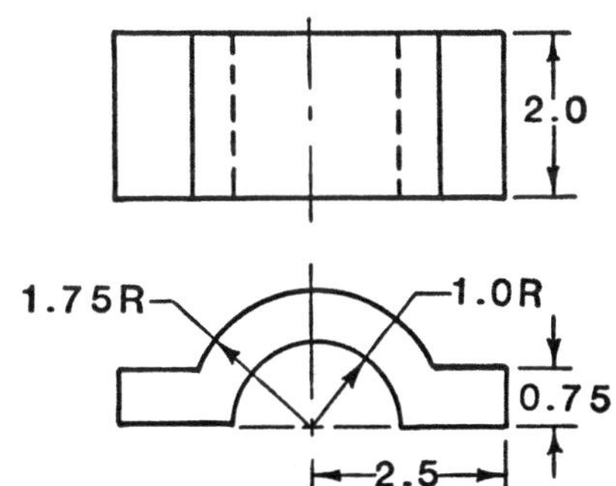

(4) STRAP

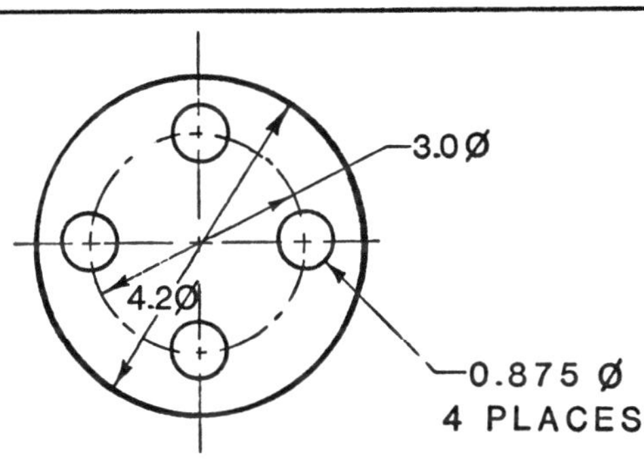

(5) FACEPLATE

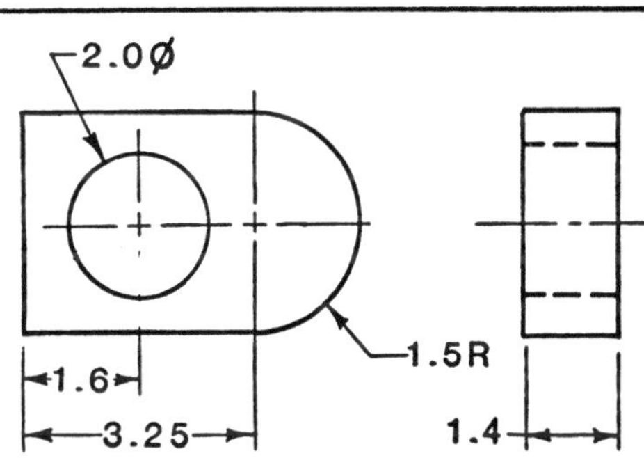

(6) GUIDE BLOCK

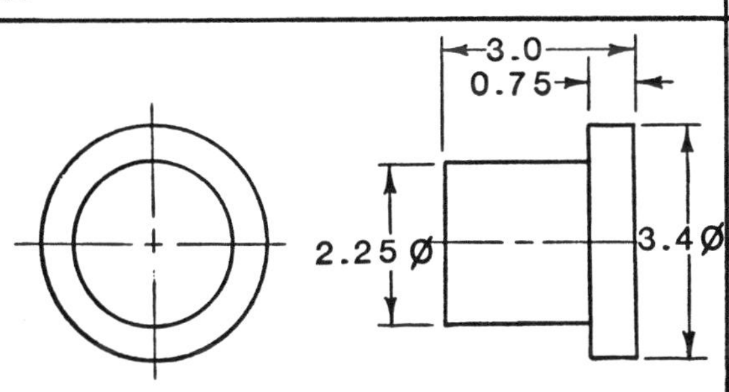

(7) RIVET

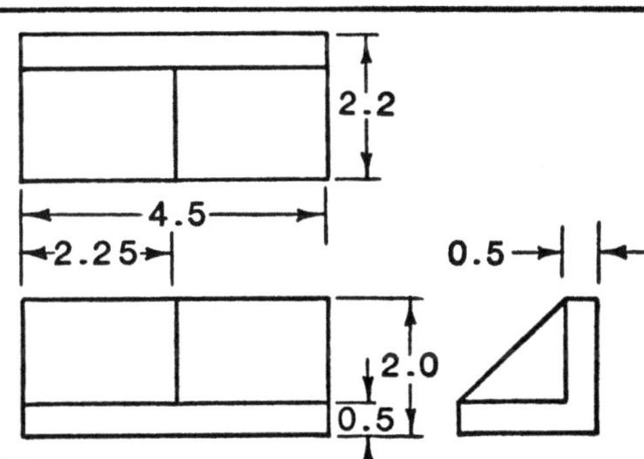

(8) BRACE

ASSIGNMENT

Draw the isometric and the three-view orthographic projection of one of these drawings.

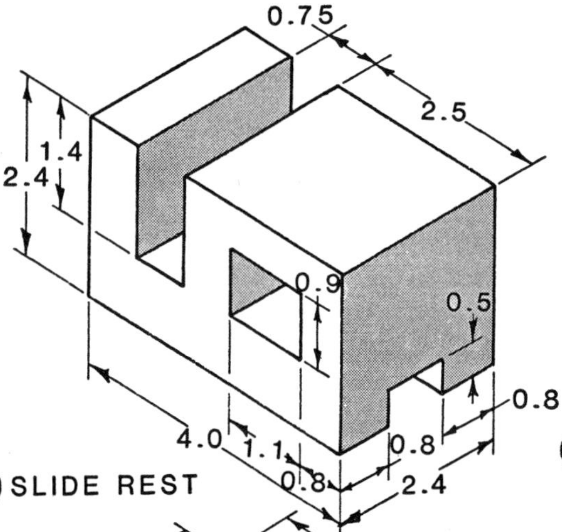

(1) SLIDE REST

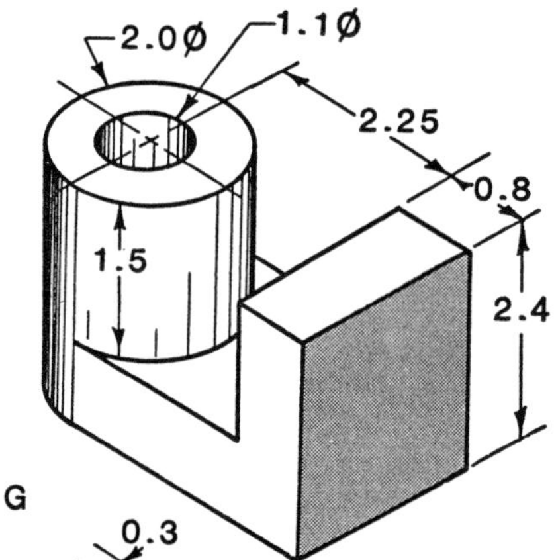

(2) BEARING

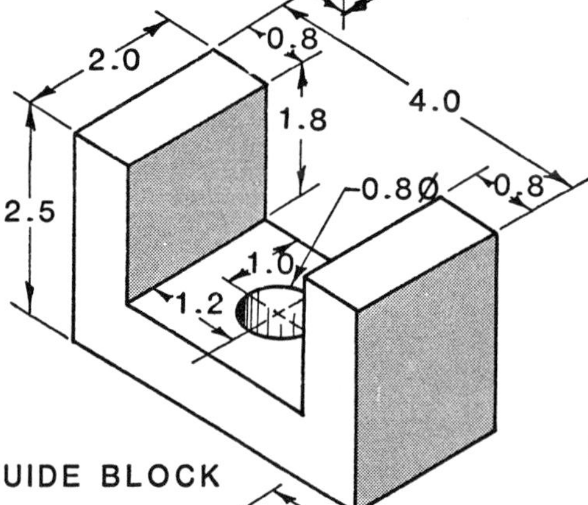

(3) GUIDE BLOCK

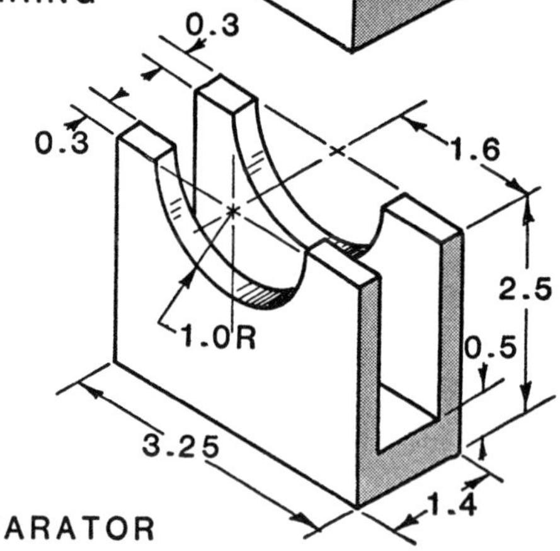

(4) SEPARATOR

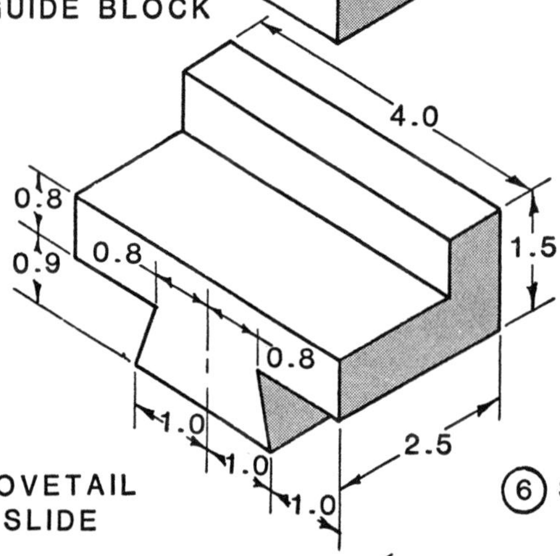

(5) DOVETAIL SLIDE

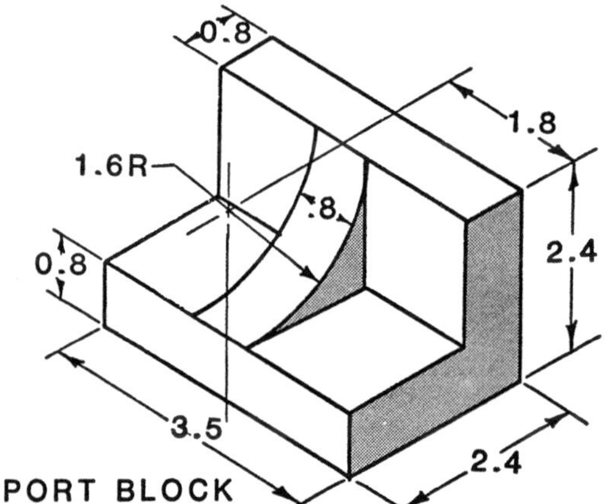

(6) SUPPORT BLOCK

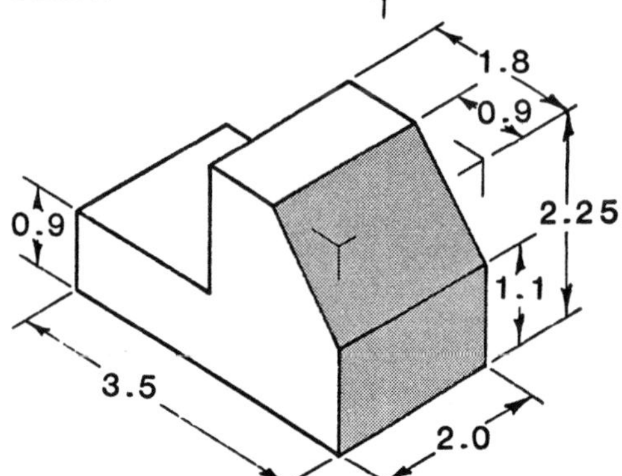

(7) CHAMFERED BLOCK

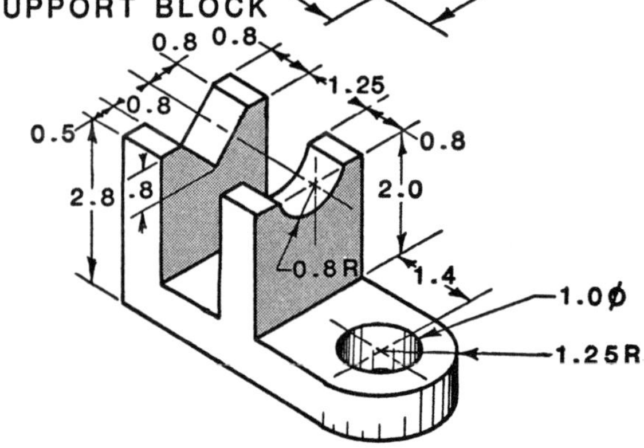

(8) JIG

UNIT 12

SECTIONS

Hidden edges are usually shown by means of broken lines known as **hidden edge lines**. If the interior of an object is simple, the drawing may be easily read. If the interior of an object is complex, especially if it consists of several pieces, the use of hidden edge lines results in a drawing that is almost impossible to read.

The drafter has devised a method of drawing that enables him or her to show the interior of an object as though it were exposed to view. Hidden edges are practically eliminated.

This type of drawing is known as **section drawing**. Views drawn in this manner are called **sectional views**.

Sectional views of an object may be made as cabinet drawings, oblique drawings, isometric drawings, or as orthographic projections. In other words, section drawings may be made either as pictorial drawings or as orthographic projections.

The basic idea behind the making of sectional views is a simple one. The drafter uses an imaginary cutting plane in order to cut the object apart. When the front portion of the object has been removed, the interior of the object is exposed to view. The resulting illustration is called a **sectional view.**

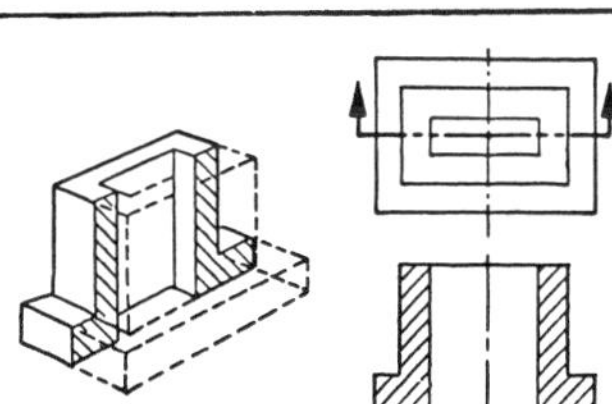

FULL SECTION

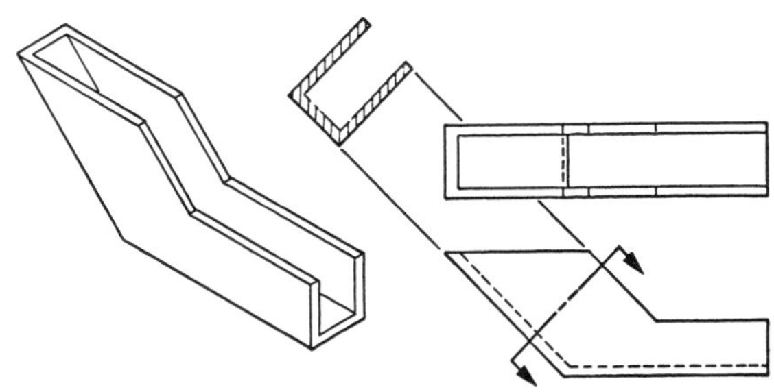

AUXILIARY SECTION

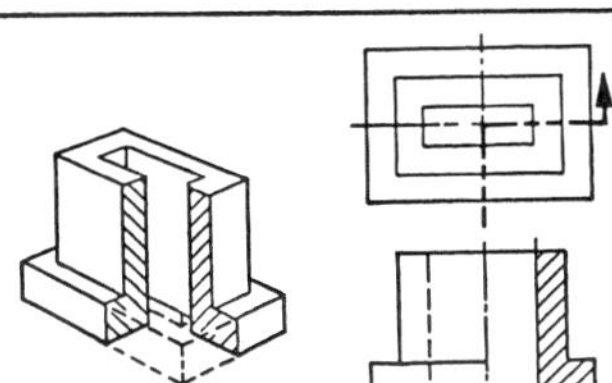

HALF SECTION

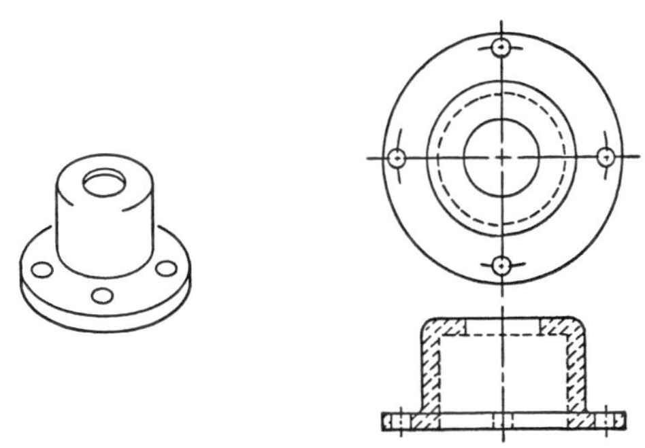

PHANTOM SECTION

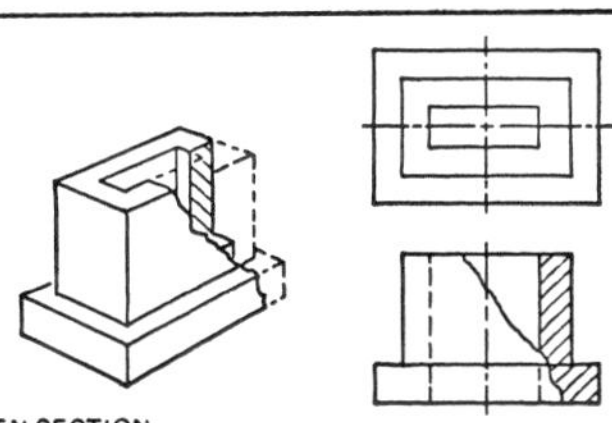

BROKEN SECTION

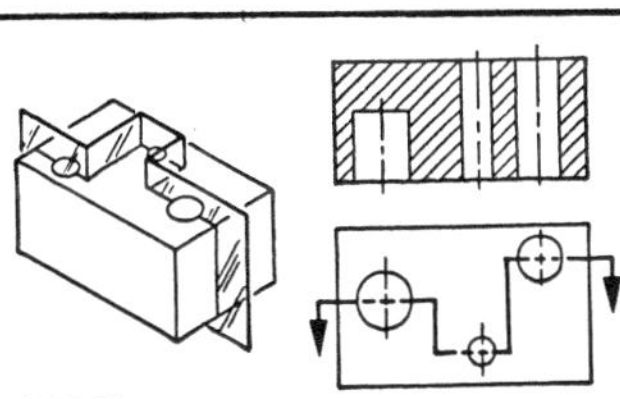

OFFSET SECTION

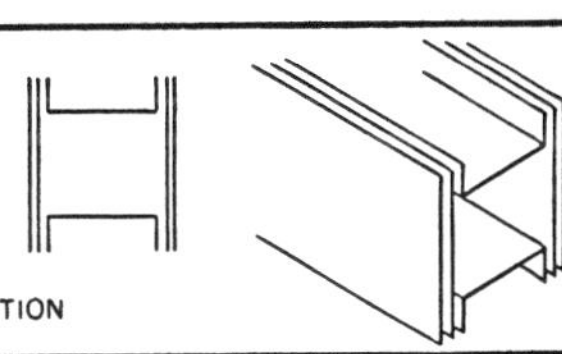

THIN MATERIAL SECTION

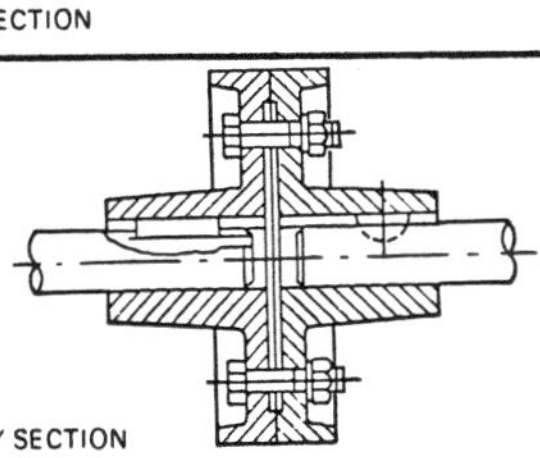

ASSEMBLY SECTION

REMOVED SECTION

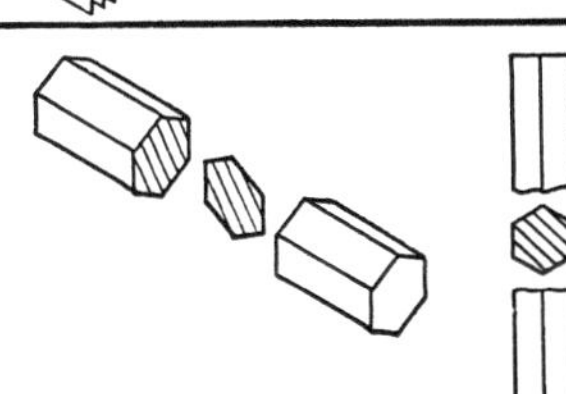

REVOLVED SECTION

THE CUTTING PLANE AND THE CUTTING PLANE LINE

It is important to remember that the cutting plane is imaginary. In common practice, when one view of an orthographic projection is shown as a sectional view, the other view or views are not affected by the cutting but are drawn as normal views.

THE CUTTING PLANE AND FULL SECTIONS

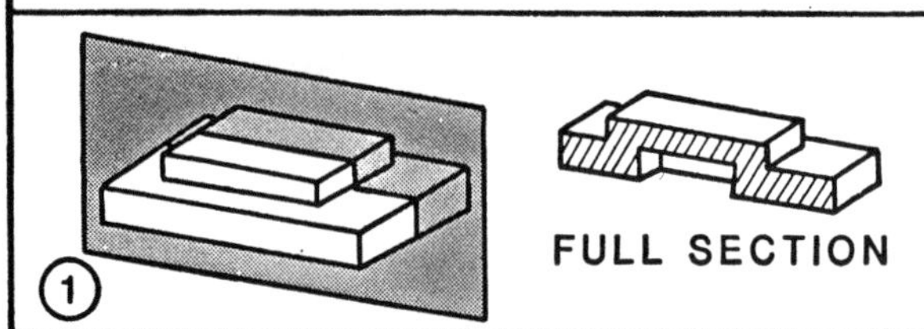

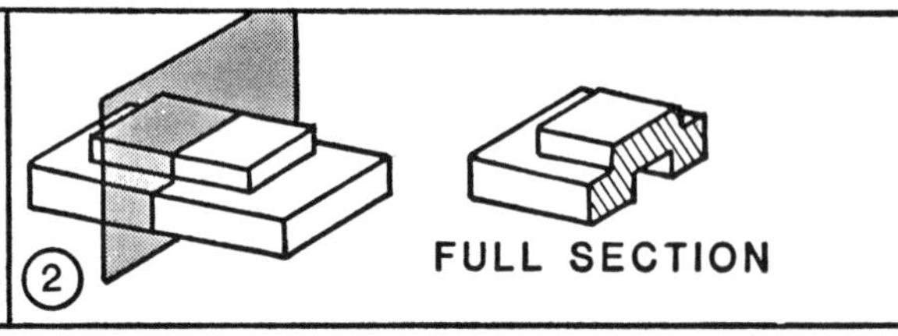

The cutting plane is an imaginary line that divides the object into two parts. When the object is cut entirely across, the resulting sectional drawing is called a full section.

THE CUTTING PLANE LINE

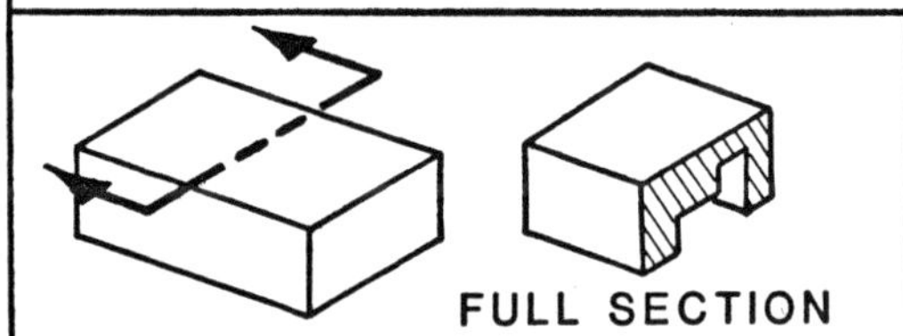

(1) Drafters use a symbol called a **cutting plane line** instead of using the cutting plane itself.

(2) Compare the above lines:
A. Cutting plane line.
B. Object line.
C. Center line.

THE THREE PURPOSES OF THE CUTTING PLANE LINE

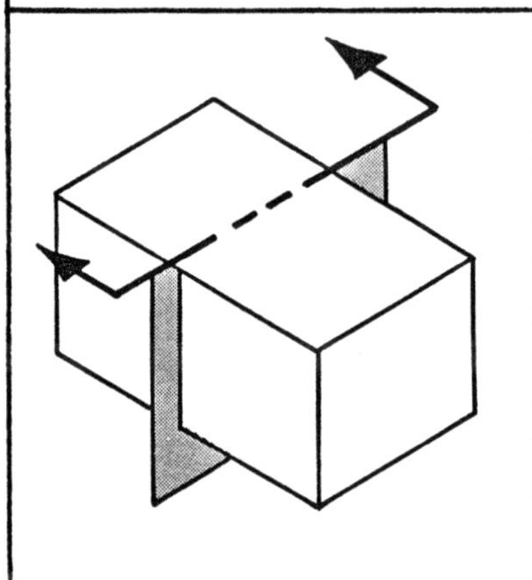

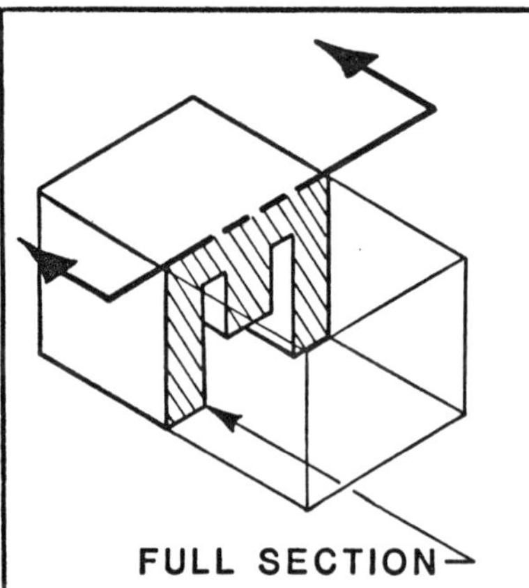

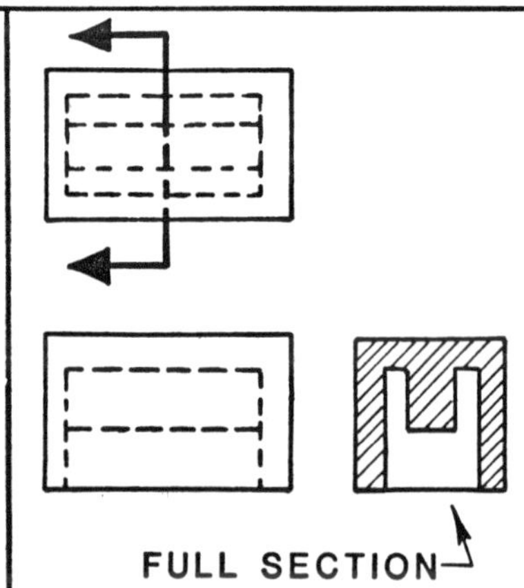

(1) The heavy line represents the edge of the cutting plane and indicates the direction of the cut.

(2) The arrows show which portion of the object is to be retained.

(3) The location of the cutting plane line on the object determines in which view the section drawing must appear.

For ordinary working drawings, section lining consists of thin, evenly spaced lines drawn about 1/16" apart. They are drawn at an angle of 45°. These lines are used to indicate the imaginary cut surface of the object.

VISUALIZING FULL SECTIONS

It is extremely important to study the direction of the cutting plane line before attempting to make a sectional view.

HOW TO VISUALIZE OR "SEE" A SECTIONAL VIEW

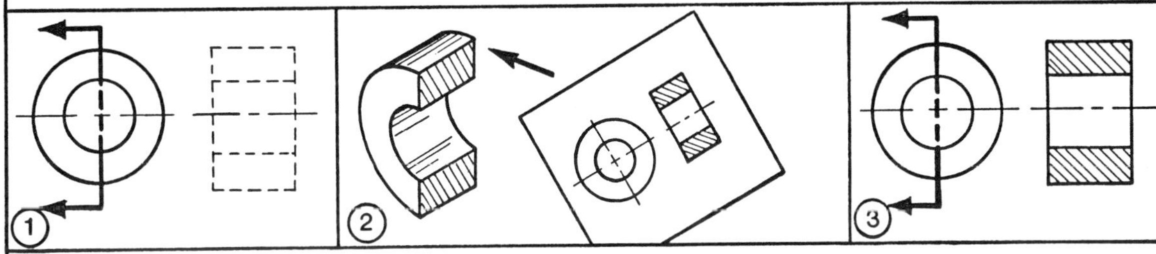

1. This cutting plane line shows three things. **First:** a full section, in a vertical position, must be drawn. **Second:** The left side of the object is to be retained. **Third:** The sectional view must be drawn as an end view (not a top view).

2. Imagine that you are looking at the cut surface of a pictorial full section when you are drawing the end view.

3. The completed drawing, showing a full sectional view.

REVIEWING A BASIC PRINCIPLE

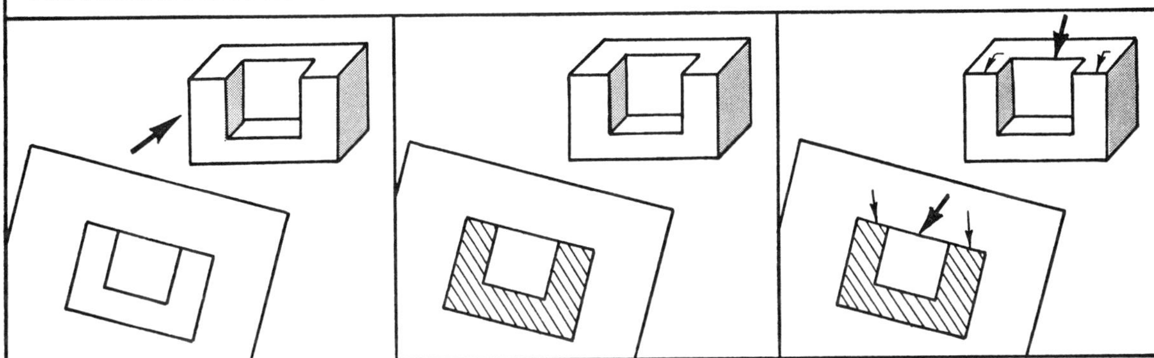

(1) You are familiar with the method of drawing the front view of an object such as this.

(2) The principle remains the same when a sectional view is to be made.

(3) It is important to remember that the top horizontal line represents three edges.

APPLYING YOUR KNOWLEDGE

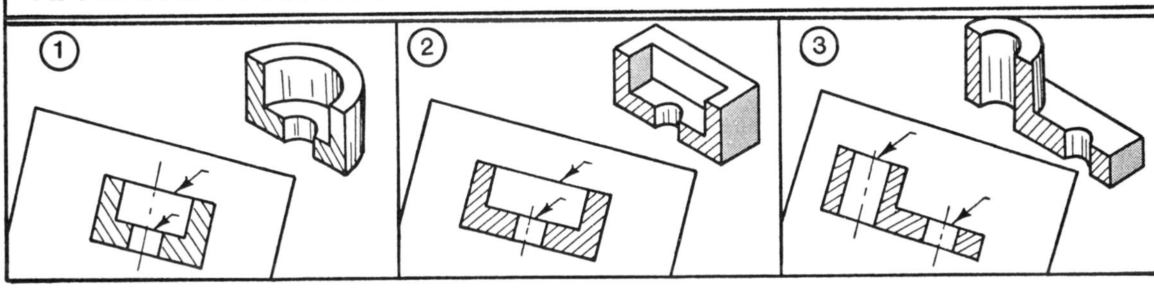

Carefully study these drawings. Be sure that you thoroughly understand the meaning of the lines indicated by the arrows.

STUDY THESE FULL SECTIONS

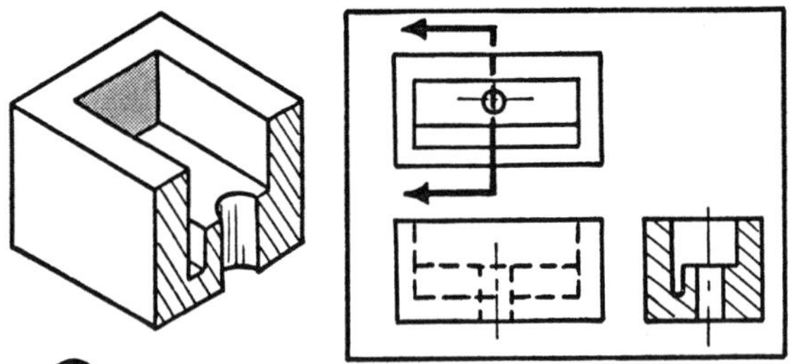

(1)

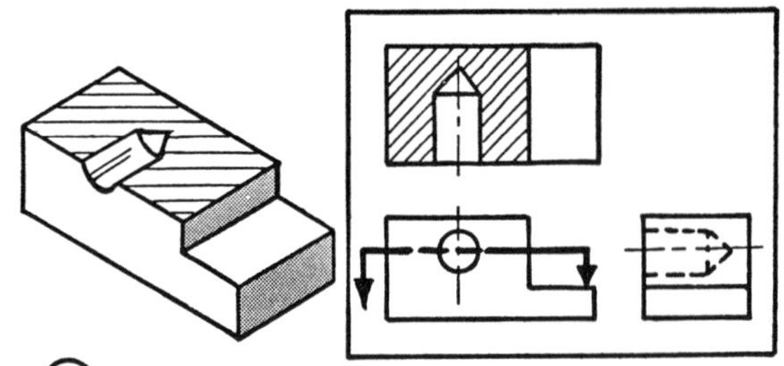

(2)

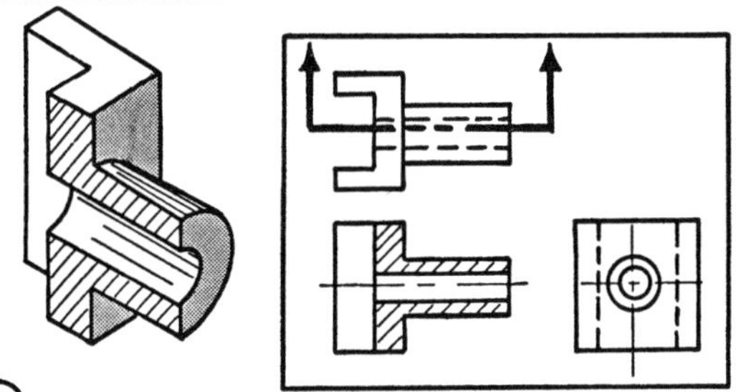

(3)

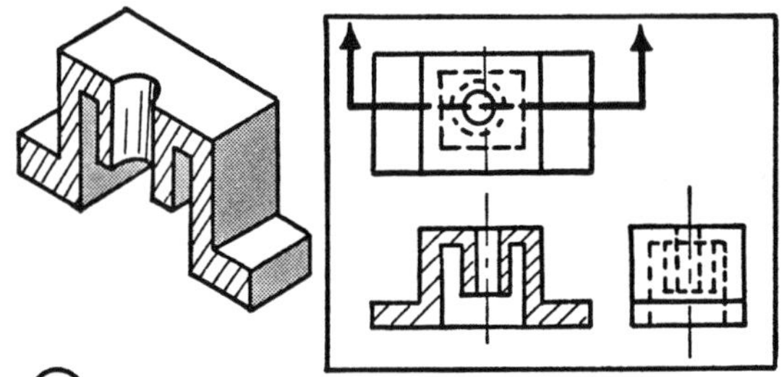

(4)

OFFSET SECTIONS

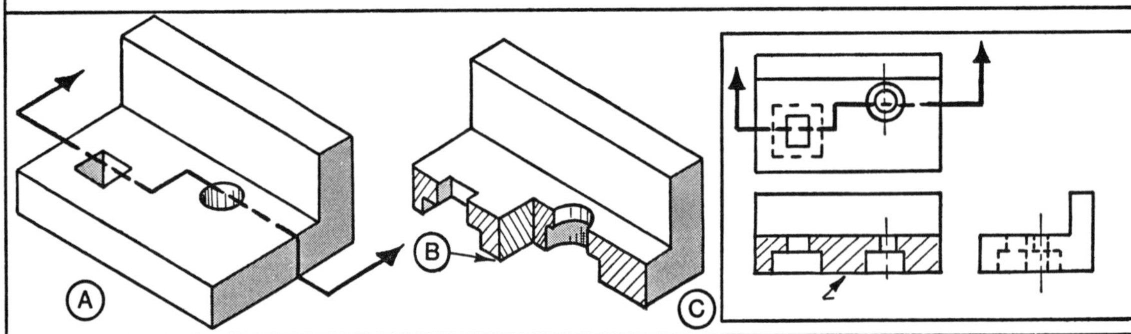

(A) The cutting plane line may be offset. Study the section lining at (B). Note at (C) the one vertical line may be omitted.

ASSIGNMENT

Make an orthographic working drawing showing a full section.

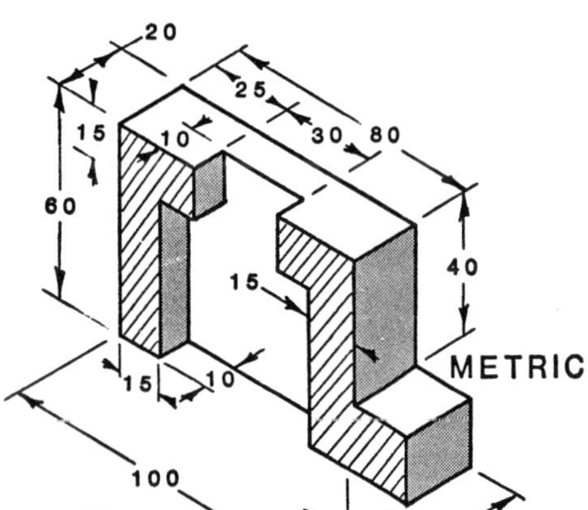

(1)

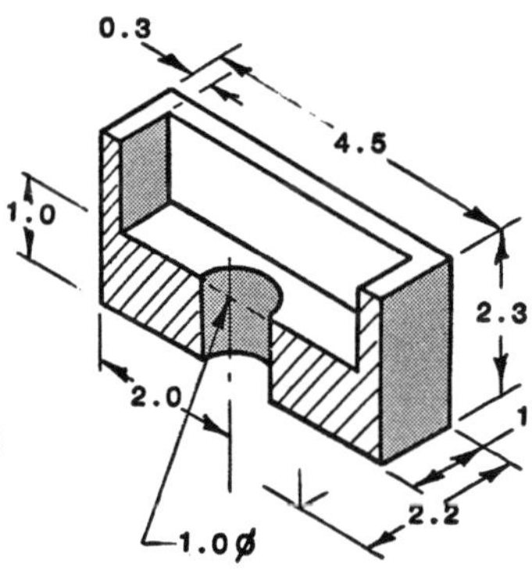

(2)

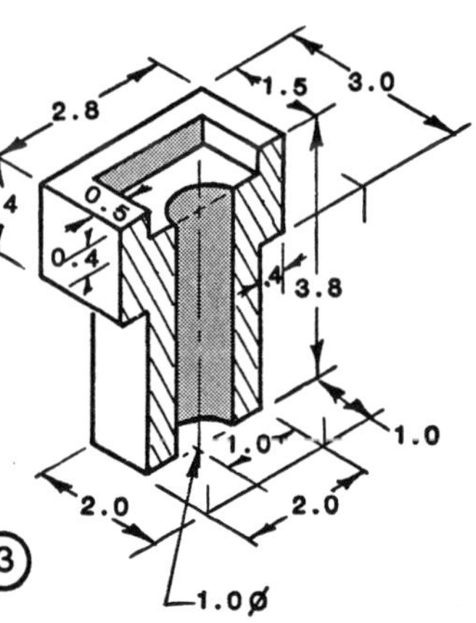

(3)

TECHNICAL POINTS

It should be remembered that the purpose of a section drawing is to show the surface that is exposed by the cutting plane. Edges that are hidden behind this exposed surface are often omitted. This practice tends to make the section drawing more understandable. However, if the hidden edge is very important, it must be drawn.

WHEN TO OMIT HIDDEN EDGE LINES

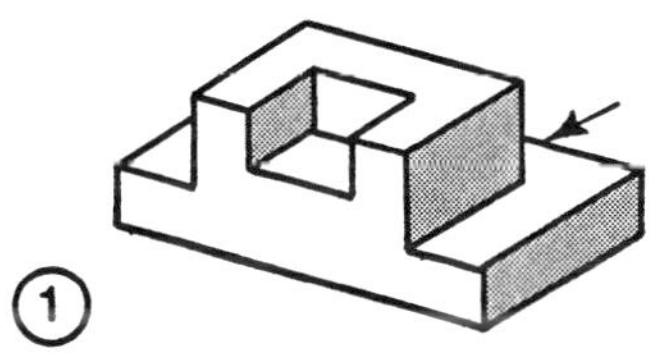

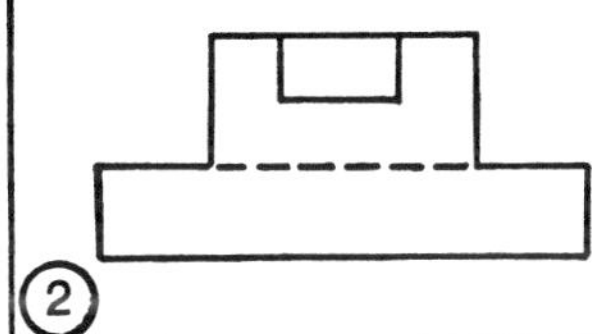

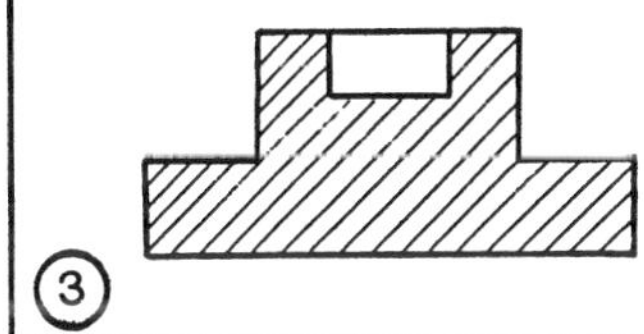

(1) The back edge pointed out by the arrow will be a hidden edge in (2).

(2) In orthographic projection, all hidden edges should be drawn with hidden edge lines.

(3) In section drawing, the hidden edge is omitted in order to give a clear picture of the cut surface.

EXAMPLES OF OMITTED HIDDEN EDGES

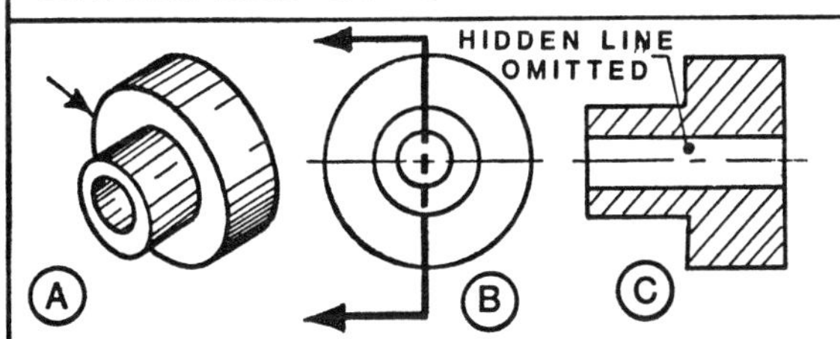

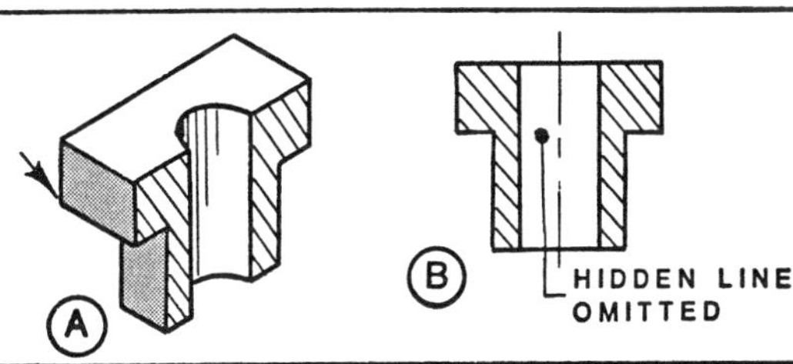

(1) The back edge shown by the arrow at (A) should be omitted in the section drawing (C).

(2) The back edge shown at (A) should be omitted as shown at (B).

MINOR DETAILS SHOULD BE OMITTED

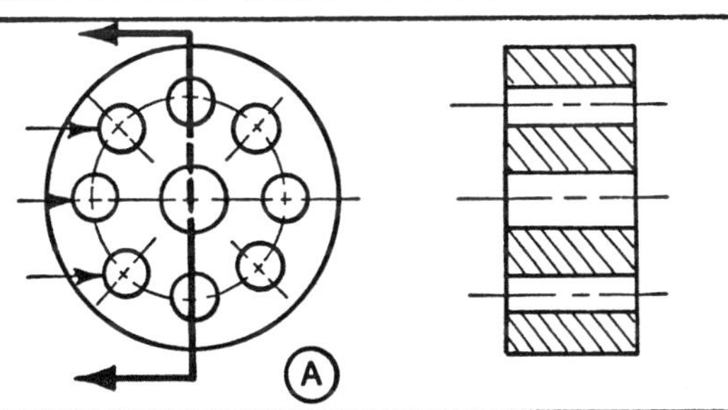

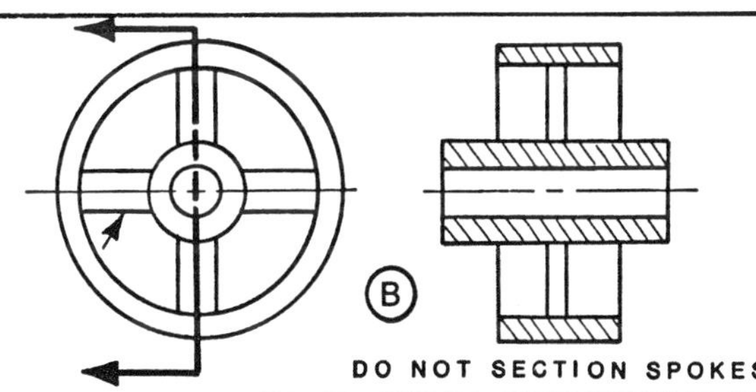

(1) Omit the three holes shown at (A).

(2) Omit the spoke shown at (B)

SOME HIDDEN EDGES ARE ESSENTIAL

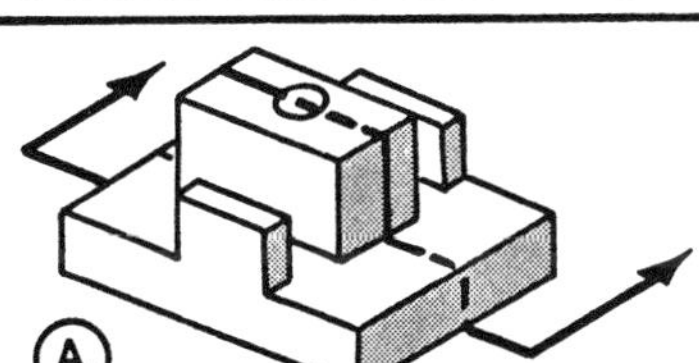

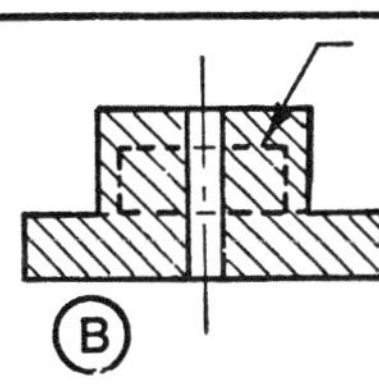

When a hidden part is considered to be important, it may be shown by using hidden edges on the section drawing as shown at (B).

LEARN TO VISUALIZE FROM TWO VIEWS

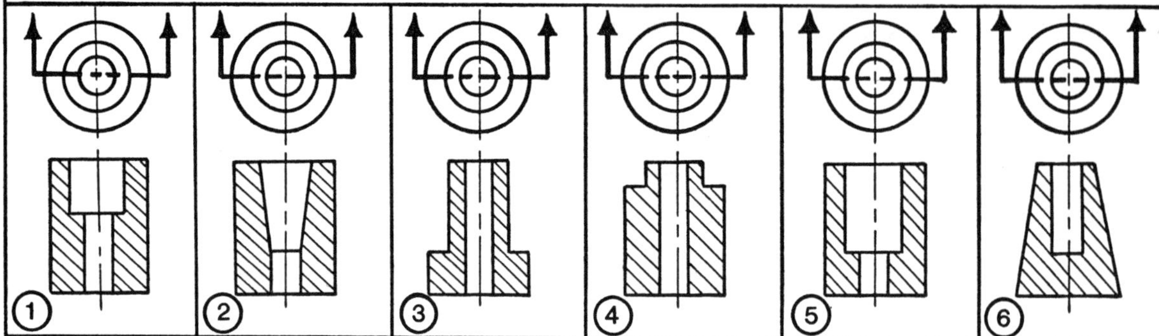

The six top views are exactly alike. The six front views are entirely different. Learn to visualize an object from two views and from three views.

STUDY THESE SECTION DRAWINGS

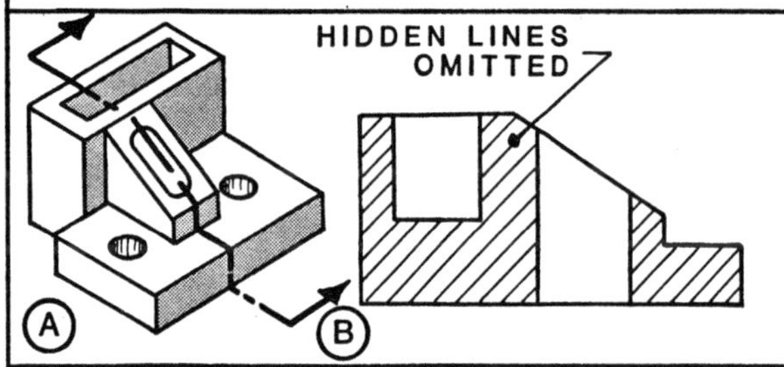

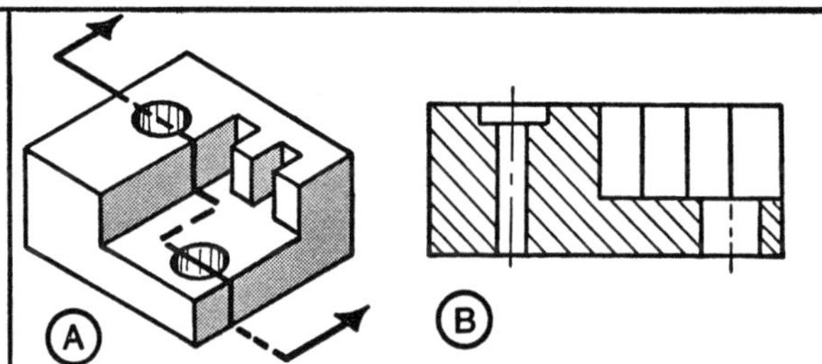

(1) The cutting plane line need not be drawn through the center of an object. Note that the two small holes have been omitted in the sectional view.

(2) This is an offset section. Study the upper right-hand area of the sectional view. Study the path of the cutting plane line.

ASSIGNMENT

Make an orthographic full section of each.

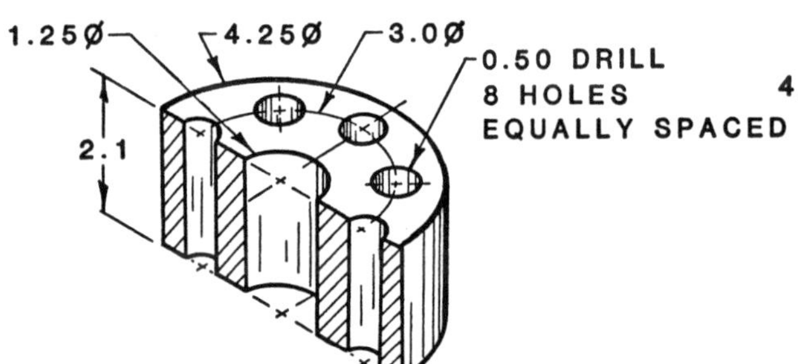

(1) PRESSURE BLOCK

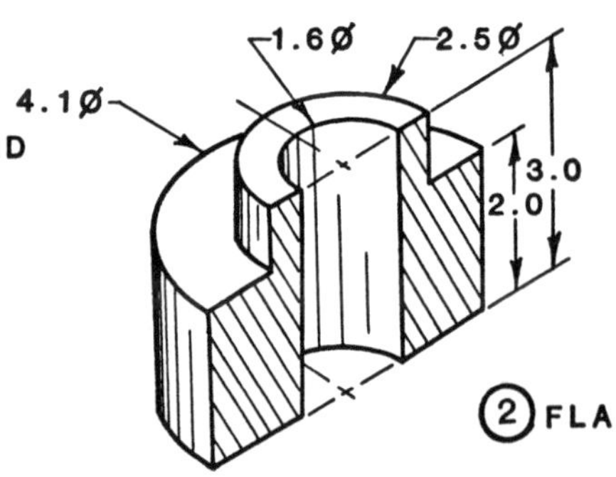

(2) FLANGE

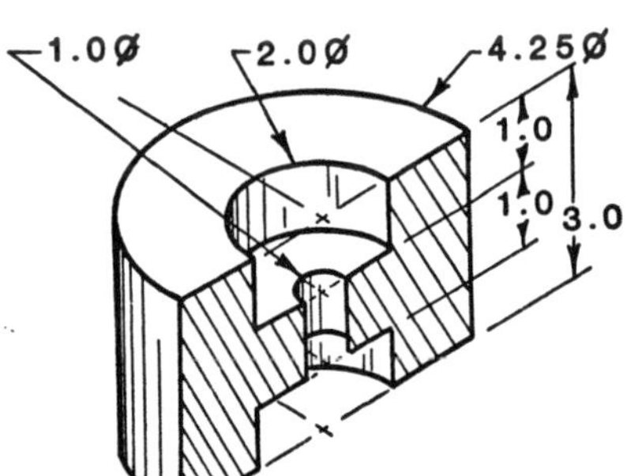

(3) VALVE SEAT

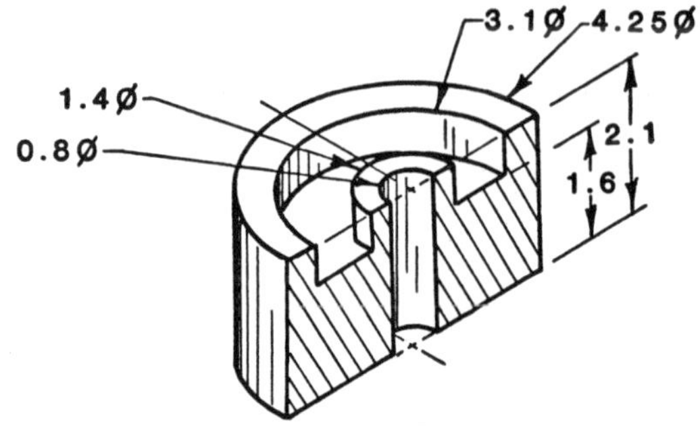

(4) FLYWHEEL

SECTION LINING

Section lining, or crosshatching, consists of fine lines, comparable to projection lines; they are used to give a shaded effect to the cut surfaces of a section drawing. Section lines are drawn about 1/16″ apart and are spaced by eye. They are usually drawn at a 45° angle.

EXAMPLES OF SECTION LINING

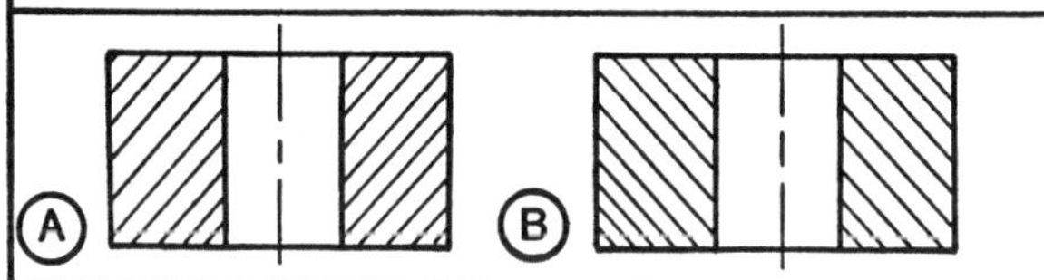

(1) Section lining is usually drawn at an angle of 45° to the left or to the right.

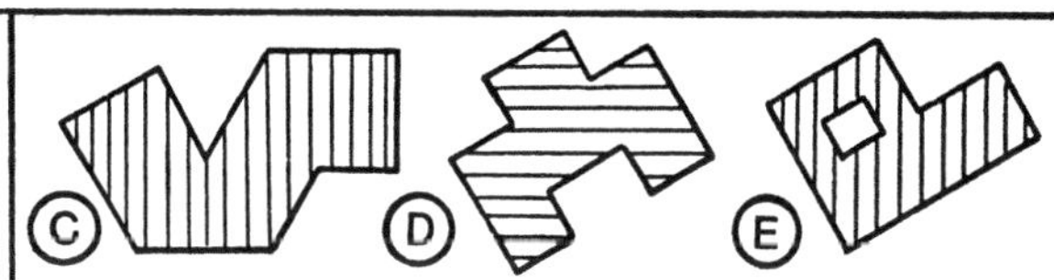

(2) Sometimes personal judgment is required concerning the direction of section lining.

ADJACENT PARTS IN SECTION

(1)

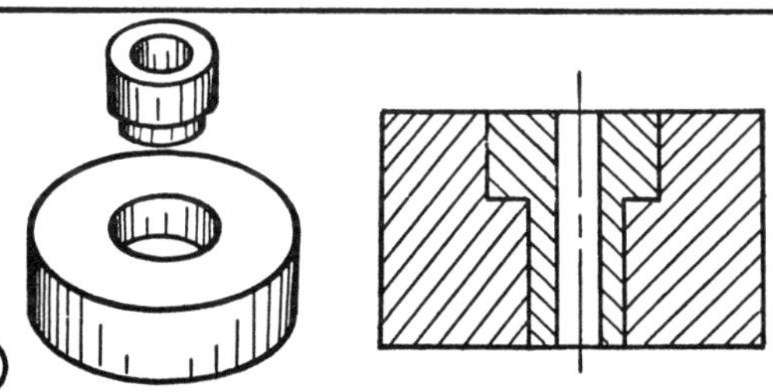

(2)

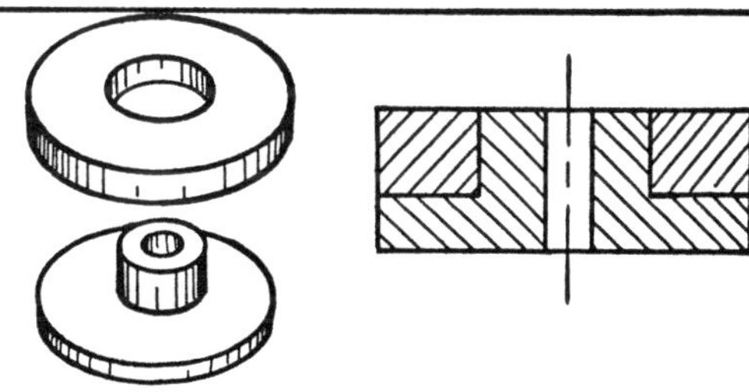

When an object consists of two or more parts, each part must be crosshatched in such a manner that it is clearly recognized. However, the angle of the section lining must remain uniform for each piece.

SYMBOLS FOR SECTION LINING

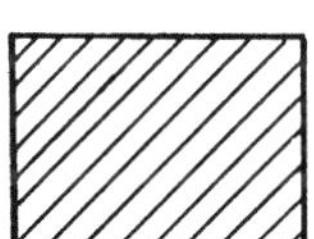

(1) Cast iron.

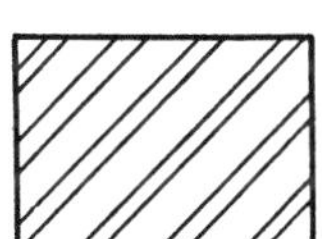

(2) Steel.

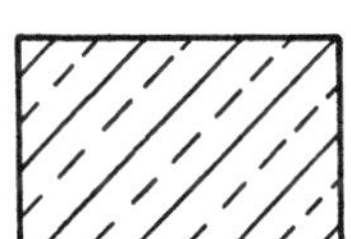

(3) Brass, bronze, compositions.

Different symbols may be used to denote various materials, but it is acceptable to use the regular section lining and indicate the kind of material by note.

ADJACENT SURFACES AND ADJACENT PARTS (In section)

(1)

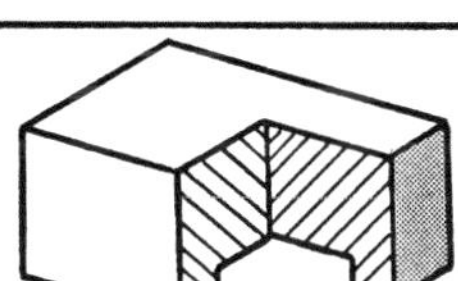

When two cut surfaces meet at right angles, the direction of the section lining must be changed for each surface. Isometric section lines are usually 60°.

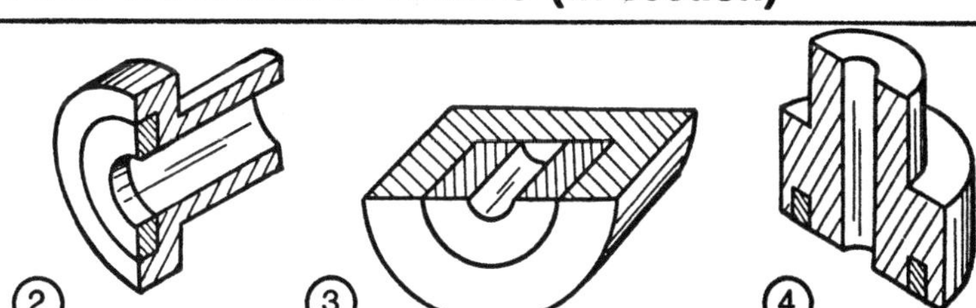

Study the method used to show adjacent parts in section.

HALF SECTIONS

When an object is symmetrical, only one half of a full section need be drawn. Such a drawing is called a **half section.** A half section enables one to see the inside and the outside in one view.

ORTHOGRAPHIC AND PICTORIAL HALF SECTIONS

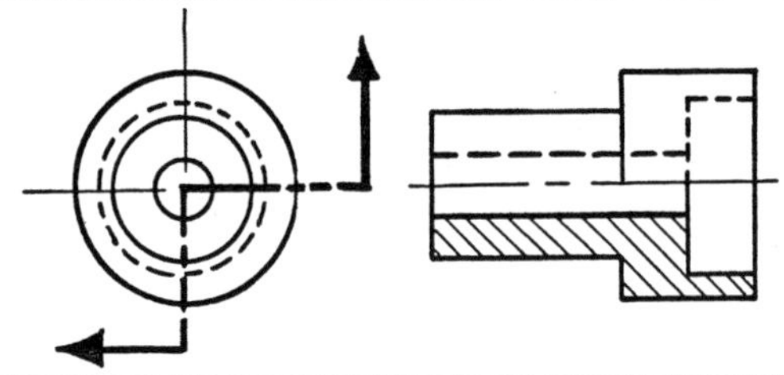

① Consider the three purposes of the cutting plane line. Note that the end view shows both the interior and the exterior.

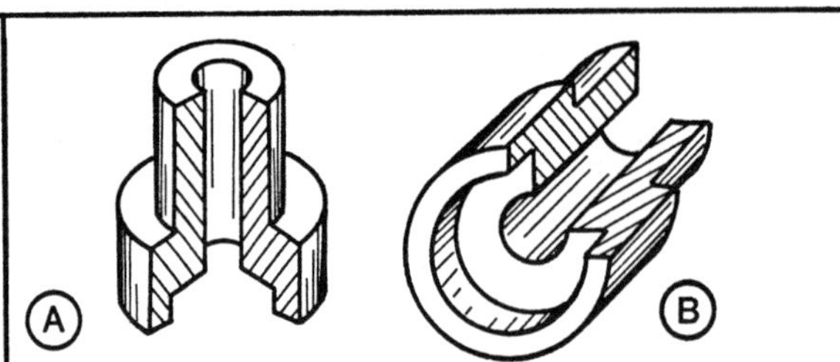

② Ⓐ is an isometric half section. Ⓑ is a cabinet half section. Both are pictorial half sections.

STUDY THESE HALF SECTIONS

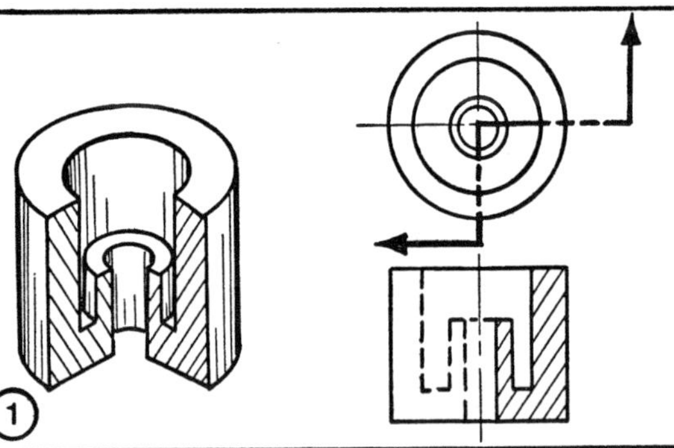

①

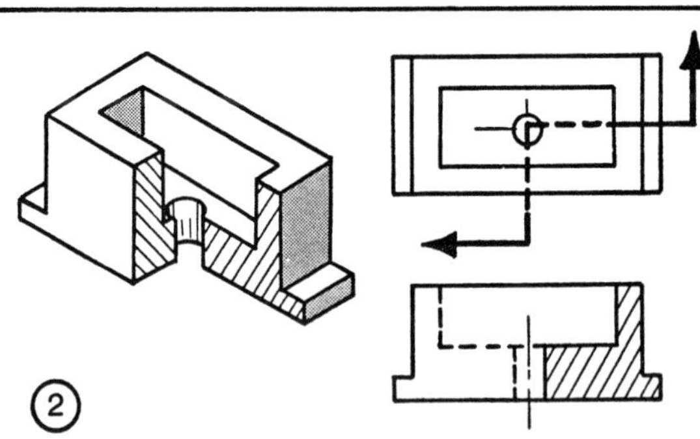

②

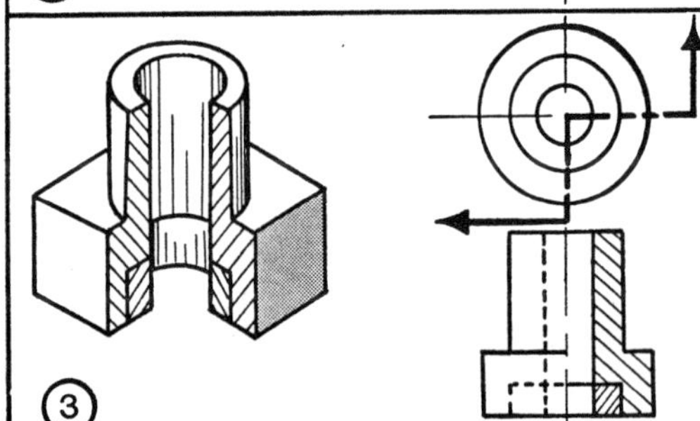

③

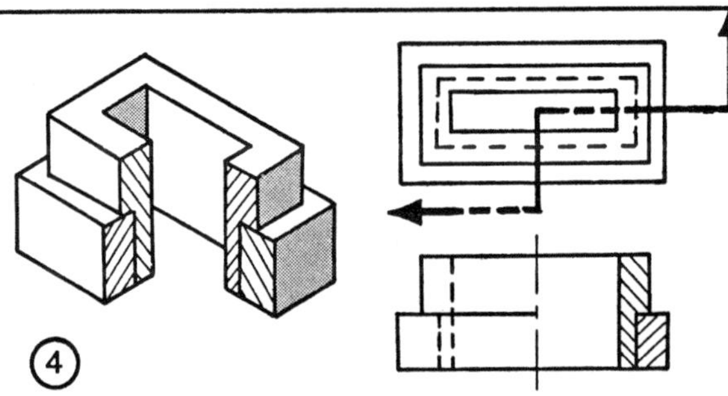

④

ASSIGNMENT

Make an orthographic half section.

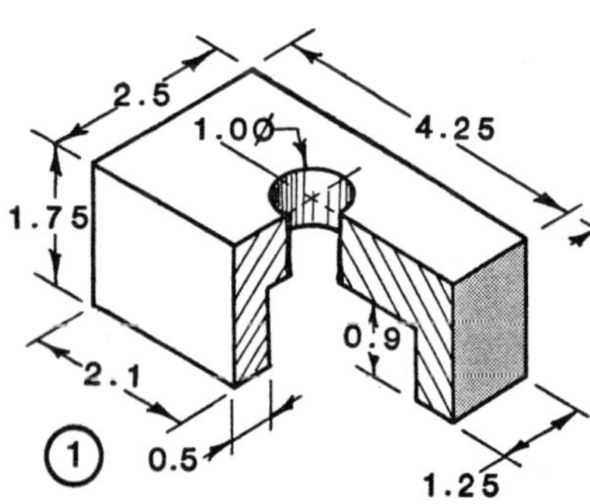

①

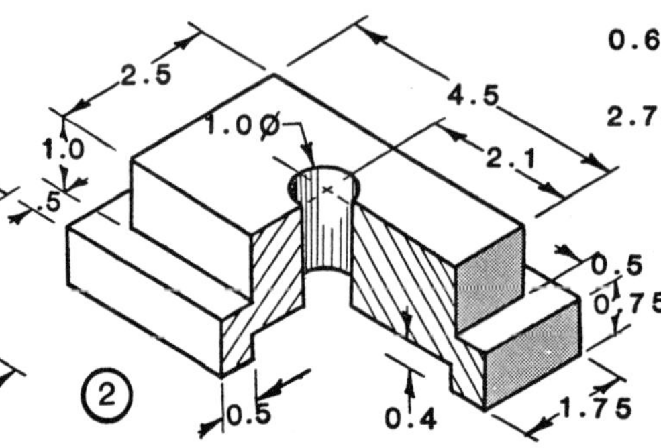

②

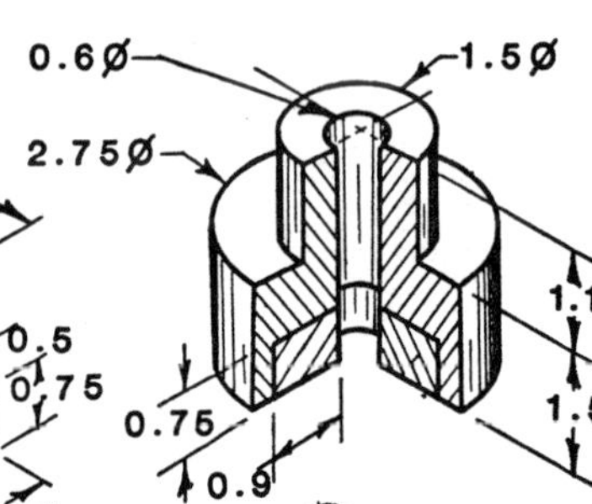

③

OTHER TYPES OF SECTION DRAWINGS

Broken-out sections, **revolved** sections, and **removed** sections are used to great advantage by the drafter.

BROKEN-OUT SECTIONS (No cutting plane line is needed.)

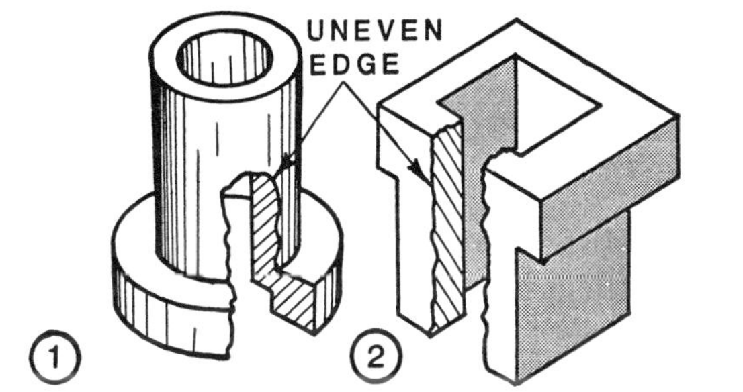

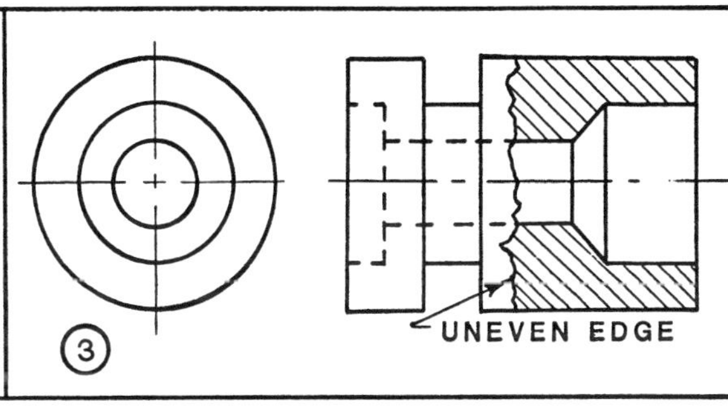

Occasionally only a small piece of an object need be "broken out" in order to show a detail more clearly. Note the use of the uneven edge to denote a broken-out section.

REVOLVED SECTIONS (No cutting plane line is needed.)

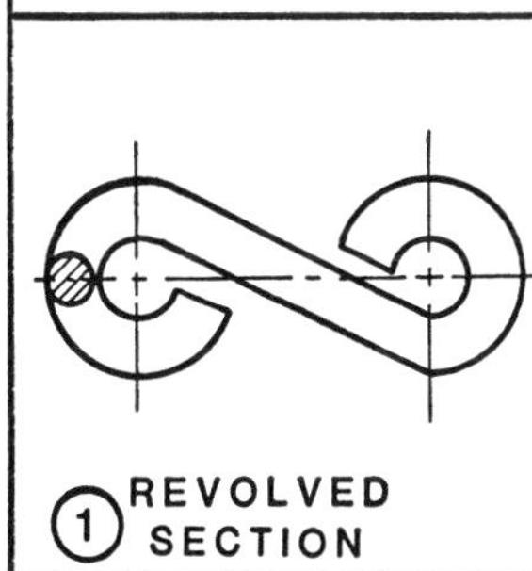

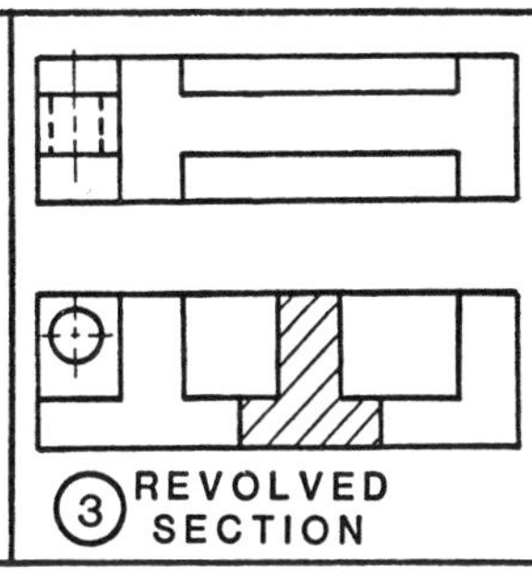

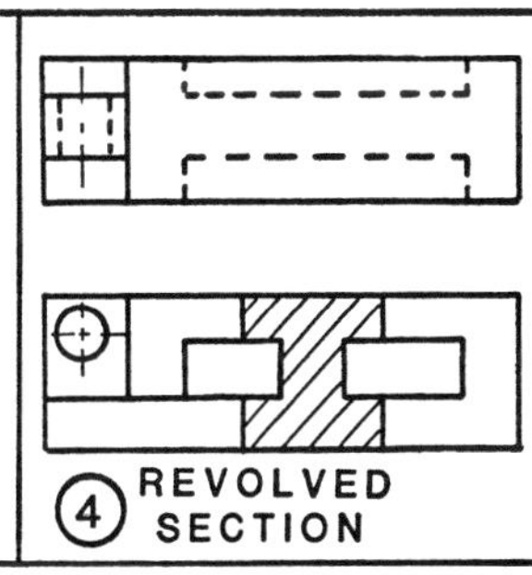

A revolved section shows the exact shape of the object at the point where the revolved section is drawn.

It would be difficult to describe the form in each case above if it were not for the use of revolved-section drawing.

REMOVED SECTIONS DESCRIBE CHANGING SHAPES

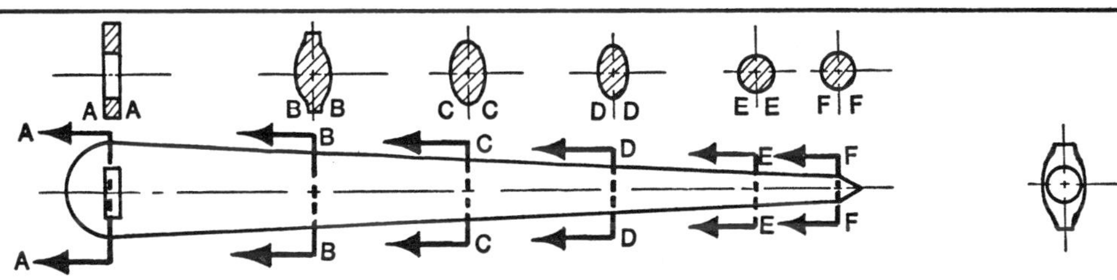

Removed sections serve the same purpose as do revolved sections in that they show the shape of the object at the exact point indicated. The sections are usually removed from the drawing in order to see and dimension them more clearly.

The needle shown above would be difficult to describe without the aid of removed-section drawings.

BOLTS, RIVETS, AND SHAFTS IN SECTIONAL VIEWS

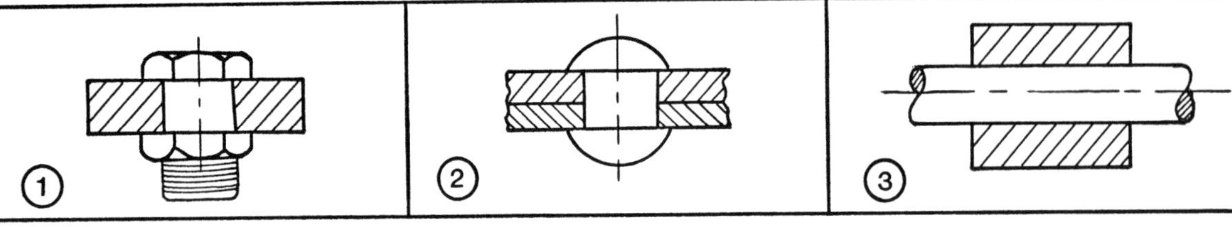

Bolts, nuts, rivets, keyways, shafts, etc., are not shown in section on section drawings. They are solid objects and need no further detailing.

RODS, PIPES, TUBES, ETC.

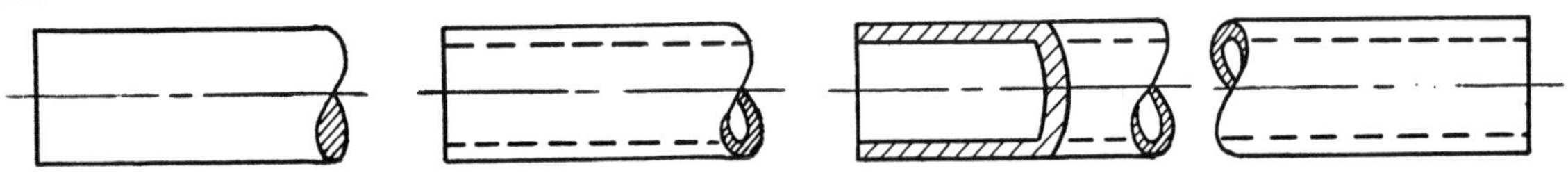

(A) ROUND ROD (B) PIPE OR TUBING (C) BREAK IN LENGTH OF PIPE

The symbols used to designate cylindrical rods, pipes, and tubes may be considered as a type of section drawing since they indicate an imaginary cut in an object.

MAKING THE SYMBOL FOR CYLINDRICAL ROD, ETC.

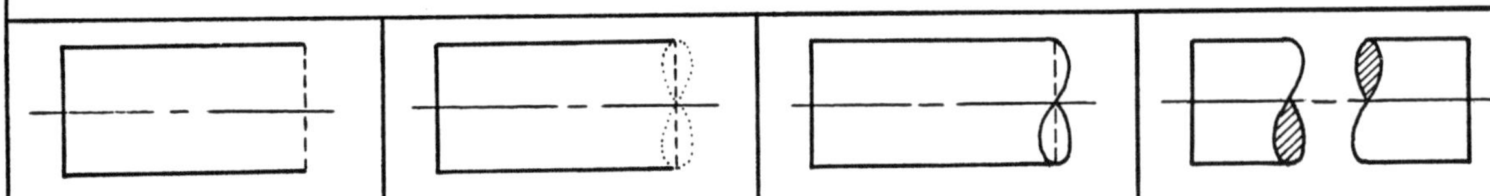

(1) Draw the layout for the rod.

(2) Sketch a flattened "8" as shown.

(3) Darken the required part. Draw the section lining.

(4) The symbols must appear as reverse duplicates.

VIOLATION OF RULES TO OBTAIN CLARITY

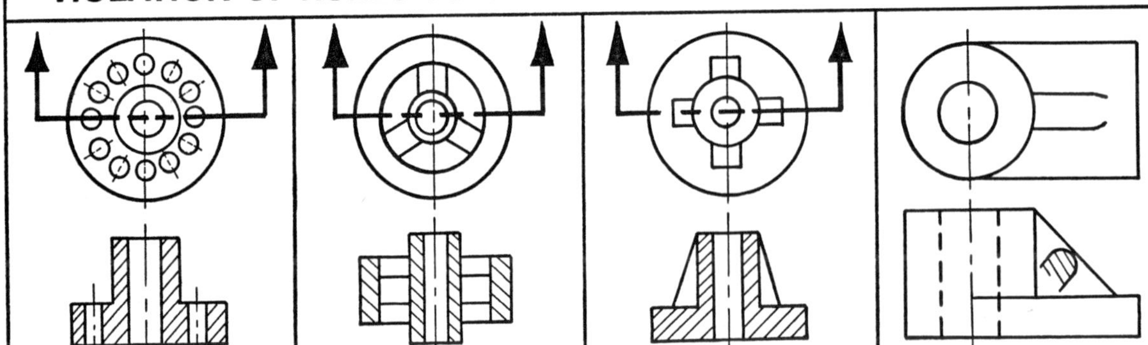

(1) Holes may sometimes be omitted for clarity on a section drawing.

(2) Spokes may be moved when required for clarity. Do not use section lines on the spokes.

(3) Section lining of supports may be omitted when necessary.

(4) Partial sections may be drawn for clarity.

PICTORIAL SECTION DRAWINGS

Since you know how to make pictorial drawings and understand full sections and half sections, you will find no difficulty in making pictorial section drawings.

AN ISOMETRIC FULL SECTION (PRISM)

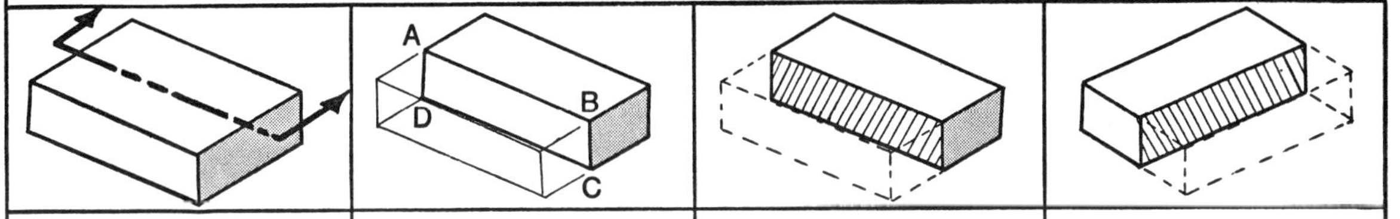

(1) The cutting plane line runs at 30° through the center of the isometric drawing.	(2) Drop a vertical line from "A" and from "B." Draw "CD" at 30°.	(3) Draw the section lining at 60°.	(4) The drawing could have been turned in this direction.

AN ISOMETRIC FULL SECTION (CYLINDER)

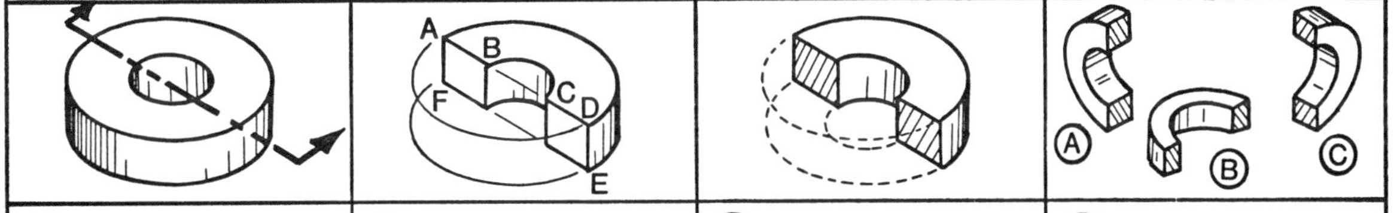

(1) The cutting plane line is drawn at 30°.	(2) Drop vertical lines from "A," "B," "C," and "D." Draw "EF" at 30°.	(3) Draw the section lining at 60°.	(4) The object may be revolved.

AN ISOMETRIC HALF SECTION (PRISM)

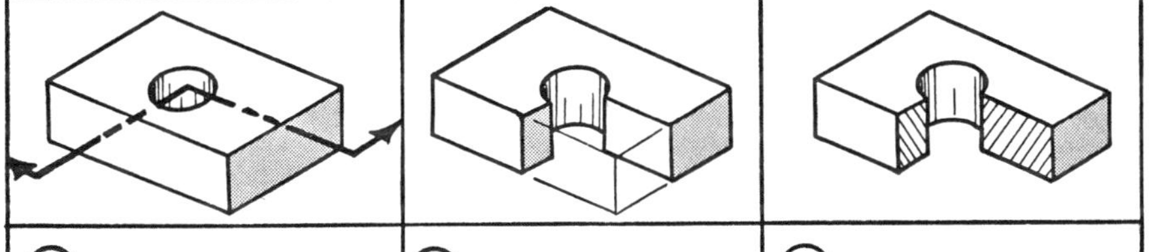

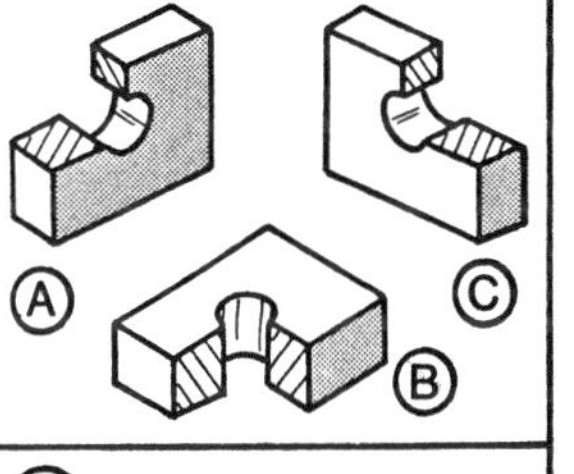

(1) The cutting plane line starts at the middle and goes to the left and right at 30°.	(2) Drop the necessary vertical lines from the top surface.	(3) Draw the section lining at 60°. Study the direction of the section lining.	(4) The object could have been revolved.

ISOMETRIC HALF SECTIONS (CYLINDERS)

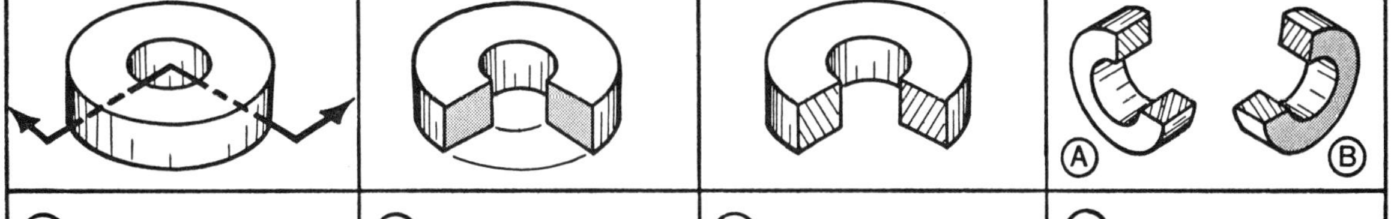

(1) Study the cutting plane line.	(2) Drop the necessary vertical lines.	(3) Draw the section lining at 60°.	(4) The object could have been revolved.

CABINET AND CAVALIER SECTION DRAWINGS

The principles of pictorial drawing and section drawing must be followed when making cabinet or cavalier section drawings.

OBLIQUE (OR CABINET) FULL SECTIONS (CYLINDERS)

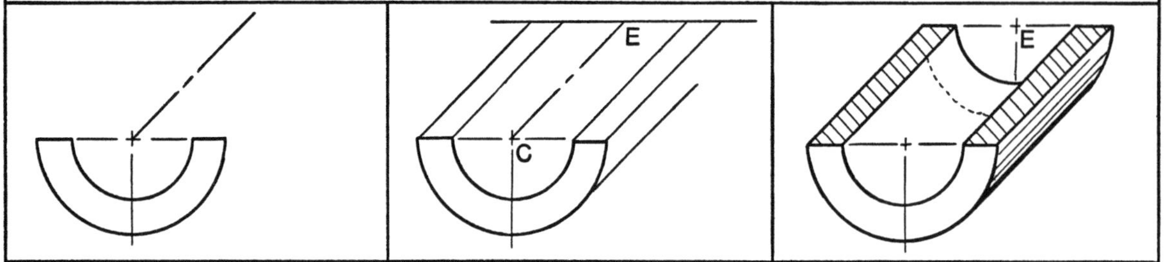

(1) Start by drawing the half circles. Draw the center line back at 45°.

(2) Measure the required length of the cylinder. Extend the necessary edges backward at 45°.

(3) Using "E" as a center, draw the arcs in the back. Complete the drawing.

OBLIQUE (OR CABINET) FULL SECTIONS (PRISMS)

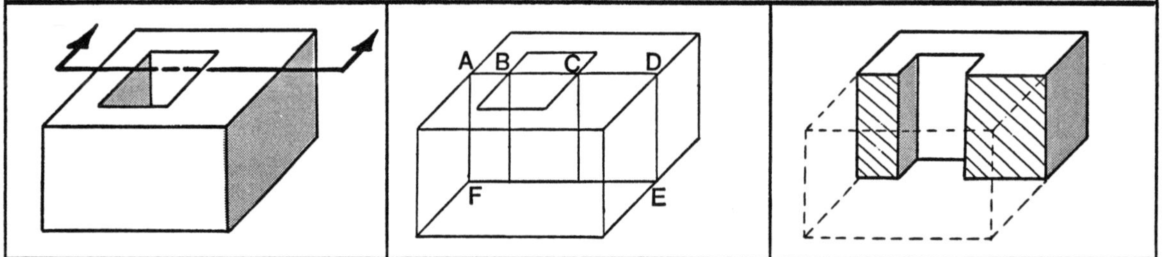

(1) Draw the cutting plane line in a horizontal position.

(2) Drop verticals from "A," "B," "C," and "D" to line "EF."

(3) Complete the full section by adding the 45° section lining.

SECTION DRAWINGS MAY BE REVOLVED (Study the sectioning lining.)

ASSIGNMENT

Make a pictorial section drawing of any object shown in this unit. Draw either a full section or a half section. Revolve the object if you wish. Draw the orthographic section of each pictorial section you draw.

DIMENSIONING SECTION DRAWINGS

Section drawings are dimensioned in the same manner as regular pictorial drawings and orthographic projections with one exception: Make every effort to avoid placing numerals and arrows within the areas that have section lining.

AVOID DIMENSIONING IN THE SHADED AREAS IF POSSIBLE

When it is absolutely necessary that you dimension in a shaded area, break the section lining so that the numerals may be read easily and accurately.

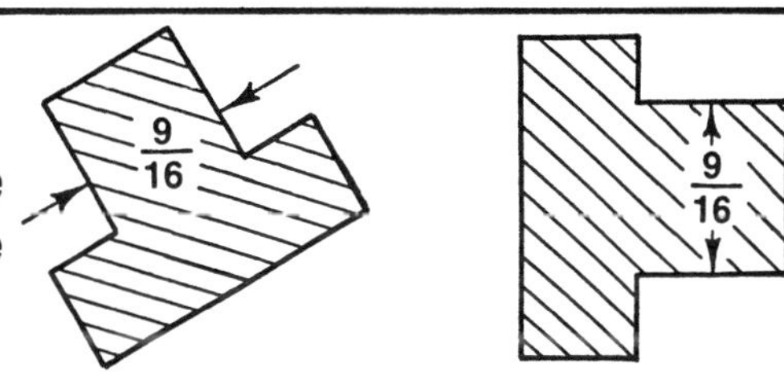

STUDY THE DIMENSIONING OF THESE SECTION DRAWINGS

(1)

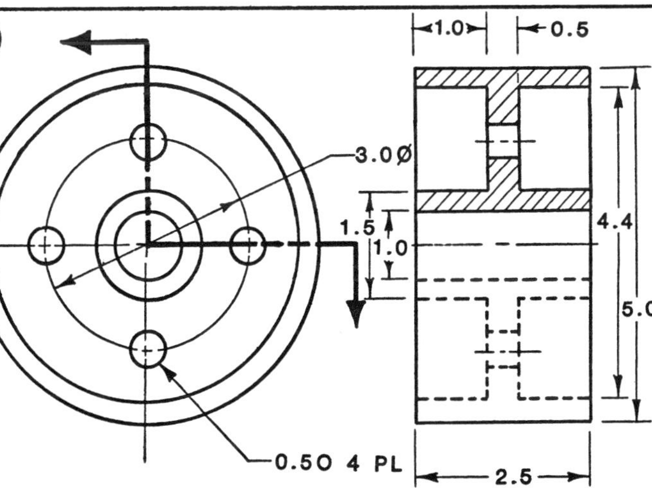

(2)

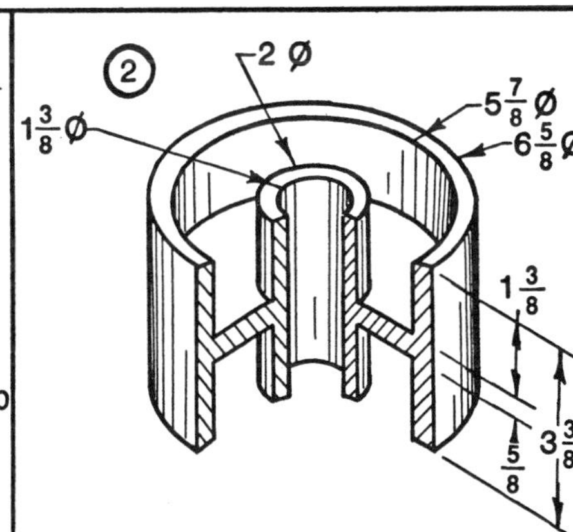

(3)

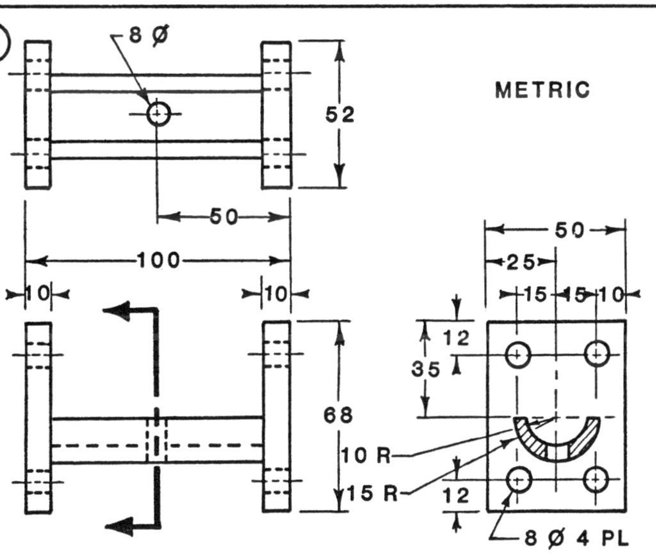

(4)

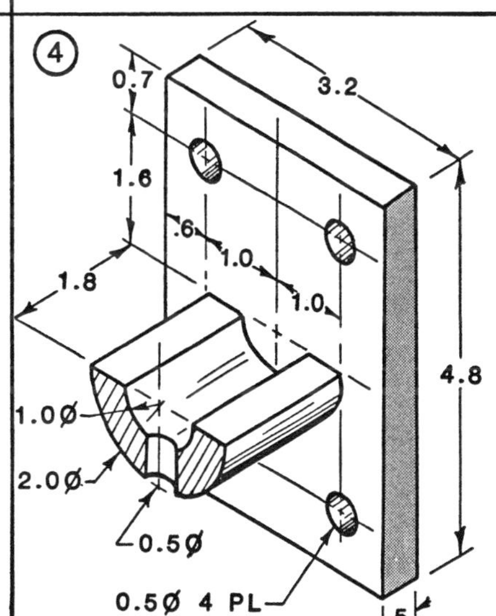

ASSIGNMENT

Dimension the section drawings that you have made to date.

PROBLEMS IN SECTION DRAWING

At this point in your training, you should realize that it is possible to make a section drawing using any of the four types of drawing that you have learned (isometric, oblique, cabinet, or orthographic).

Make a section drawing of one of the objects shown below. Use any of the four types of drawing mentioned above.

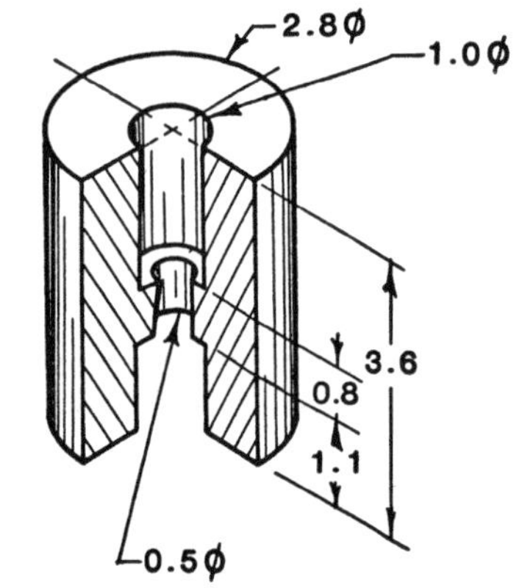

(1) DIE

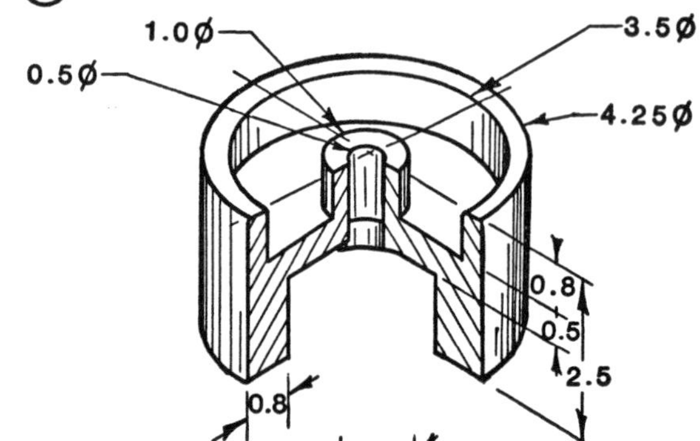

(2) FLYWHEEL

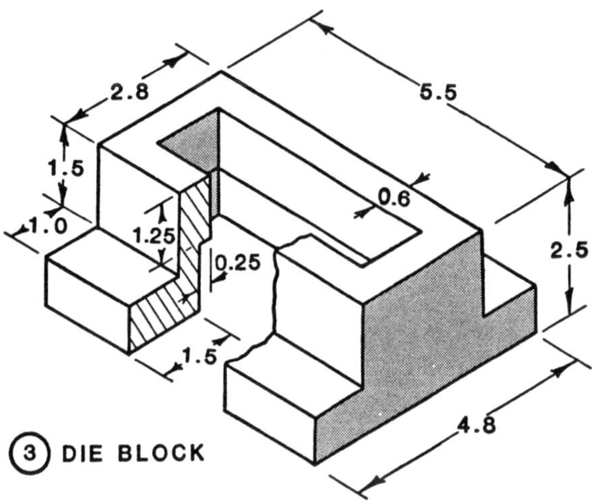

(3) DIE BLOCK

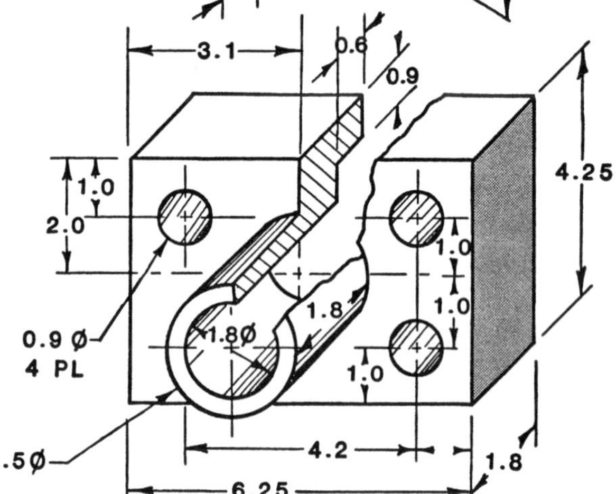

(4) BEARING

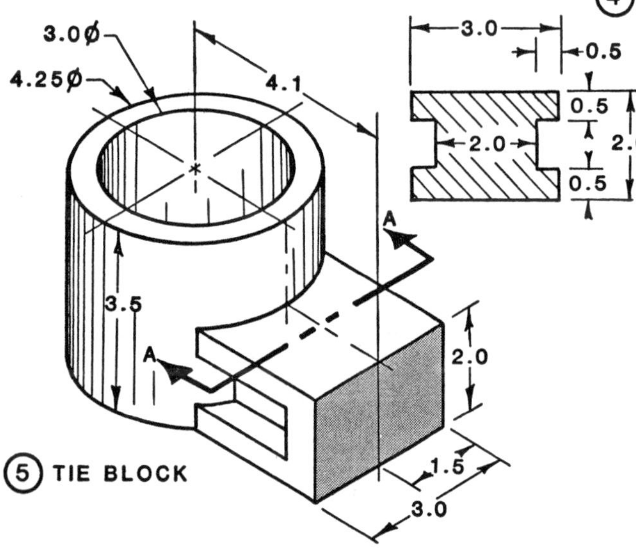

(5) TIE BLOCK

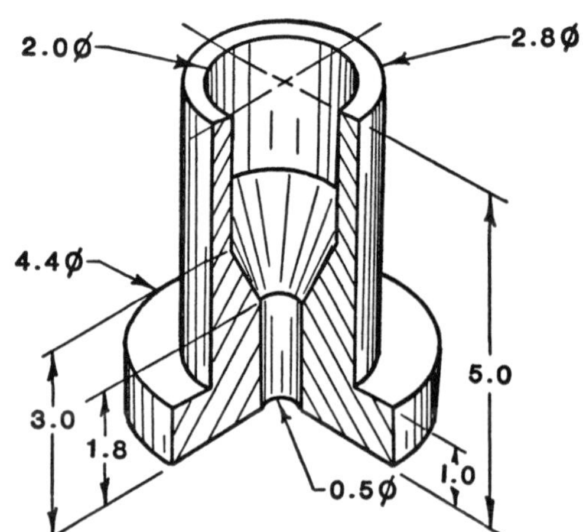

(6) BUSHING

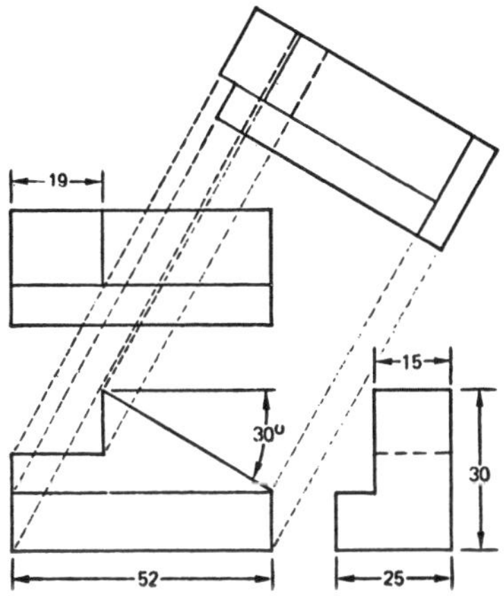

UNIT 13

AUXILIARIES

A mechanical drawing must represent the true shape of an object. Each surface must show all the necessary information needed to build the object. The orthographic drawing is most often used for this information. An auxiliary view is another kind of orthographic view used to give additional information.

Any number of auxiliary views can be projected for an orthographic drawing. The main purpose of an auxiliary view is to show the true size of a surface that cannot be shown on an orthographic drawing.

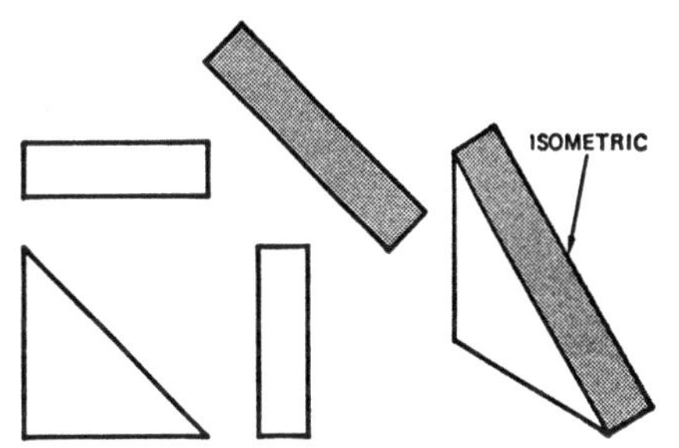

PRIMARY ORTHOGRAPHIC VIEWS

PRIMARY VIEWS
THIRD-ANGLE
PROJECTION

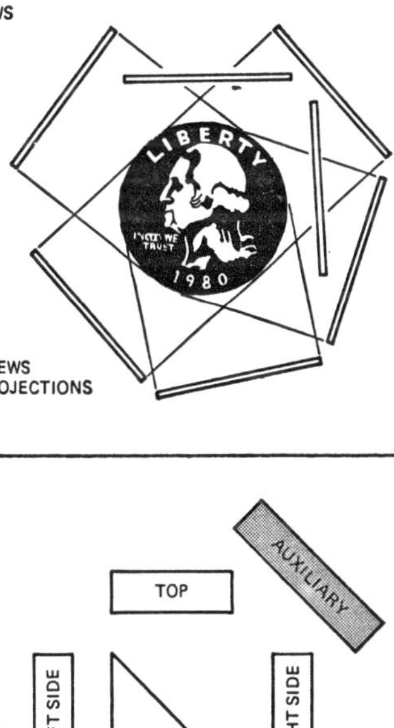

AUXILIARY VIEWS
UNLIMITED PROJECTIONS

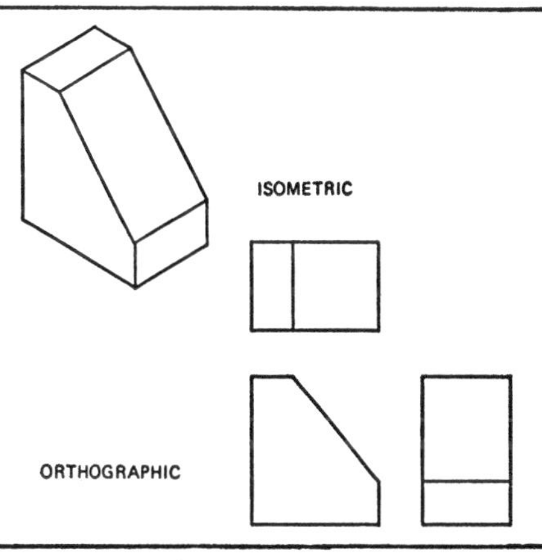

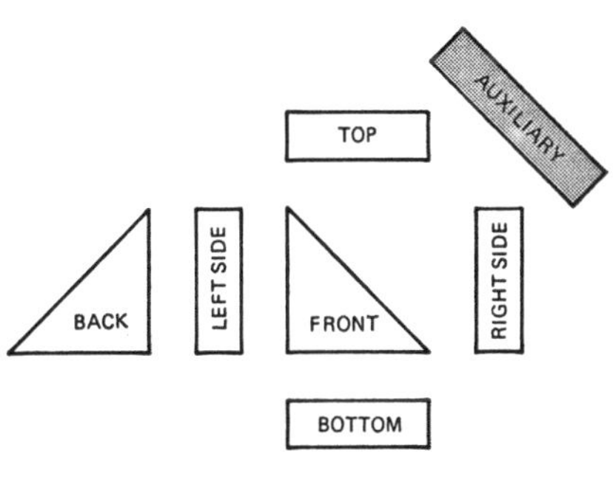

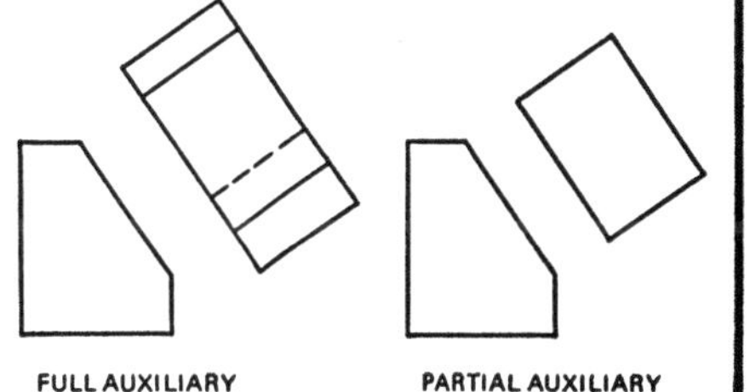

FULL AUXILIARY

PARTIAL AUXILIARY

STEPS IN MAKING AN AUXILIARY VIEW

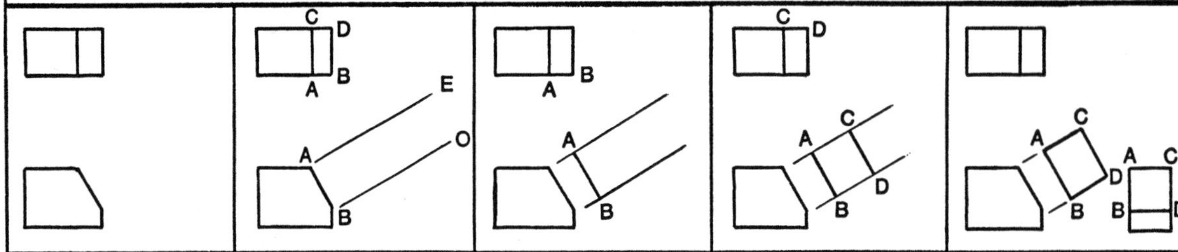

1. Draw the front view and the top view.
2. Draw "AE" and "BO" at right angles to "AB."*
3. Draw "AB" parallel to "AB."*
4. Draw "CD" equal to "AB."
5. Draw "AC" and "BD" to complete the auxiliary view.

***Special Note:** An auxiliary view can be projected from any orthographic view at any convenient angle.

STUDY THESE PARTIAL AUXILIARY VIEWS

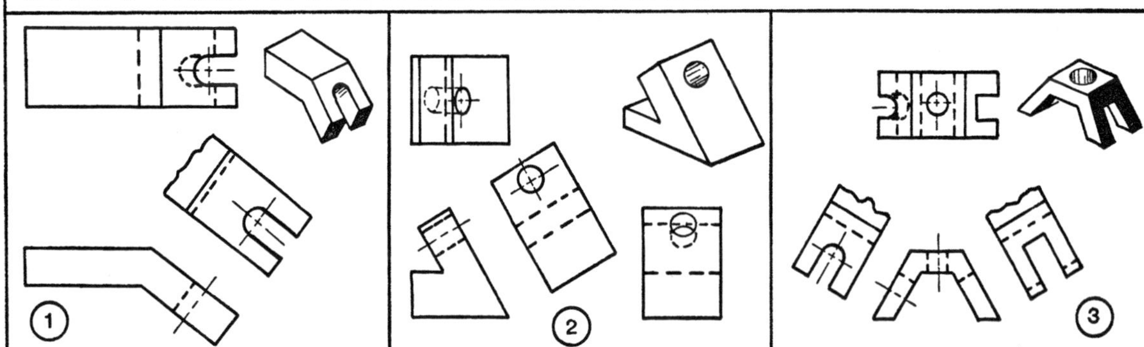

A COMPLETE AUXILIARY VIEW

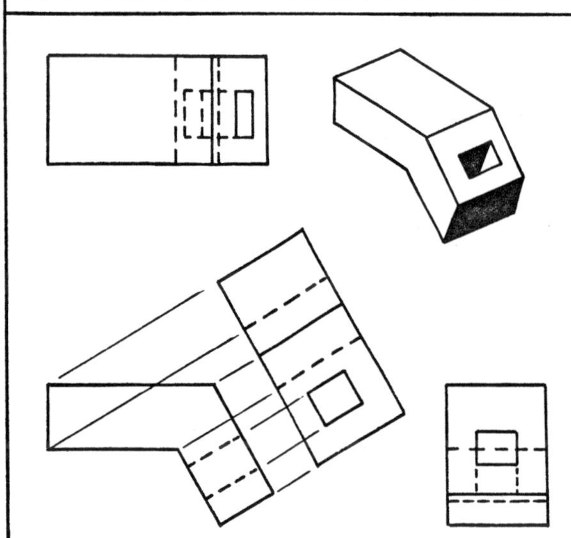

A PARTIAL AUXILIARY VIEW

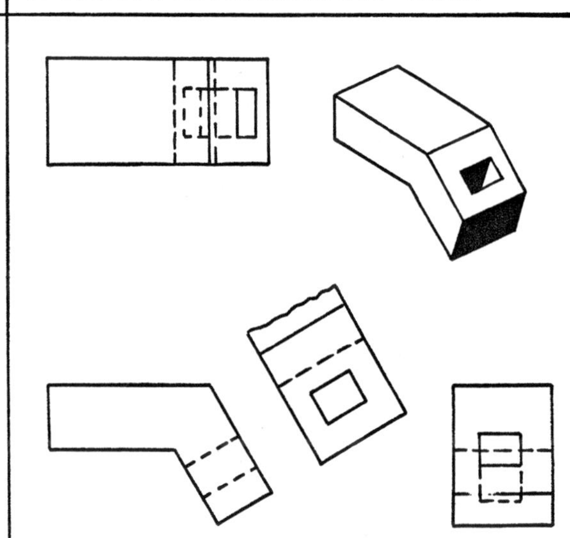

PAGE LAYOUTS

Careful planning is required in laying out a drawing where room must be left between the three views for an auxiliary view. When an auxiliary view is necessary, plan your layout with this in mind.

STUDY THIS AUXILIARY DRAWING AND SKETCH IT

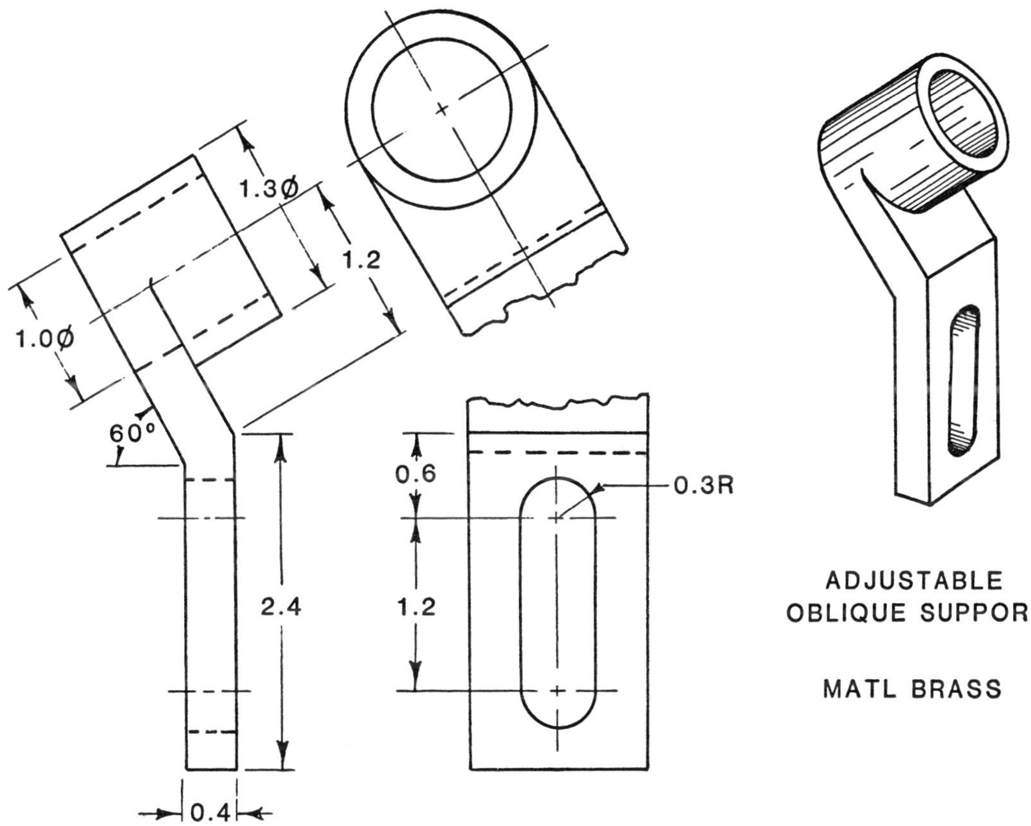

ADJUSTABLE
OBLIQUE SUPPORT

MATL BRASS

ASSIGNMENT

Make a complete auxiliary drawing of one of these objects.

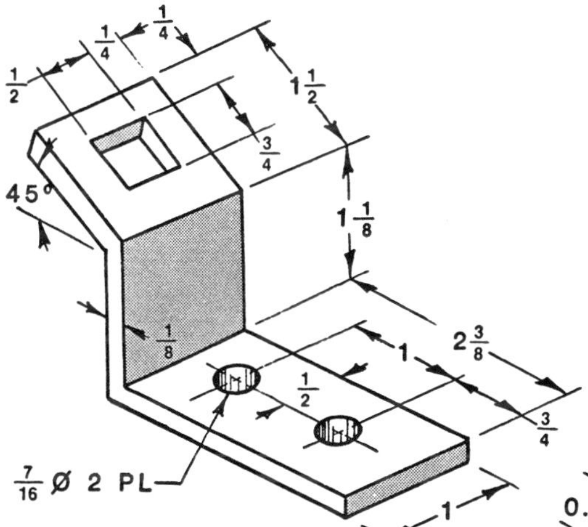

(1) ANGLE SUPPORT BRACKET

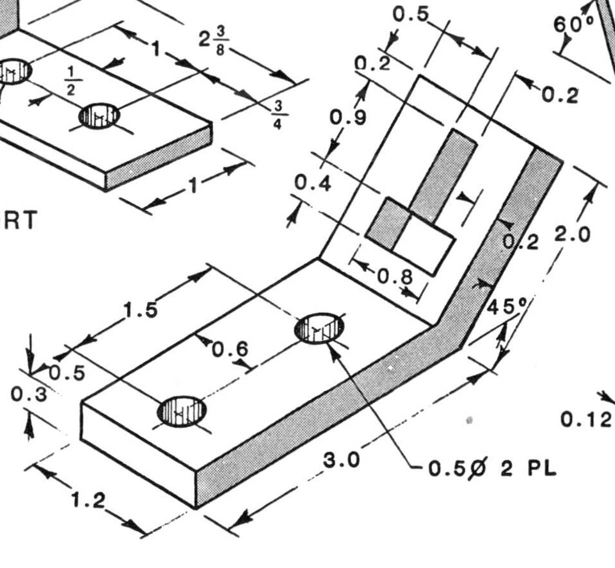

(2) ANGLED HANGER

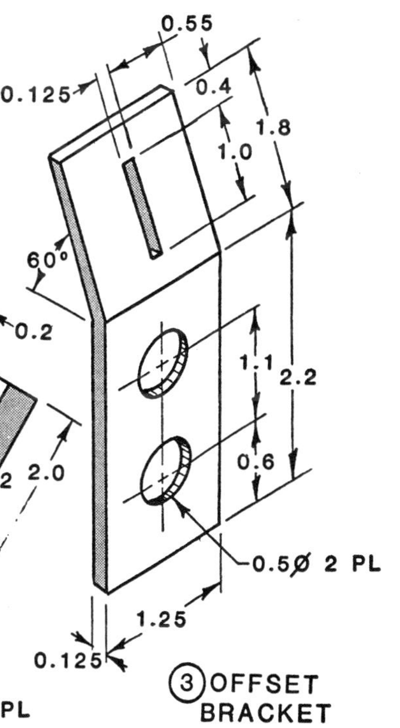

(3) OFFSET BRACKET

DRAW THE ORTHOGRAPHIC, AUXILIARY, AND ISOMETRIC DRAWINGS.

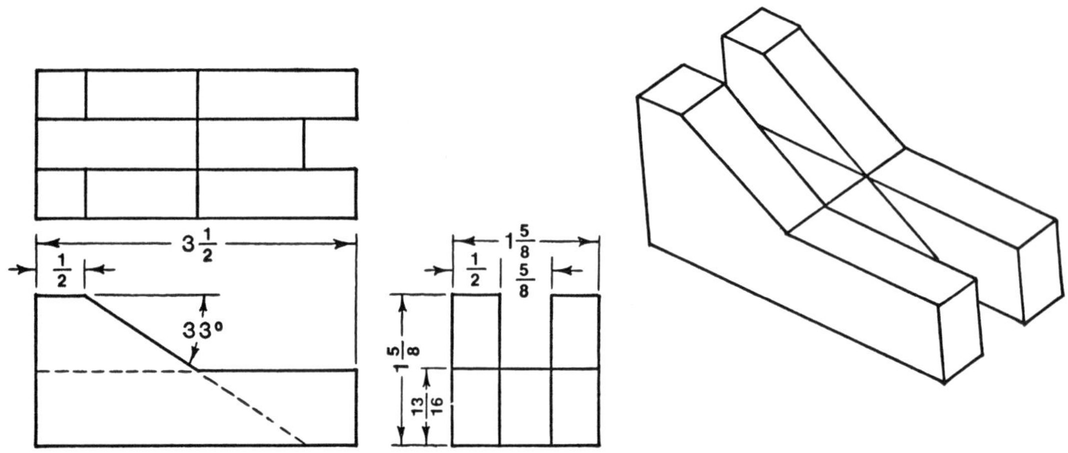

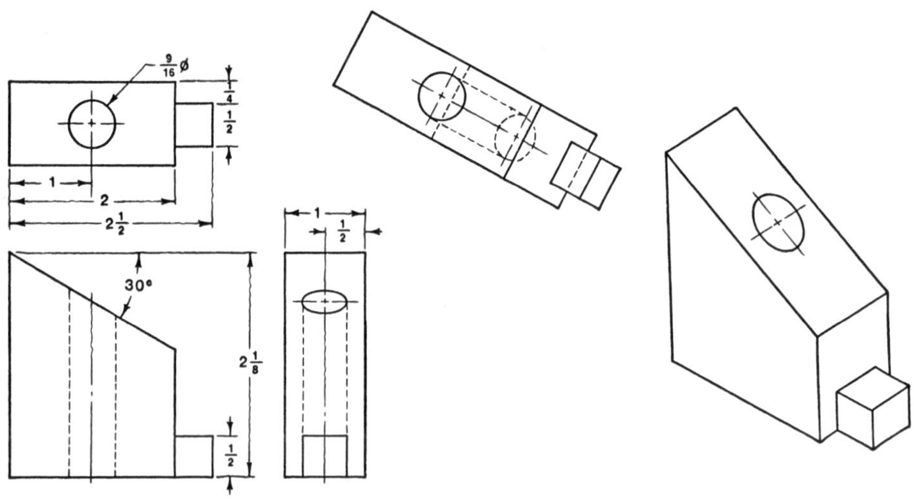

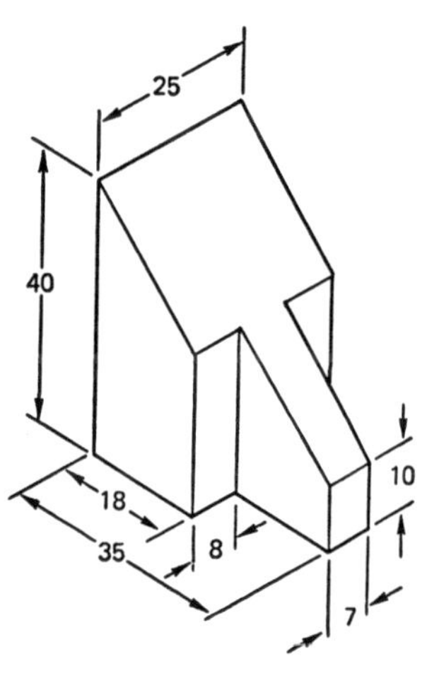

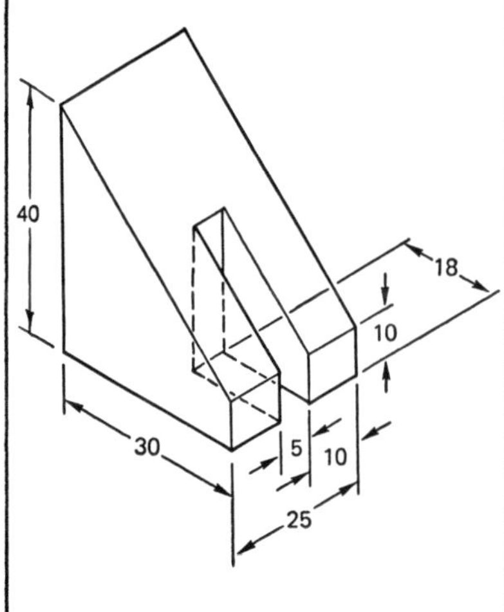

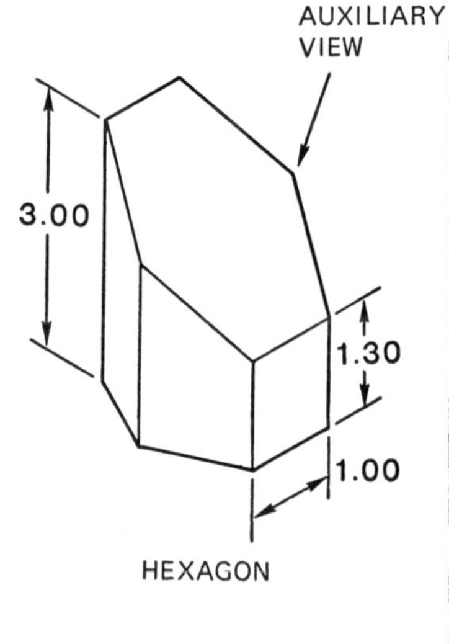

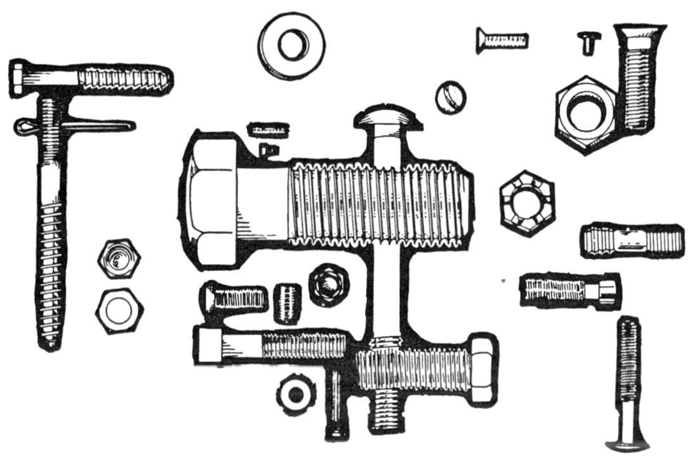

UNIT 14

THREAD CONVENTIONS

Many machine parts are threaded in order that the parts may be assembled. It is important to know the basic methods used by the drafter to indicate machine threads.

While there are many types of threads, we will confine our study to the "V" thread and to the drawing of the symbols that may be used to designate any type of thread.

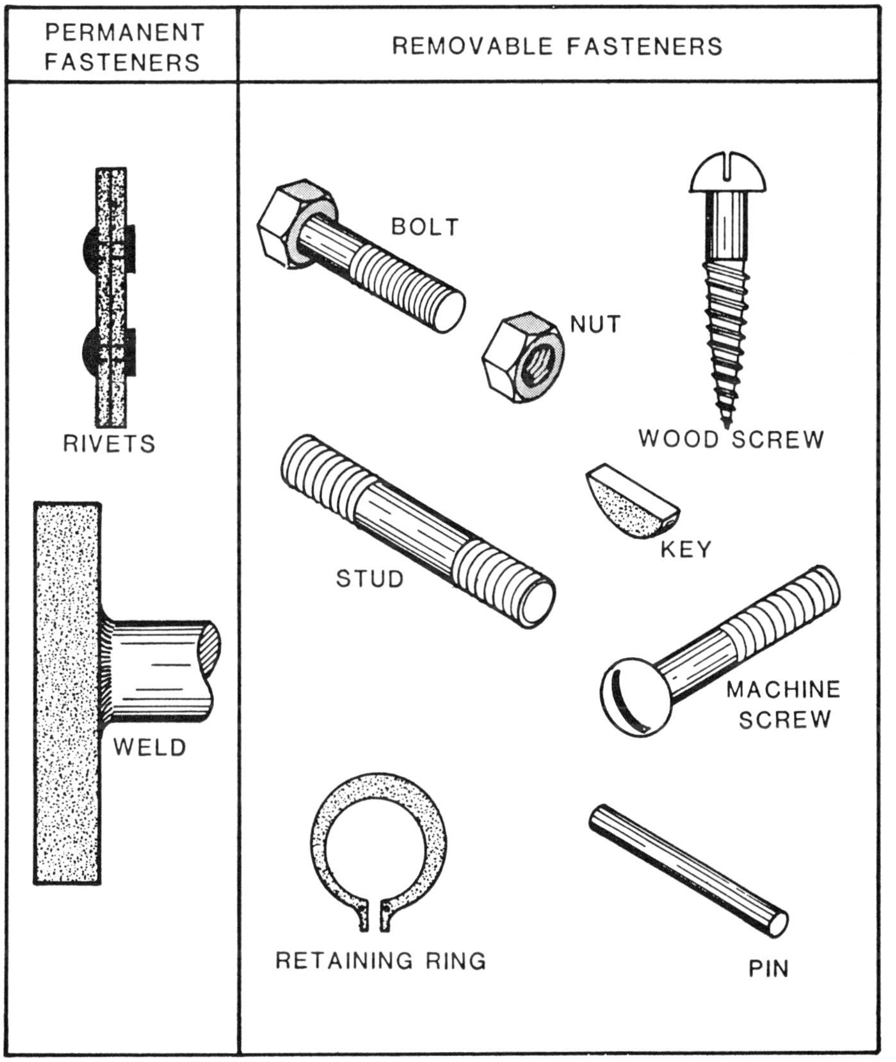

Parts of External Threads

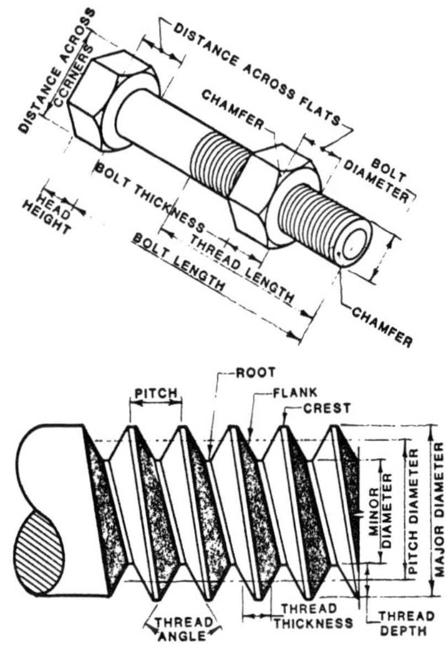

Parts of a threaded fastener

External Thread Symbols

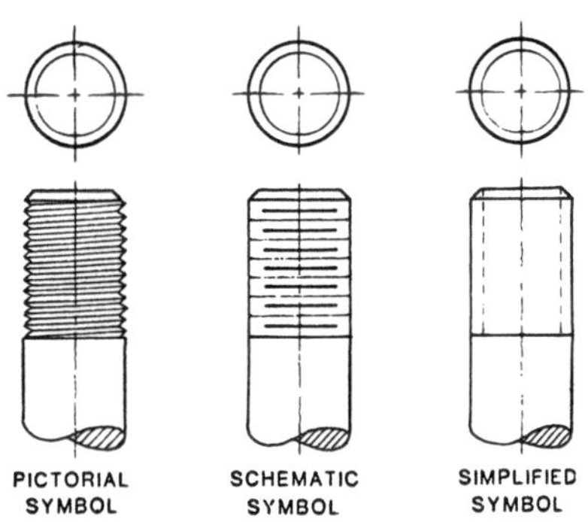

Three types of thread symbols. The simplified symbol is preferred.

A basic rule in drafting is to pick the quickest drawing method to communicate the ideas and concepts to the fabricator. All three methods of thread representation are acceptable on a drawing, but the simplified version of drawing threads is recommended because it will save time.

Types of External Threads

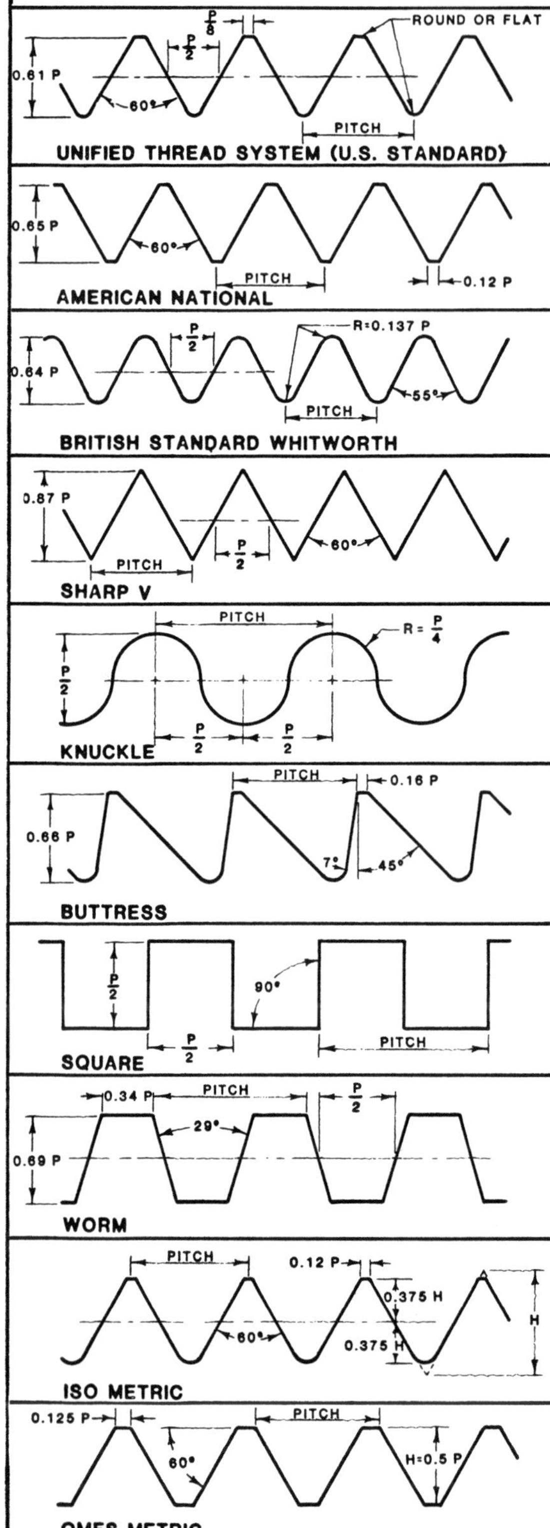

Profiles of ten types of threads

THE PROFILE OF THE SHARP "V" THREAD

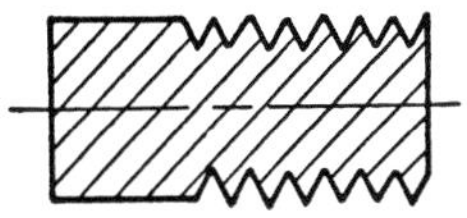

(A) EXTERNAL THREAD

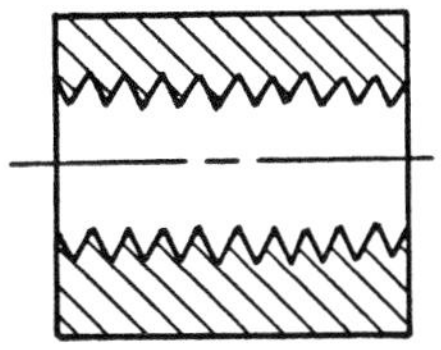

(B) INTERNAL THREAD

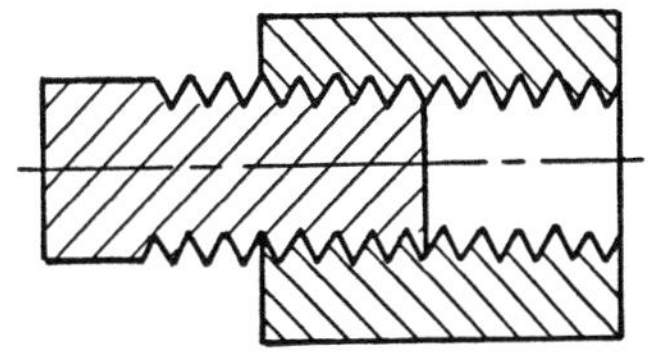

(C) THREADS ENGAGED

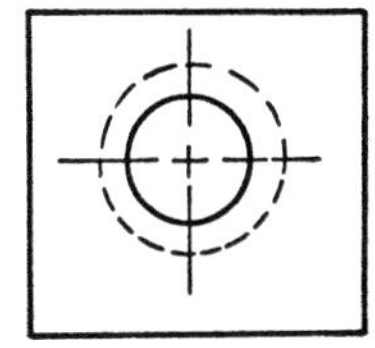

(D) END VIEW

A sectional view of a thread presents an excellent profile. An external thread is one that has been made on the outside of the object. An internal thread is one that has been made on the inside of an object.

THREE METHODS OF DRAWING THREADS

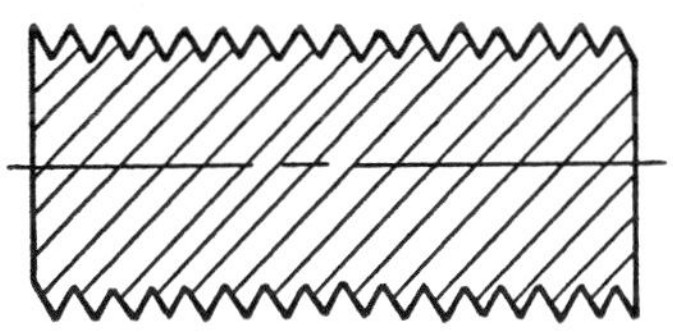

THREAD PROFILE

A profile is an outline of a thread. The profile is easily seen on a sectional view.

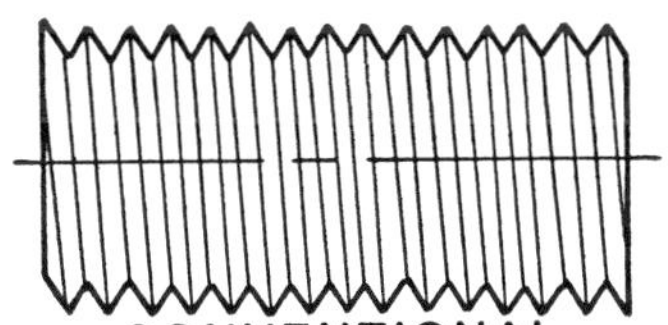

CONVENTIONAL REPRESENTATION

A conventional representation is a modified drawing that presents a likeness of a thread.

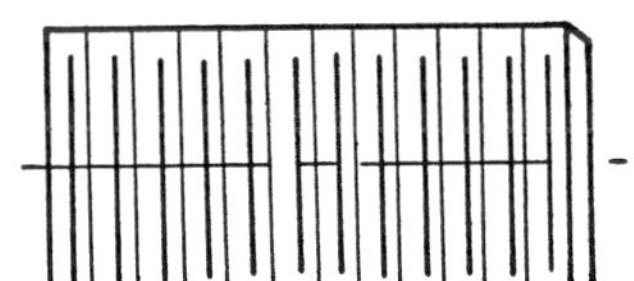

SIMPLIFIED THREAD SYMBOL

A symbol of a thread is a sign used to designate a thread.

TERMS USED IN DESCRIBING THREADS

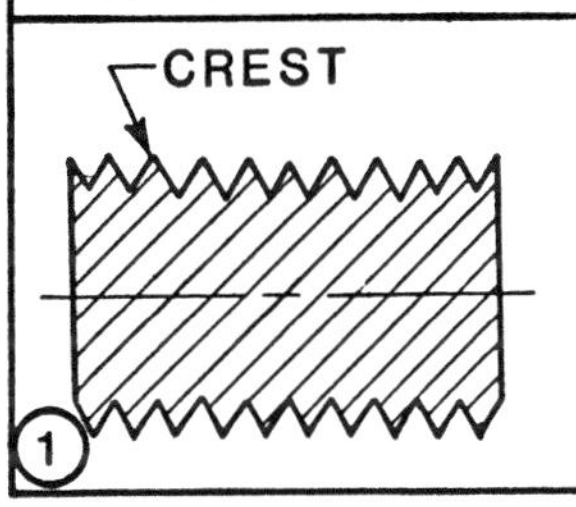

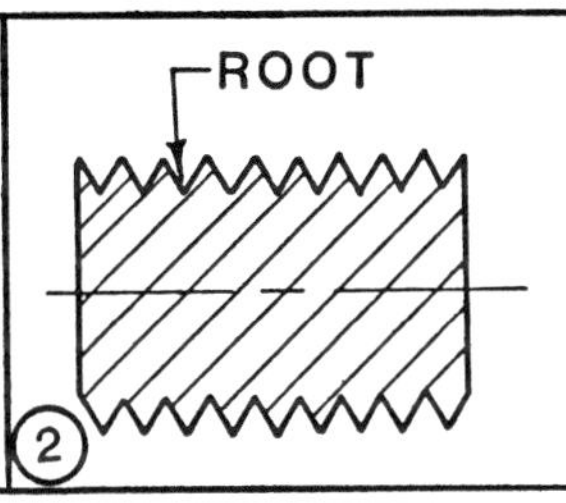

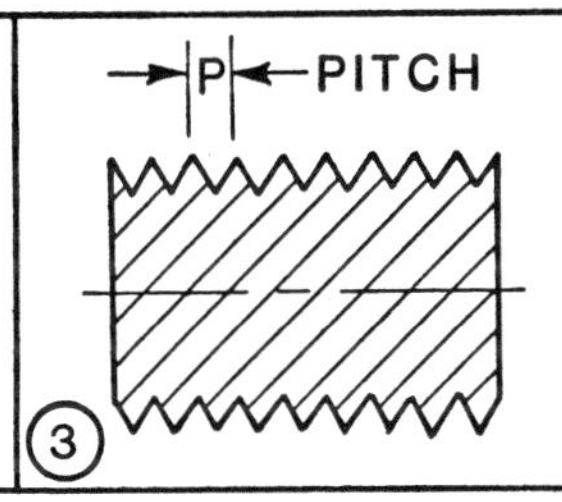

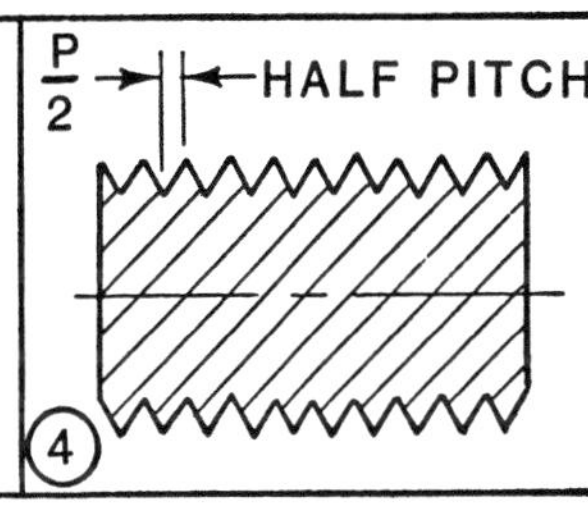

(1) The highest point of a thread is called the **crest.** (2) The lowest point of a thread is called the **root.**
(3) **Pitch** is the horizontal distance between one crest and the next crest.
(4) **Half pitch** is half of the pitch distance.
Study the above terms. You will need them in the lesson that follows.

INTERNAL THREADS Schematic Simplified

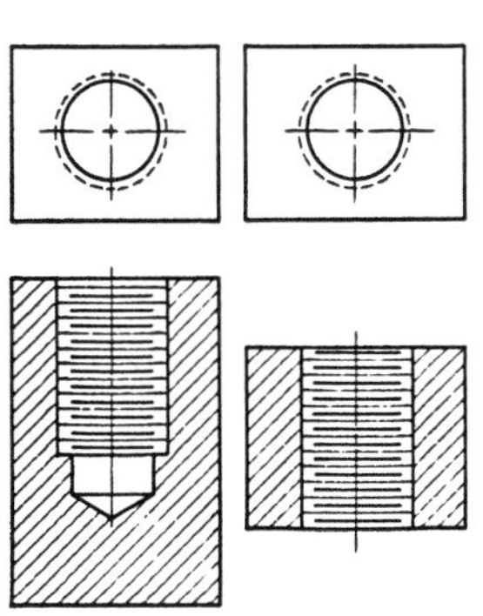

SCHEMATIC SYMBOL IN SECTION

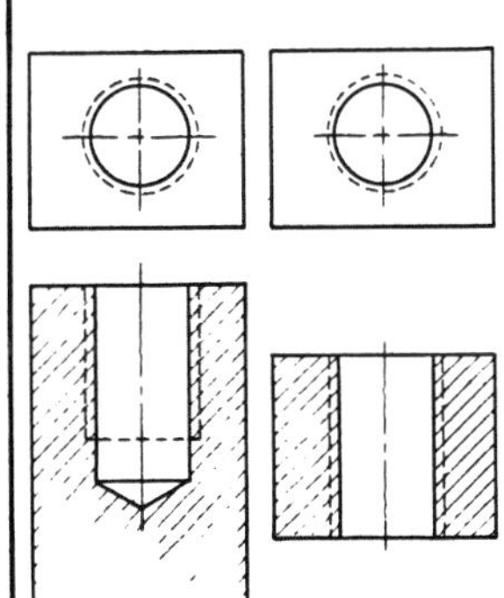

SIMPLIFIED SYMBOL IN SECTION

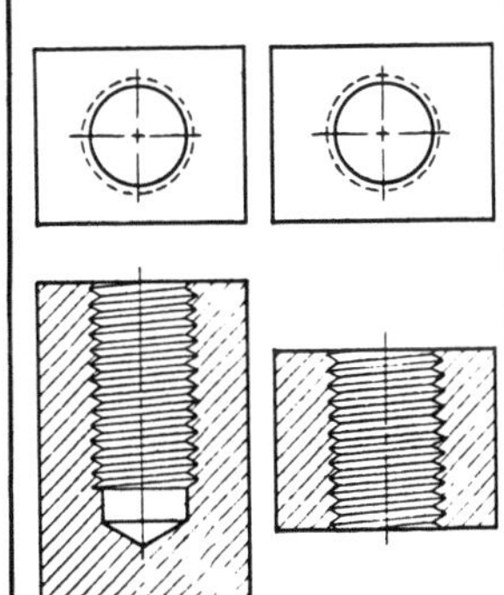

PICTORIAL THREADS IN SECTION

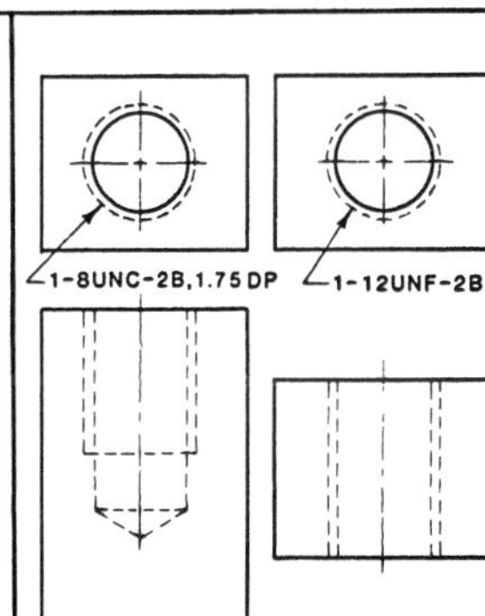

SIMPLIFIED SYMBOL

HOW A CONVENTIONAL REPRESENTATION OF A THREAD IS MADE

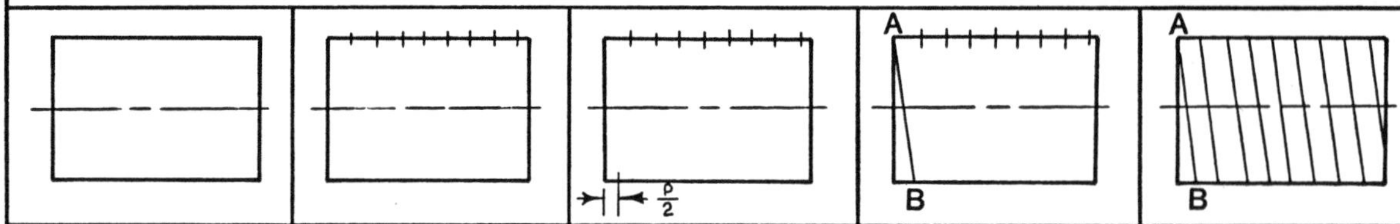

(1) Draw the contour view of the cylinder.

(2) Mark off the pitch units on the top edge.

(3) Mark off one one-half pitch unit on the bottom.

(4) Draw a line from "B" to "A."

(5) Draw the other lines parallel to "AB."

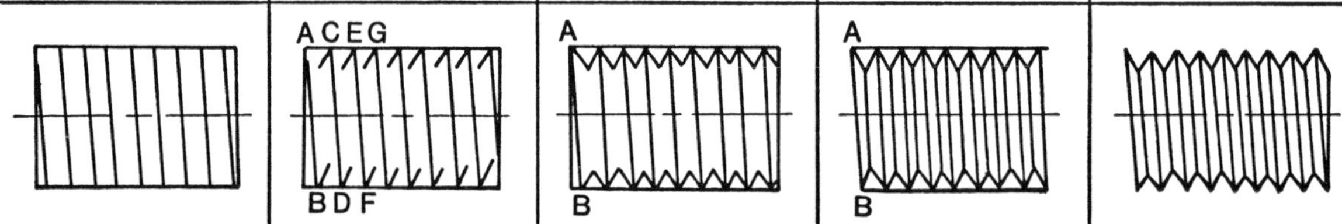

(6) These lines represent the crests of the threads.

(7) Draw 60° lines from "B," "D," "F," and "C," "E," "G," etc., as shown.

(8) Draw 60° lines in the opposite direction.

(9) Connect the roots by drawing parallel to "AB."

(10) The completed conventional representation.

It is not necessary that you draw the above conventional representation at this point in your training. However, you must study the method in order to apply this knowledge of thread drawing to the required drawings of thread symbols that follow.

TWO THREAD SYMBOLS

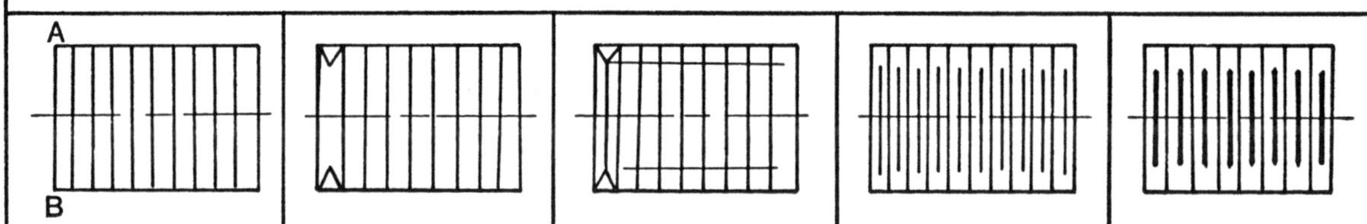

(1) Prepare the drawing as in steps 1 to 5 above.

(2) Draw one thread with 60° lines (lightly).

(3) Project a light horizontal line from each root.

(4) Draw heavy lines parallel to "AB" and ending at the root lines.

(5) A thread symbol may be drawn with lines that do not slope.

OTHER THREAD SYMBOLS

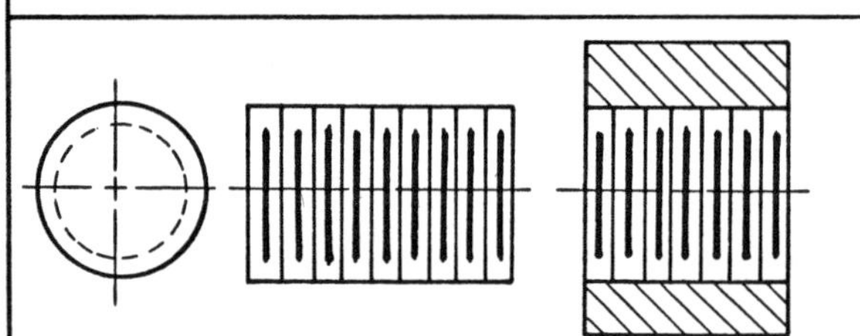

(1) The symbol for threads as used on external and internal threads.

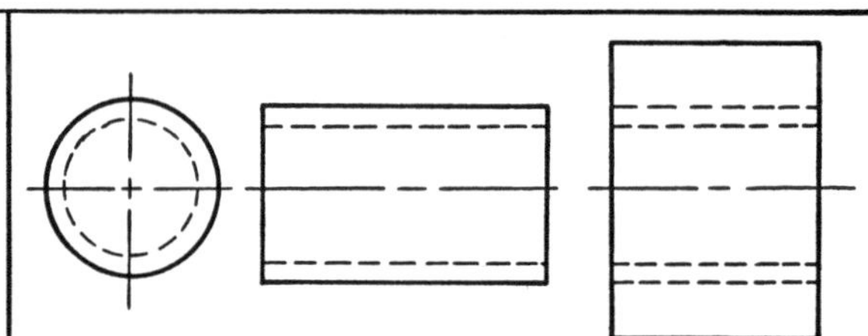

(2) This is a simplified symbol often used on detail drawings.

THREAD SYMBOLS AND THEIR APPLICATION

It must be remembered that the spacing of the crest and root lines on a thread symbol is not an exact representation of the thread wanted, and that the spacing is done entirely by eye. The drafter must prepare a lettered note describing to the fabricator the exact thread to be made.

THREAD DESIGNATIONS (AMERICAN NATIONAL STANDARDS)

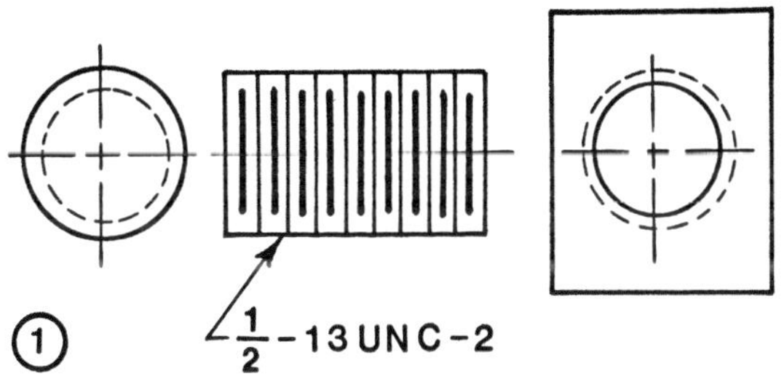

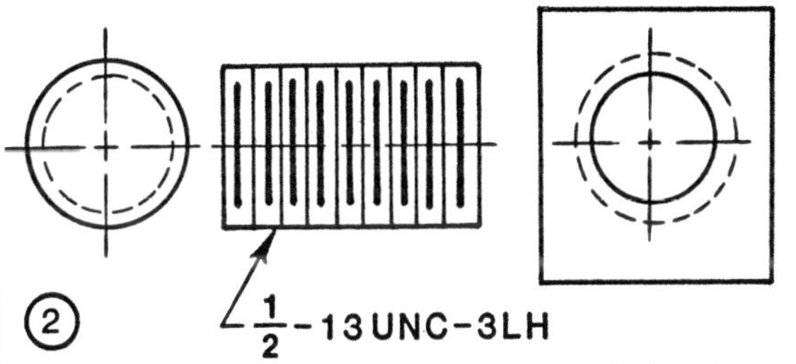

There are many types of threads that could be made on a 1/2" shaft. The drafter makes all of the thread drawings alike and depends upon the lettered note to inform the machinist as to the exact thread to be made on a 1/2" shaft.

HOW TO READ THREAD DESIGNATIONS

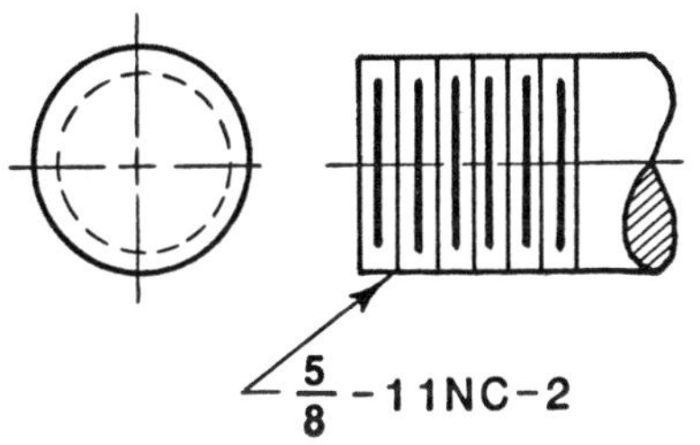

1. The "5/8" refers to the diameter of the shaft.
2. The "11" designates eleven threads per inch.
3. The "N" refers to American "National."
4. The "C" refers to "coarse series" as compared to "F" for "fine series."
5. The "2" refers to the "class" of fit. (There are four classes of fits, numbered 1 to 4.)
6. A left-handed thread must be marked "LH."

EXTERNAL AND INTERNAL THREADS

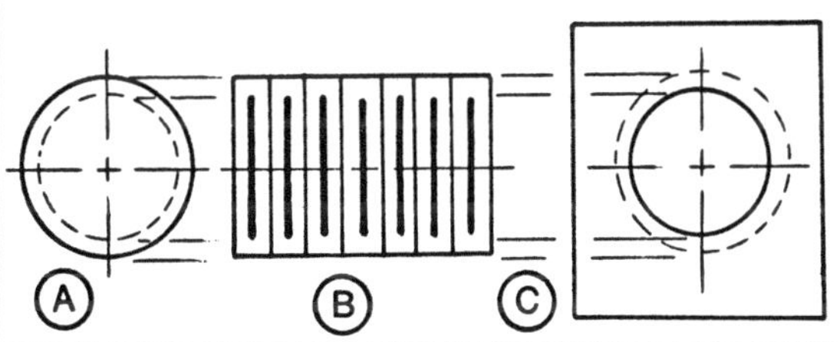

The symbol for external threads Ⓐ is drawn so that the hidden edge circle in the end view represents the root of the thread. On internal thread symbols Ⓑ, the hidden edge circle represents the crests.

"TAPPED" HOLES

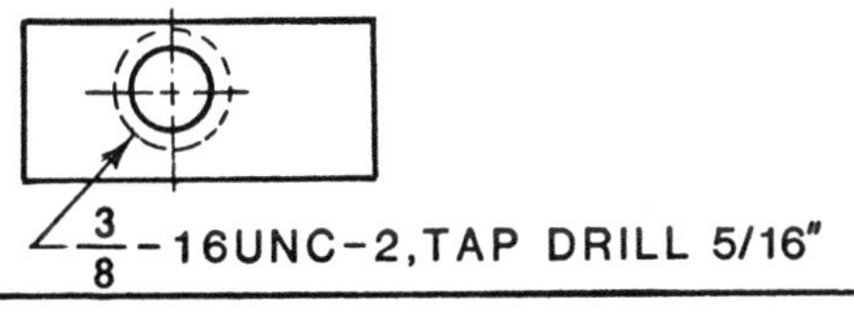

Holes of small diameter are threaded with a tool known as a "**tap**." These threaded holes are referred to as tapped holes.

The tap drill diameter, depth of the drilled hole, tap size, and length of thread must be given by note.

THREE METHODS OF DRAWING THREAD SYMBOLS

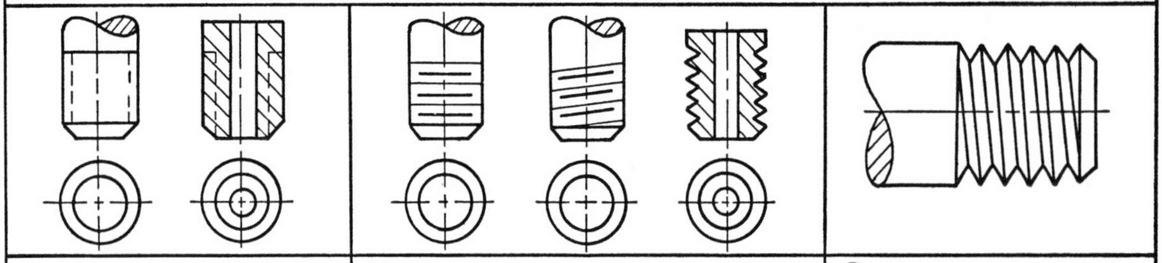

(1) A time-saving method to be used on detail drawings.

(2) More finished methods to be used on assembly drawings.

(3) The crest and root lines are the same thickness here.

SUGGESTIONS FOR DRAWING THREAD SYMBOLS

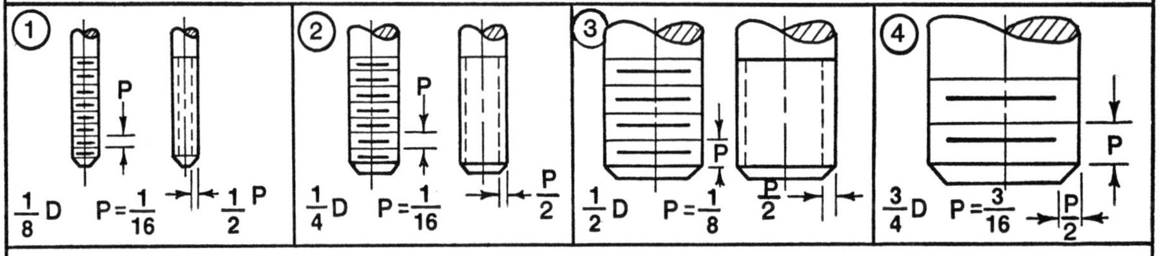

(1) For diameters of 1/8″ and 3/16″ use a pitch (P) of a scant 1/16″. Make the depth of the thread equal to one half of the pitch (1/32″).
(2), (3), (4) For diameters of 1/4″, 1/2″, and 3/4″ use the pitch and depth of thread as indicated in each of the above.

STUDY THE THREAD SYMBOLS ON THESE DRAWINGS

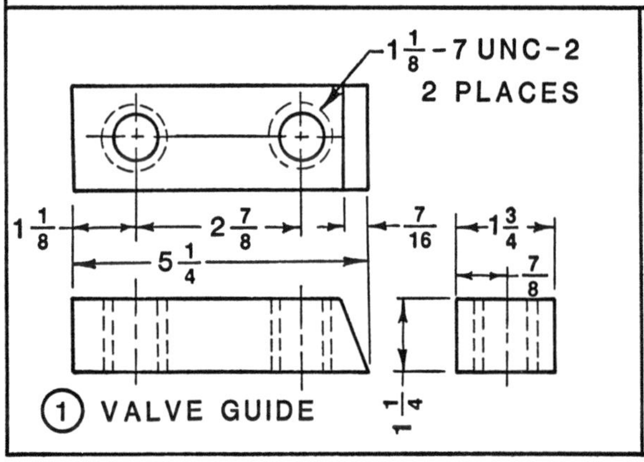

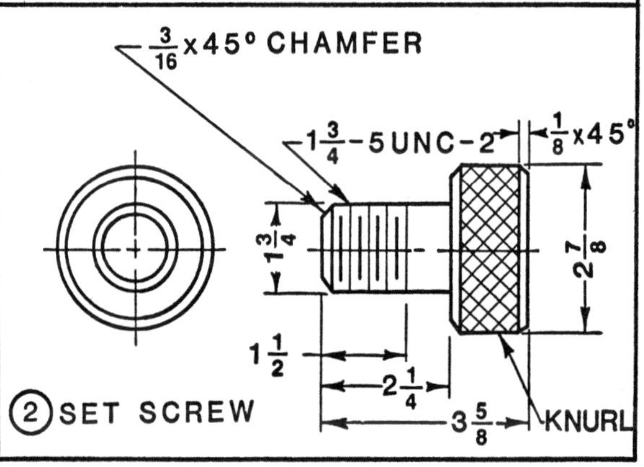

ASSIGNMENT

Make a working drawing of one of these objects.

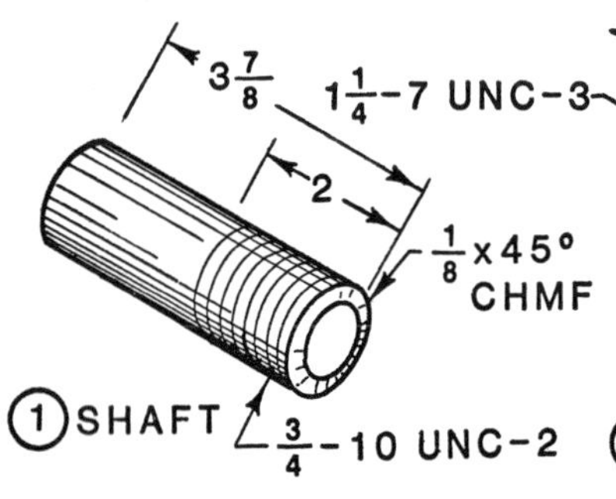

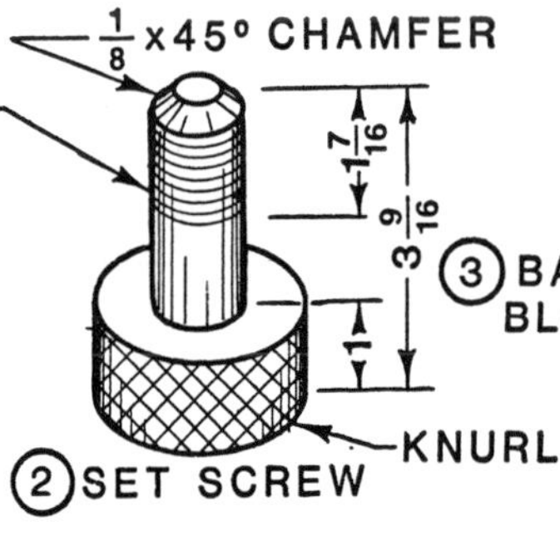

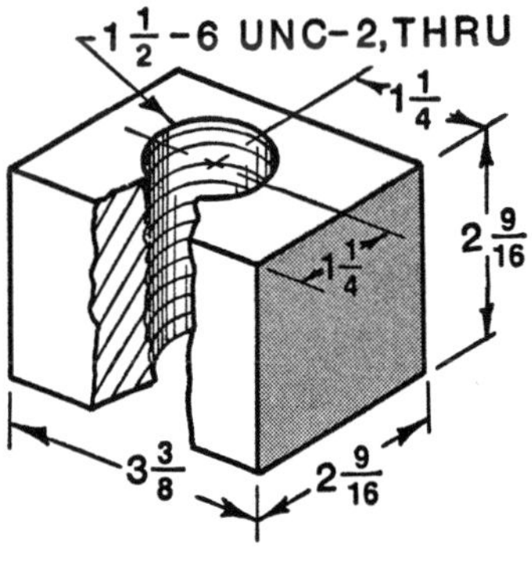

UNIT 15

WORKING DRAWINGS

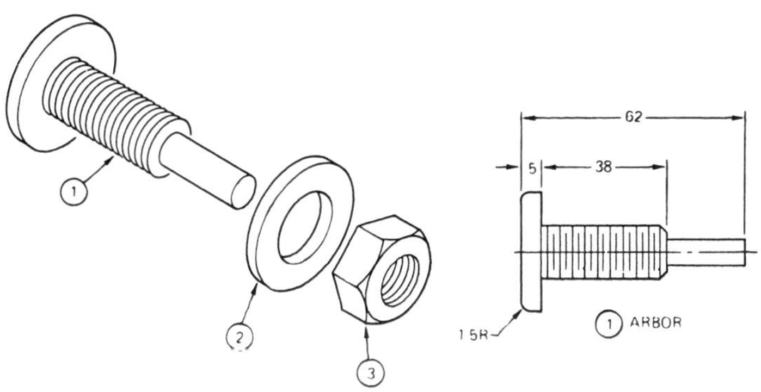

The purpose of a working drawing is to accurately describe and provide directions for the production of an item. These directions include notes, dimensions, and all the line work needed for fabricators. Working drawings may be used by a patternmaker, a foundry worker, a machinist, and many other skilled workers.

The drafter needs a working knowledge of the field of manufacturing for the item he or she is drawing. An efficient drafter must be familiar with the limitations and possibilities of each step of the manufacturing process, such as machining, casting, and welding.

This chapter will cover only the most commonly used steps in industrial fabrication. For additional information, refer to advanced texts, machinist handbooks, and experienced experts in each particular field.

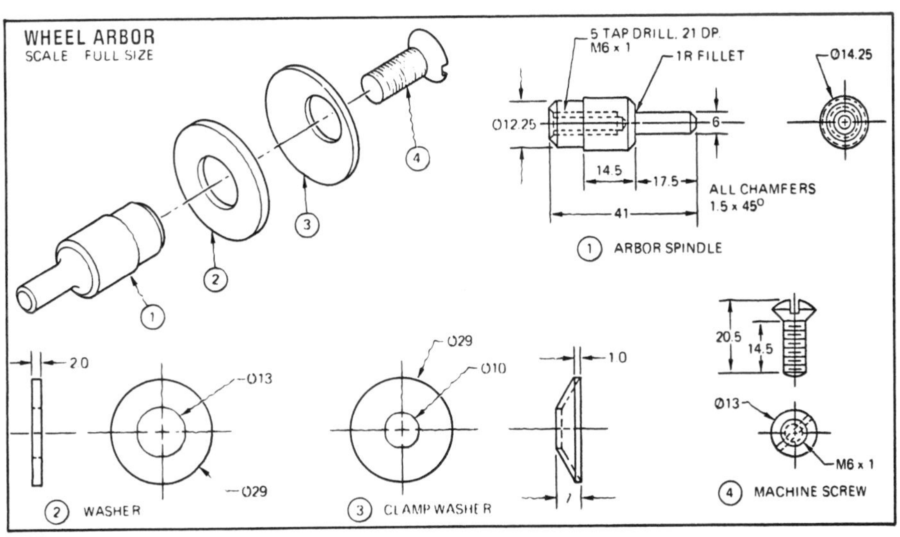

ASSEMBLY DRAWINGS

Assembly drawings show the various parts of an object arranged in their proper working relationship to each other. There are two basic types of assembly drawings. They are:

1. OUTLINE ASSEMBLY DRAWINGS

Outline assembly drawings are used to give the general appearance of an object. Dimensions and unimportant hidden edges are often omitted. However, outline assembly drawings are usually accompanied by a complete set of fully dimensioned detail drawings.

2. ASSEMBLY WORKING DRAWINGS

Assembly working drawings show how the parts of an object are assembled, and at the same time give a complete description of each of these parts. Many simple woodworking drawings are made as assembly working drawings.

DETAIL DRAWINGS

Detail drawings are separate and complete working drawings of each part of an object.

A complete set of working drawings may consist of an outline assembly drawing accompanied by a full set of detail drawings.

OUTLINE ASSEMBLY DRAWINGS

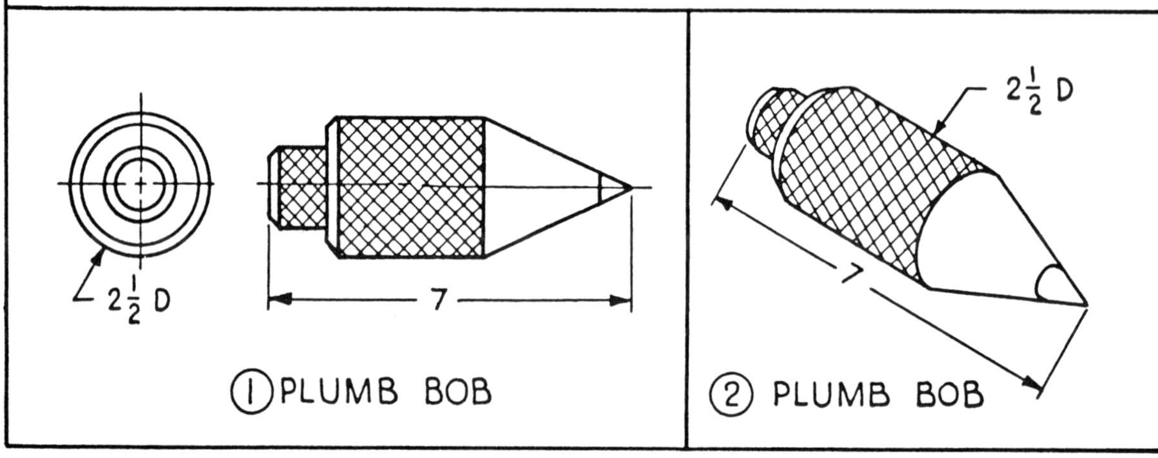

Outline assembly drawings may be made as orthographic projections or as pictorial drawings.

Only the overall dimensions are given.

Hidden parts are not shown.

Outline assembly drawings of the above type must be accompanied by a complete set of detail drawings.

A COMPLETE SET OF DRAWINGS (ON SEPARATE SHEETS)

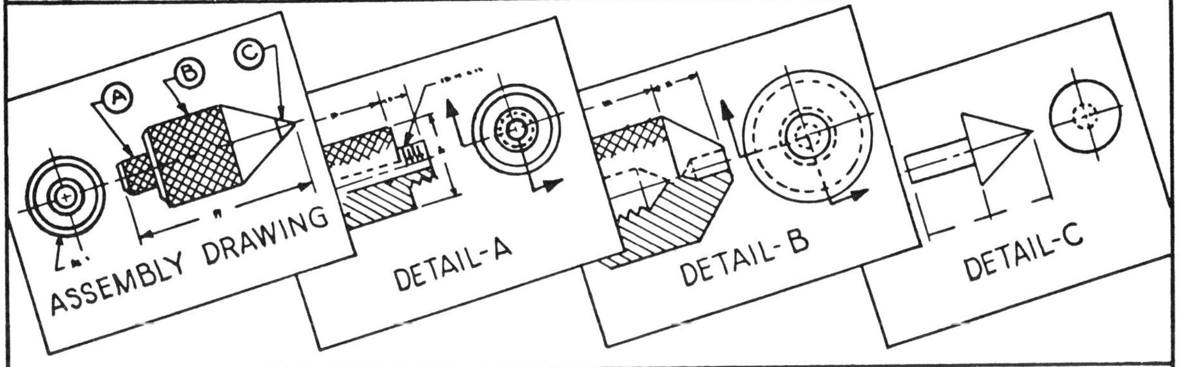

Some manufacturers must have each detail drawn on a separate sheet. These separate drawings are used in different parts of the factory, or are sent to subcontractors.

AN ASSEMBLY WORKING DRAWING

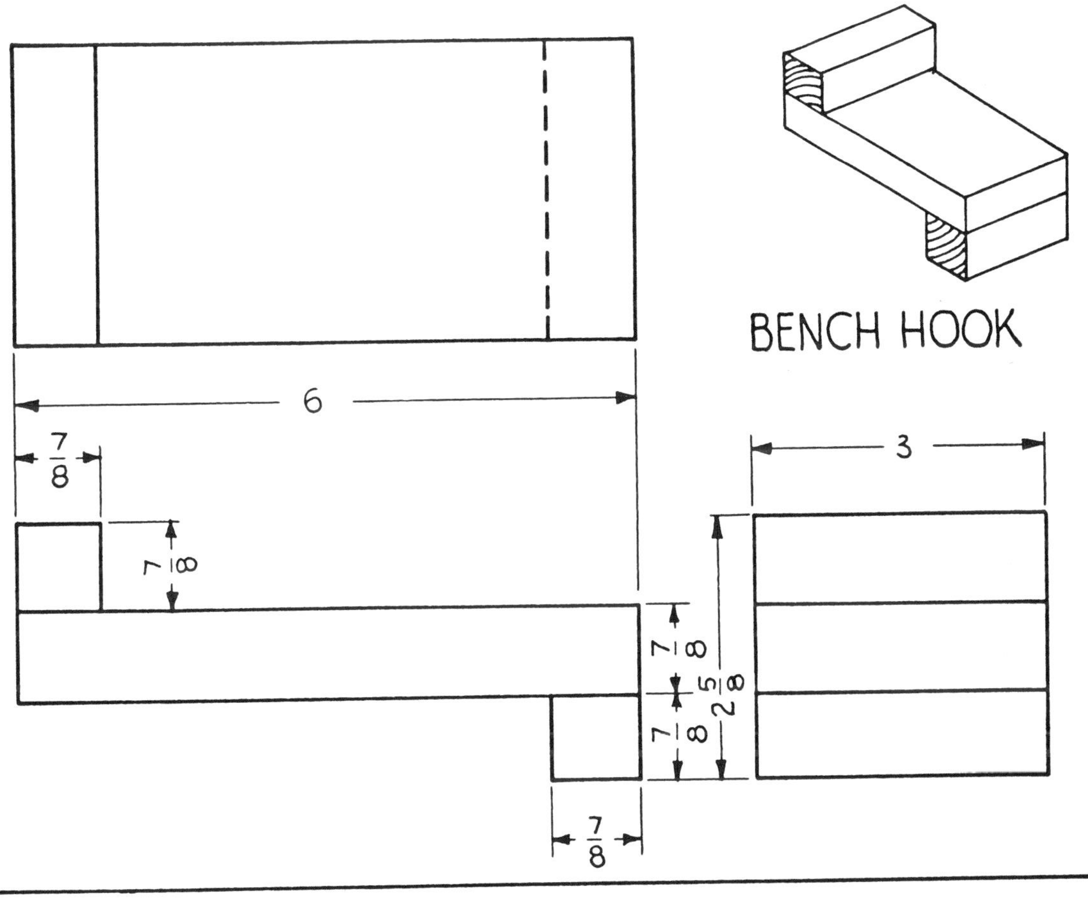

This assembly working drawing shows three pieces as they would be assembled to make a bench hook.

The sketch becomes a type of outline assembly drawing.

Many simple objects may be described by means of assembly working drawings.

ROUNDS, FILLETS, AND RUNOUTS

In order to avoid sharp edges on a casting, the patternmaker and foundryman round the edges of patterns and castings.

ROUNDS AND FILLETS

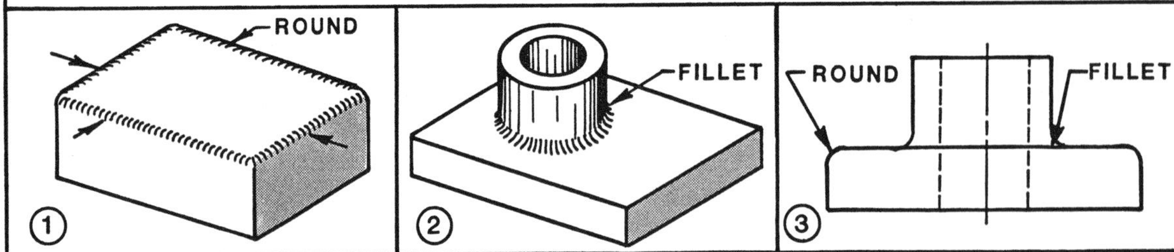

① When an outside or exterior edge is rounded, it is called a **round.**

② When two surfaces meet (as in ② above) the rounded reinforcement is called a **fillet.**

③ Rounds and fillets are made on all castings.

SHARP CORNERS AND ROUNDED CORNERS

When sharp edges appear on a finished drawing, it means one of two things:

1. The rounds and fillets are so small that they may be ignored.
2. The rounds and fillets are to be removed during machining.

STUDY THESE DRAWINGS WITH ROUNDS AND FILLETS

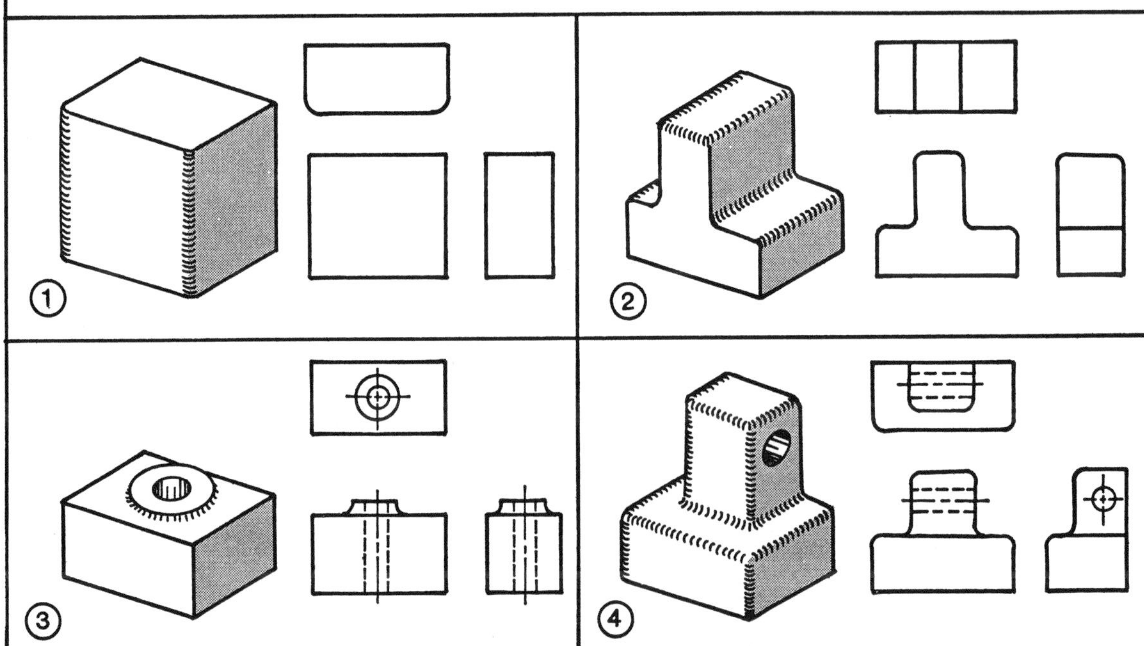

CAUTION

Beginners often have difficulty in keeping track of rounds and fillets when they are projected from view to view.

A round in one view may appear as a sharp corner in another view. Study the drawings above for examples of this.

RUNOUTS

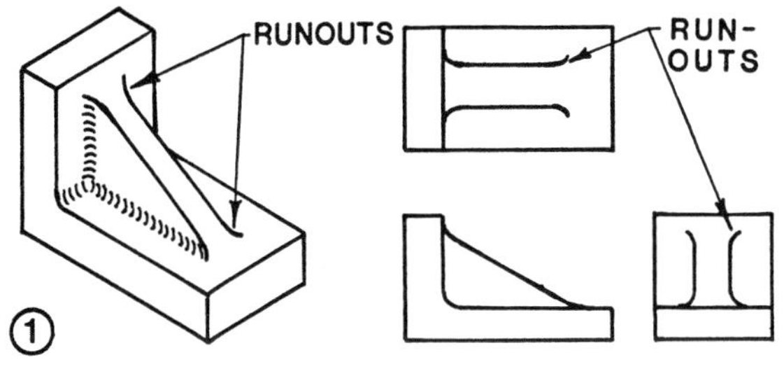

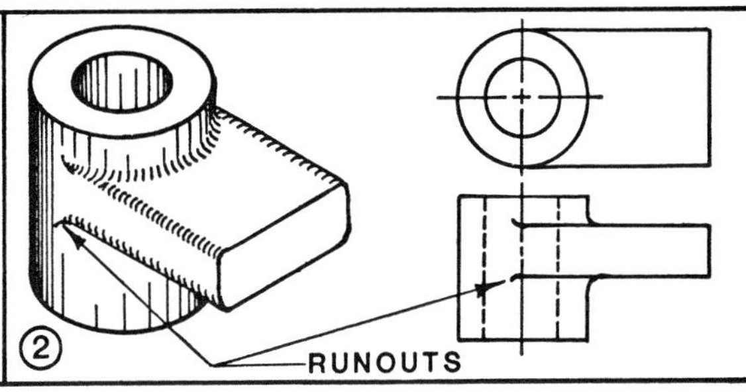

Occasionally, when a fillet occurs at an intersection, a lost edge occurs. This type of lost edge is known as a runout.

STUDY THESE ROUNDS, FILLETS, AND RUNOUTS

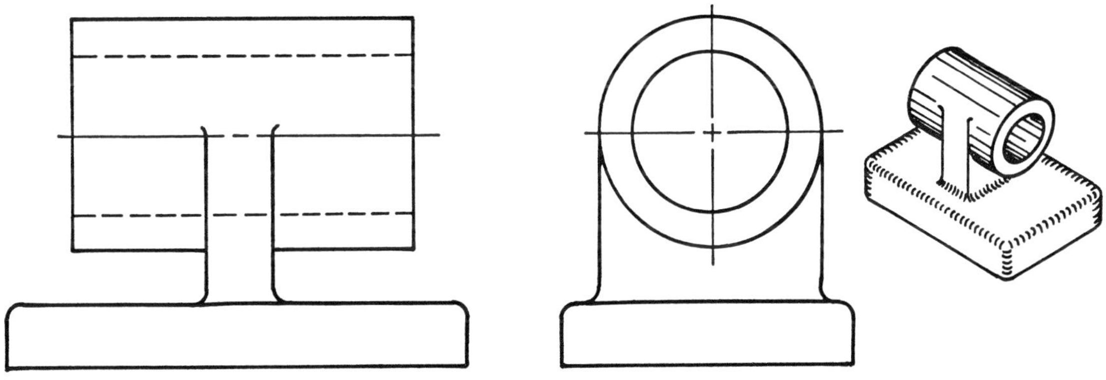

SPECIAL NOTE: When rounds, fillets, and runouts are very small (less than 1/4R) they may be drawn freehand.

ASSIGNMENT: Make a working drawing of one of these objects.

① C I SHOE

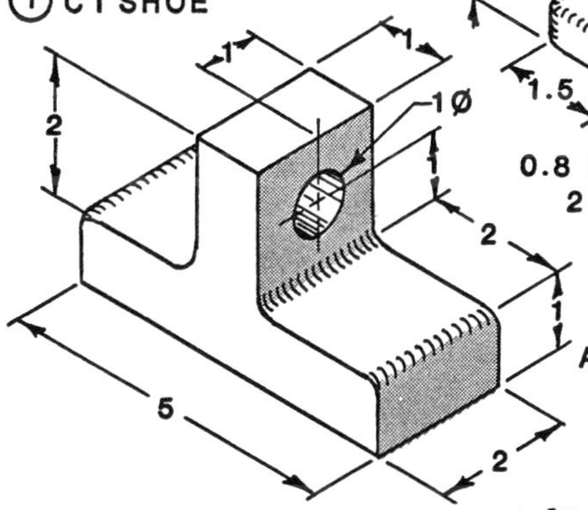

ALL ROUNDS AND FILLETS 1/8"R

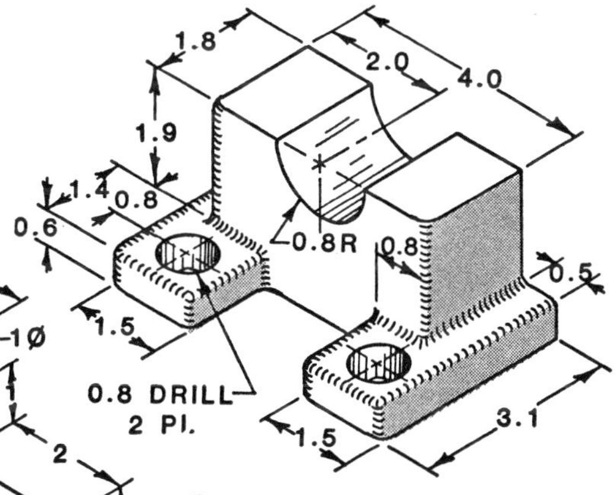

②BEARING MOUNT
ALL ROUNDS AND FILLETS 0.1R

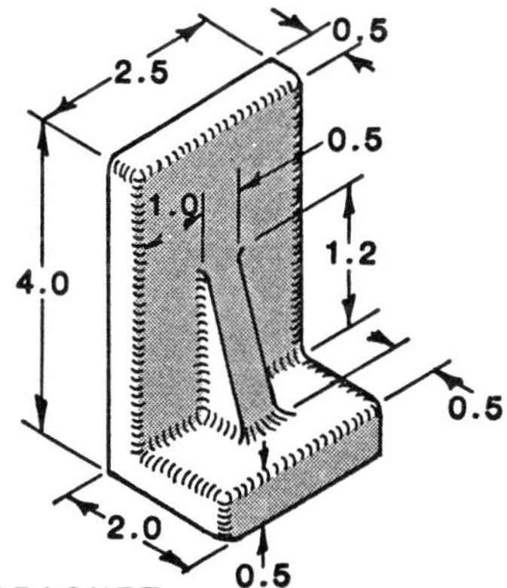

③BRACKET
ALL ROUNDS AND FILLETS 0.1R

SURFACE FINISHES

Metal surfaces are of varying degrees of smoothness. Sand castings have a rough finish when they come from the foundry. The drafter must indicate the surfaces that are to be finished and the degree of smoothness required.

INDICATING A SURFACE FINISH

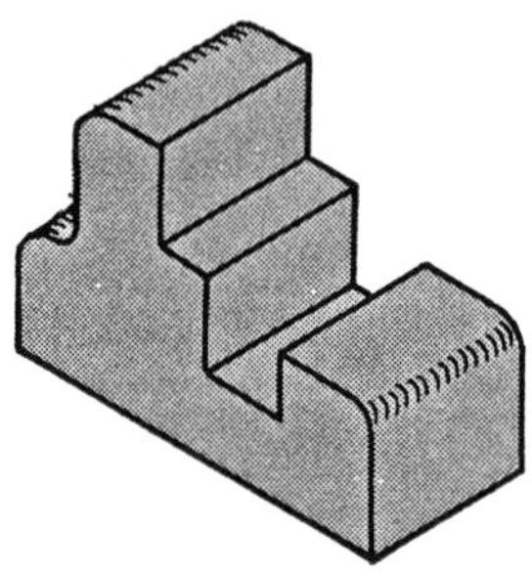

(1) Pieces as they come from the foundry have rough surfaces.

(2) The drafter must indicate the surfaces that are to be finished.

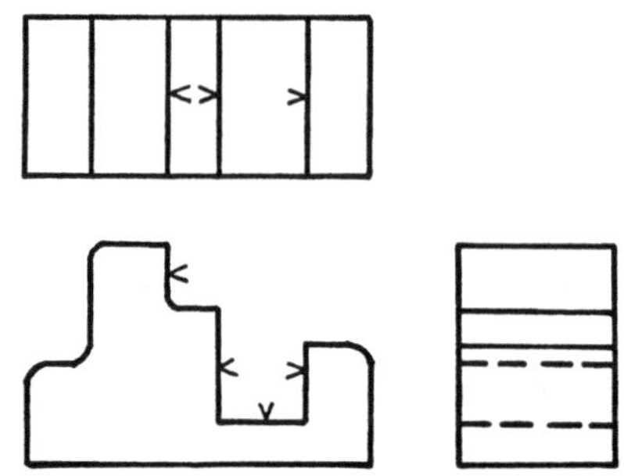

(3) The symbol used to indicate a finished surface must be used on every edge view of that surface.

SYMBOLS USED TO INDICATE SURFACE FINISHES

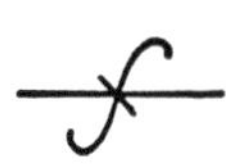

(A)

(B)

(1) Either symbol (A) or the newer symbol (B) may be used to indicate the surface to be finished.

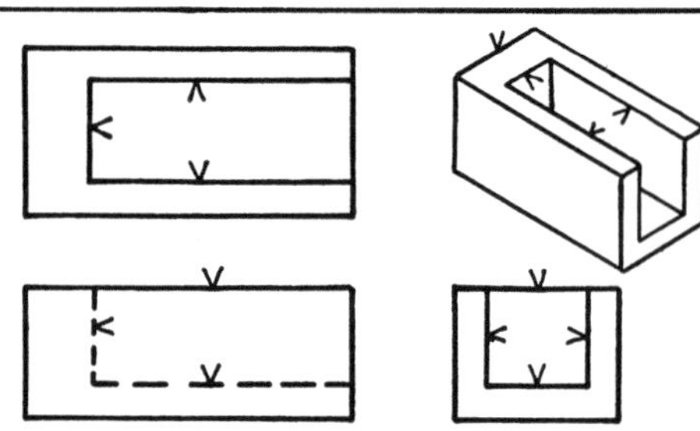

(2) Every edge, including hidden edges, of a surface must be marked with the proper symbol.

F.A.O.

(A)

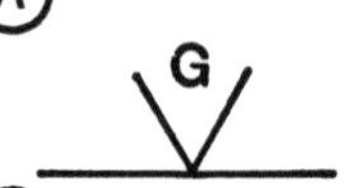

(B)

(A) "F.A.O." means "finish all over."

(B) "G" means "grind."

STUDY THE LOCATION OF THE FINISH SYMBOLS

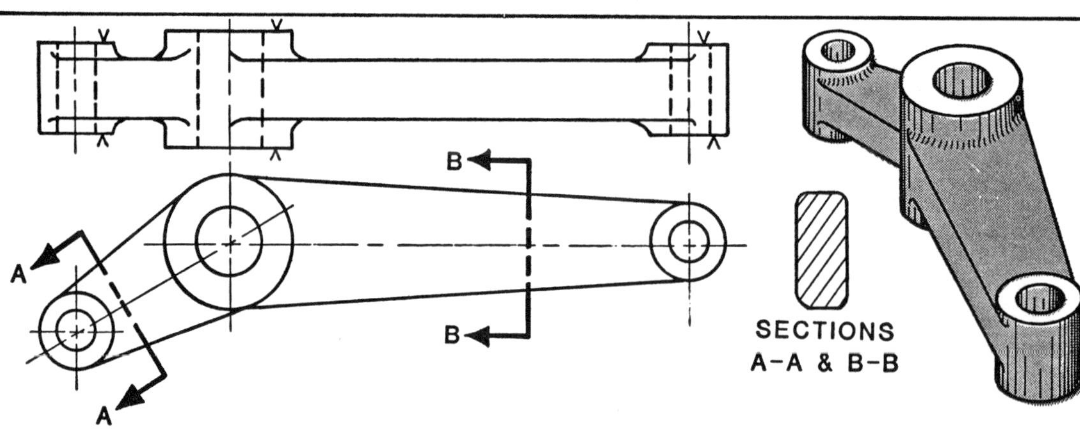

BOSS AND PAD

An area through which a bolt is to be used is often reinforced by an added thickness in the casting.

A BOSS AND A PAD

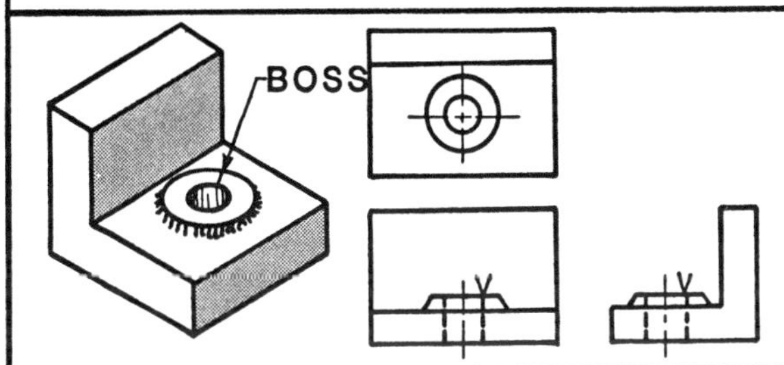

(1) A boss is a circular reinforced area through which a bolt will pass.
(Note the use of a fillet on a boss.)

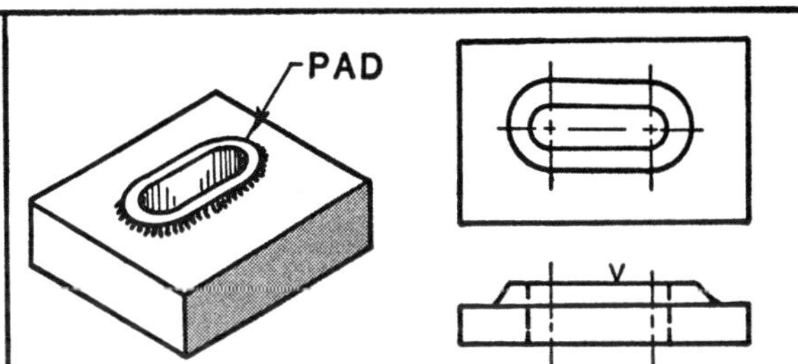

(2) An elongated boss is called a **pad.** A pad is used where pressure is to be applied over an extended area.

STUDY THESE DRAWINGS CONTAINING BOSSES AND PADS

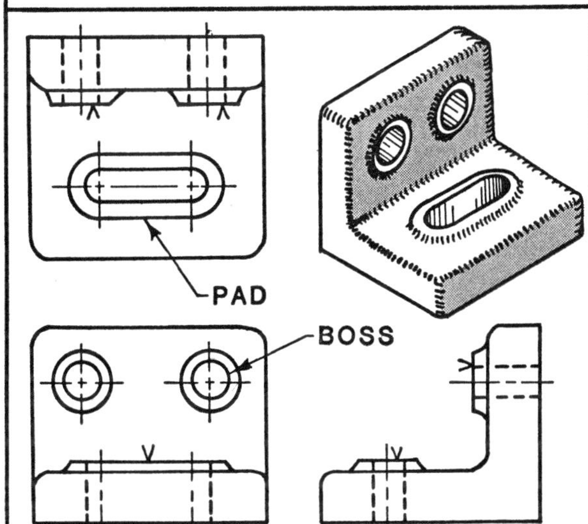

(1) ALL FILLETS AND ROUNDS 1/8"R

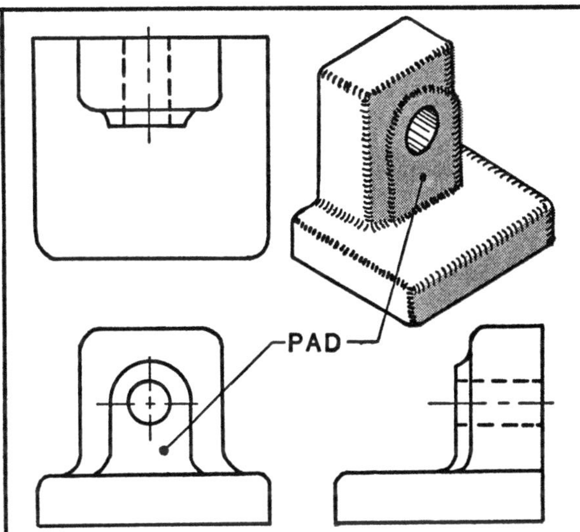

(2) ALL FILLETS AND ROUNDS 3/32"R

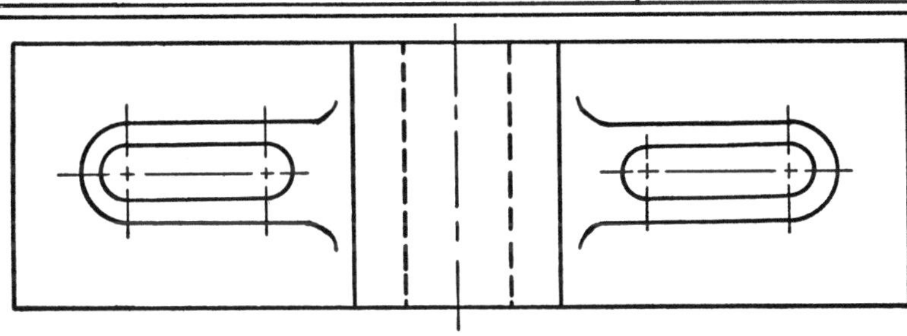

ALL FILLETS AND ROUNDS 1/4"R

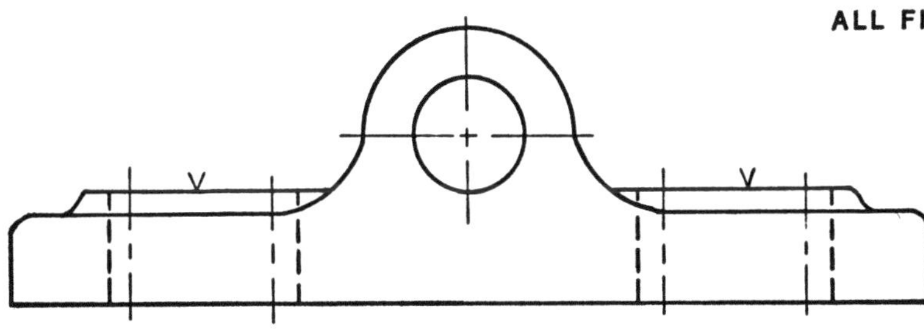

A **boss** is circular in shape.
A **pad** may be any shape.

MACHINING OPERATIONS

The drafter informs the fabricator as to the operation to be used. This is generally done in a note that appears as part of the dimensioning.

DRILL, REAM, AND BORE ARE TERMS USED FOR CIRCULAR HOLES

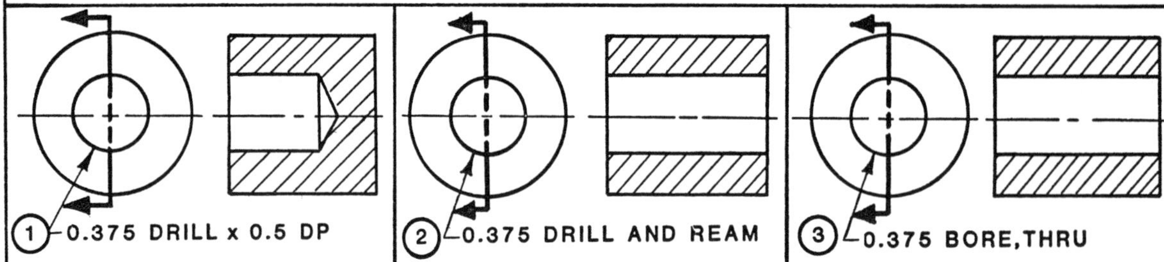

(1) A drill will make a relatively rough hole in an object.

(2) A drill slightly smaller than 3/8" is used, followed by a 3/8" reamer.

(3) Extremely accurate holes of large diameter are made by boring.

SPOTFACE

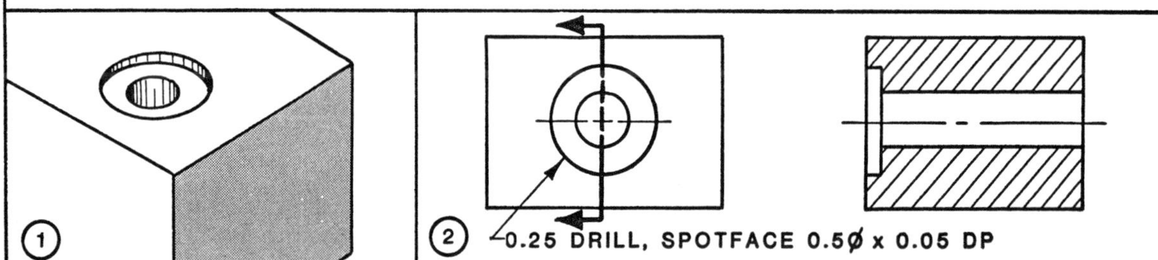

A spotface provides a smooth surface against which a bolt may be drawn up tightly. The drill diameter, spotface diameter, and depth must be given.

COUNTERBORE

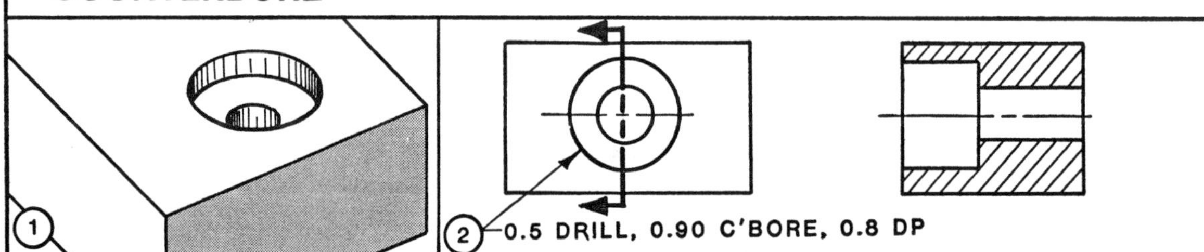

A counterbored hole provides a recessed space for a bolthead.

COUNTERSINK

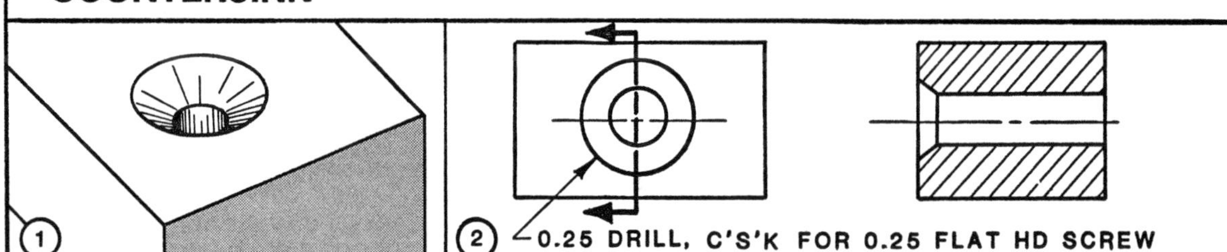

A countersink provides a cone-shaped hole for the head of a flathead screw.

BROACHING

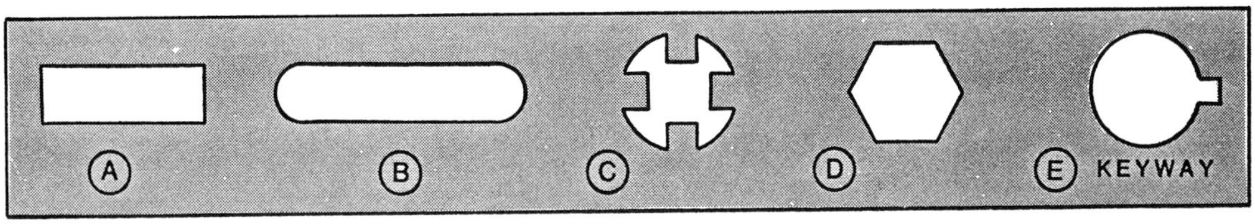

A broaching machine can produce many types of holes that would be otherwise difficult to produce.

THE NEED FOR A KEY AND A KEYWAY

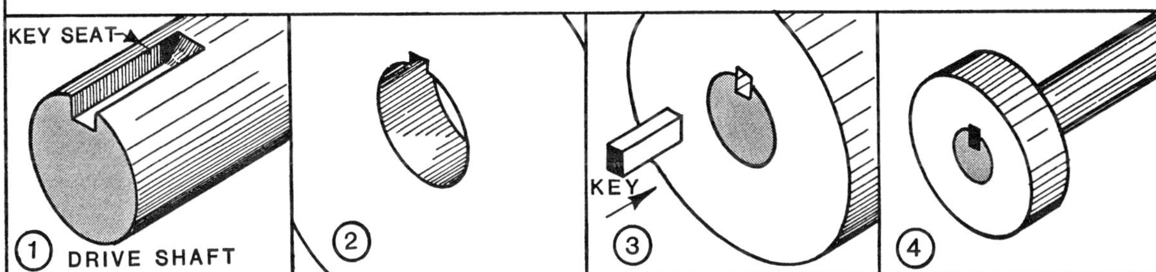

(1) A groove, called a **key seat,** is cut in the end of a drive shaft.
(2) A **keyway** is broached in the hub of a pulley.
(3) The two slots are aligned and a metal **key** is slipped into place.
(4) When the drive shaft turns, the pulley also must turn.

DRAWING AND DIMENSIONING KEYWAYS

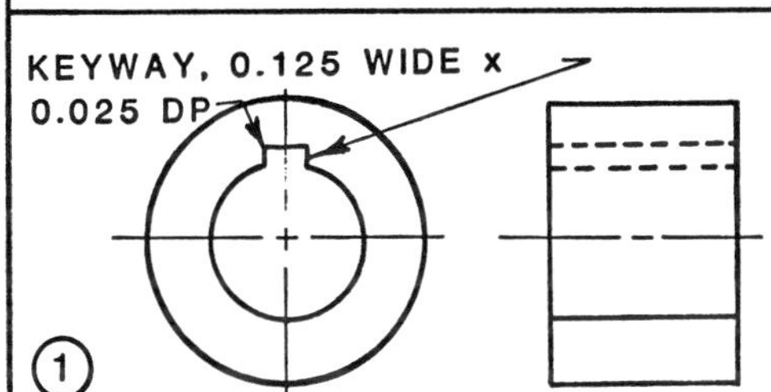

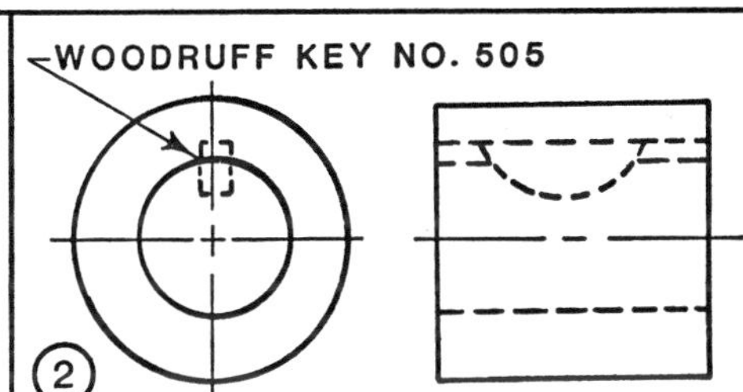

Keyways may be dimensioned directly.	Keyways may be dimensioned by reference to standard key nomenclature.

CORED HOLES AS MADE AT THE FOUNDRY

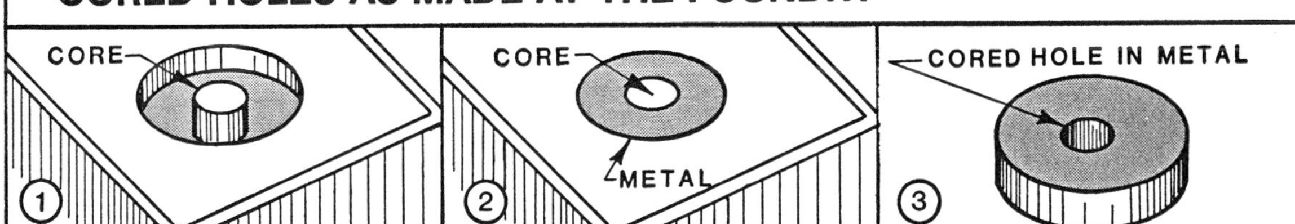

(1) A cored hole is made by placing a sand core in the mold.
(2) When the molten metal is poured, it flows around the core.
(3) A cored hole is produced in the casting.

The term **cored hole** is a specific direction to the patternmaker and to the foundryman.

KNURLING

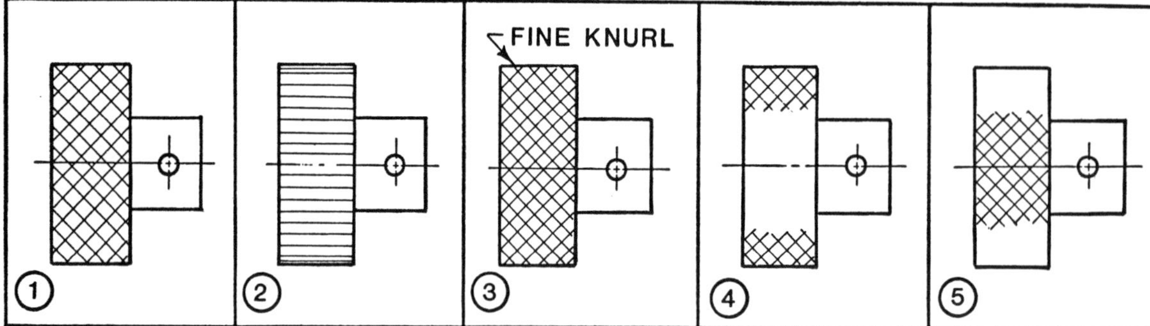

(1) Diamond knurl.	(2) Straight knurl.	(3) Knurling can be made in several grades.	(4) and (5) Two ways of suggesting knurling.

Knurling consists of small ridges made to prevent the fingers from slipping when the object is turned.

Thin lines drawn at 30° or 45° are used to represent knurling.

Lines used to represent knurling, and lines used in section lining should be made the same thickness.

A BEVEL AND A CHAMFER

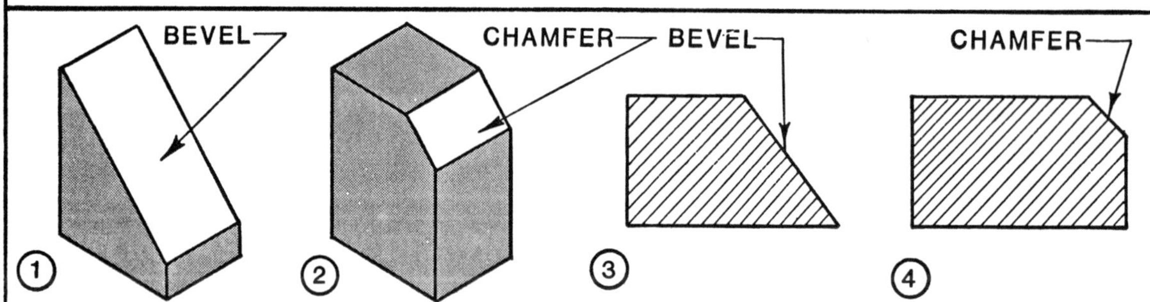

A **bevel** is a sloping cut made across the entire length or thickness of an object. A **chamfer** is a sloping cut made only partially across the edge of a surface.

DIMENSIONING BEVELS AND CHAMFERS

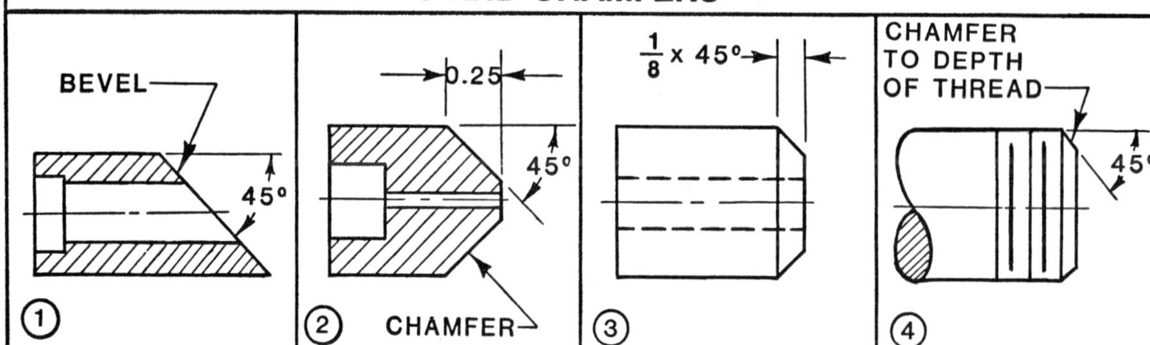

Only the angle of a bevel need be given.

Two dimensions are needed when dimensioning a chamfer.

HOW TO USE NOTES ON A DRAWING

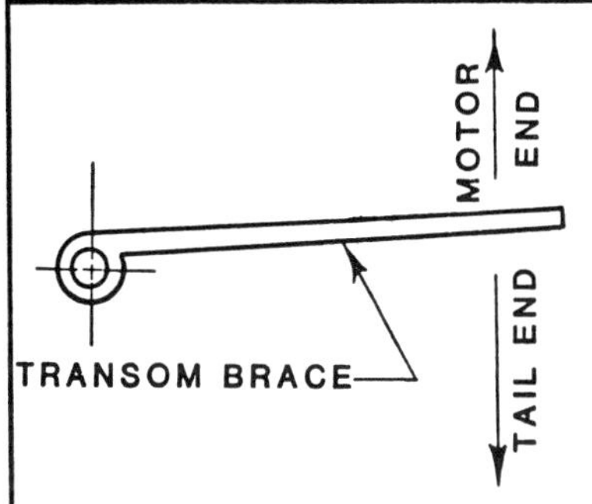

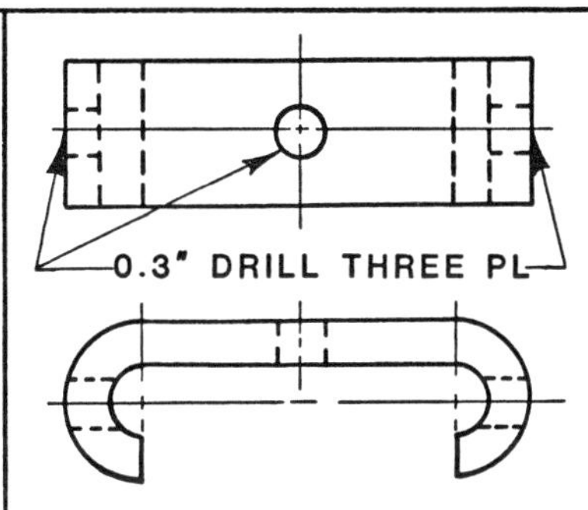

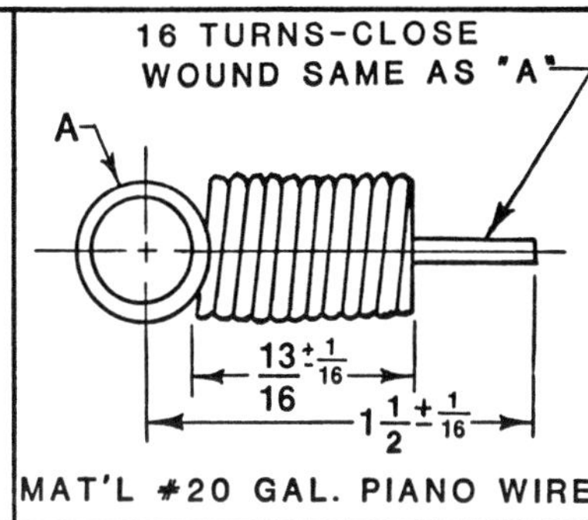

(1) Letter notes 1/8" high in either a horizontal or a vertical direction.	(2) Notes may be connected to the drawing by means of a leader.	(3) Keep notes near the view for which they are intended.

STUDY THE USE OF NOTES ON THIS DRAWING

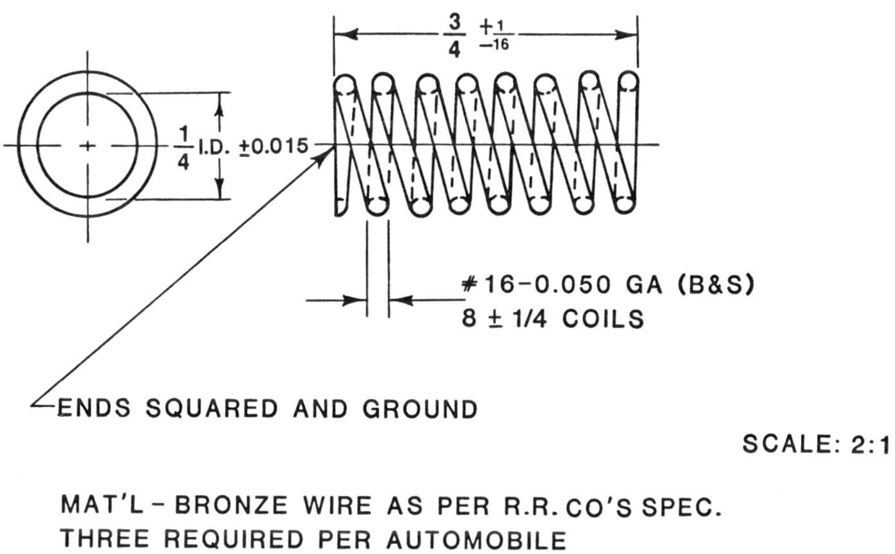

ASSIGNMENT

Make a working drawing of these objects.

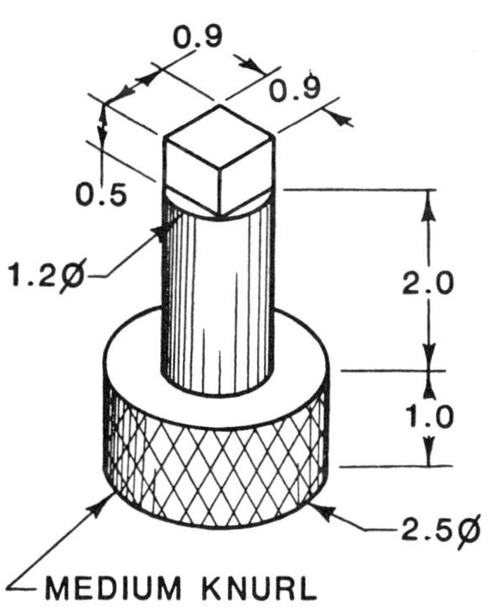

(1) KEY - 2 REQUIRED

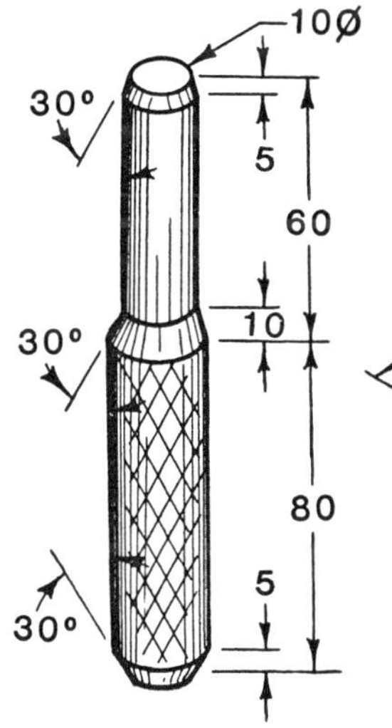

(2) METRIC PUNCH

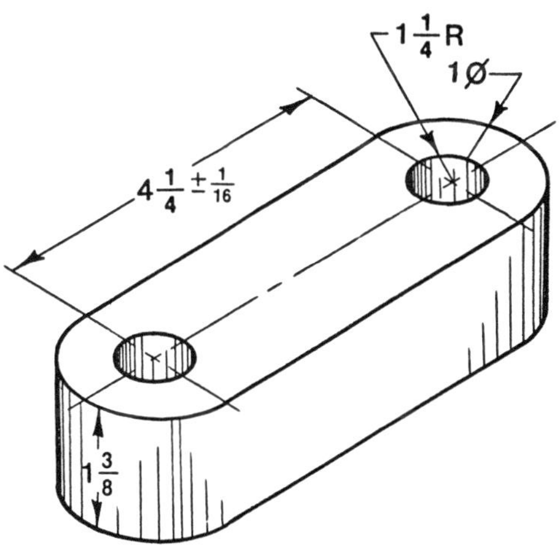

(3) LINK - 20 REQUIRED

AN ASSEMBLY WORKING DRAWING OF AN AWL

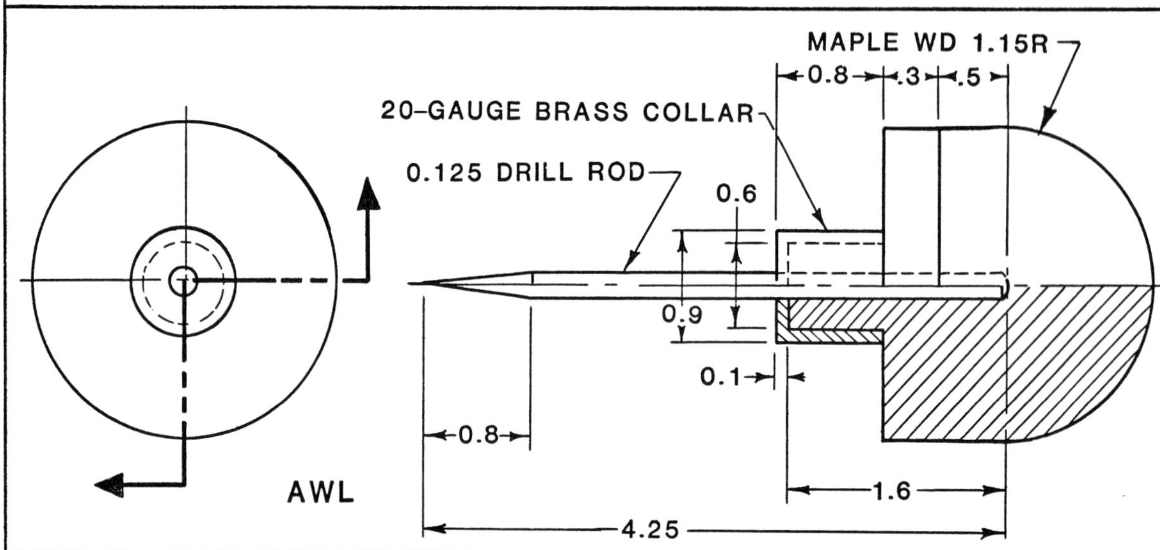

A WORKING DRAWING OF A HAMMER (ASSEMBLY AND DETAILS)

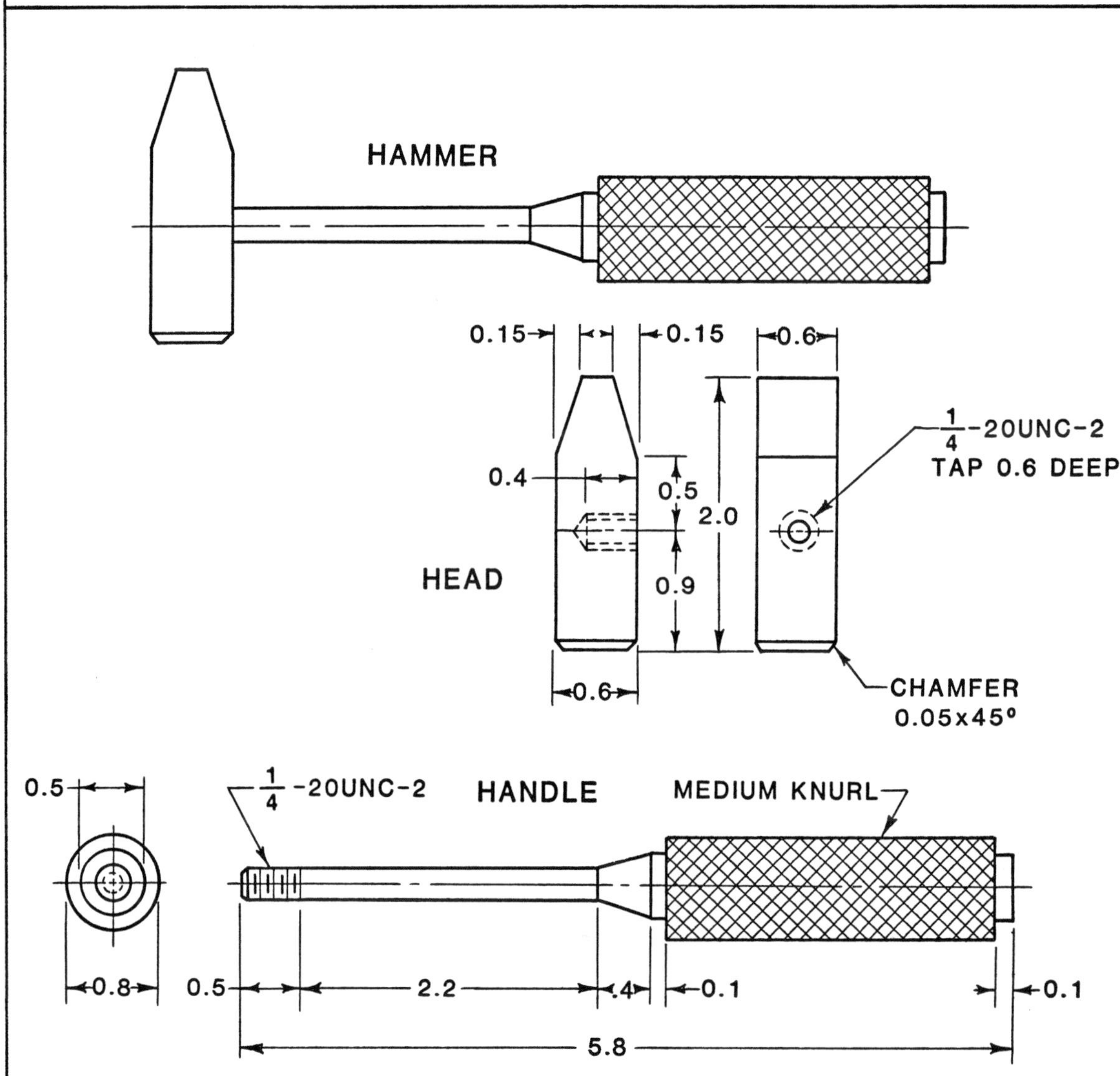

LADDER LOCK

Here is a simple device with which you are probably familiar. There are many other such devices in your home and school. Study each one and decide how you would make a working drawing of it. Then, make the necessary drawings of some of them.

The drawings shown below will serve as a model for other similar drawings.

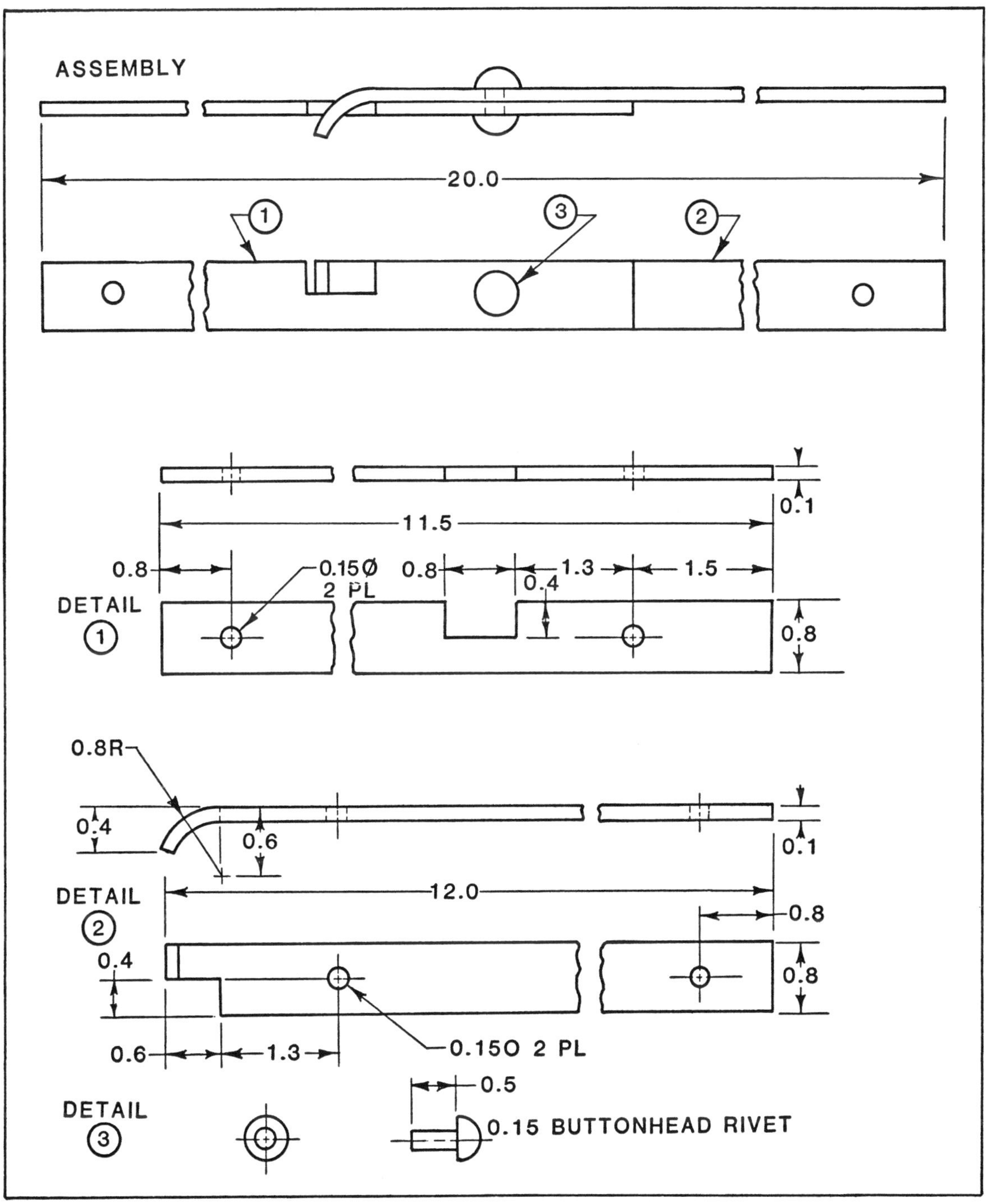

ASSIGNMENT

Make a complete working drawing of any type that you think necessary to fully describe one of the objects below.

It should be noted that when an object has two or more parts that are exactly alike, only one part need be drawn. A note such as **"two required"** will tell the fabricator of the duplicate parts (see ① and ② below).

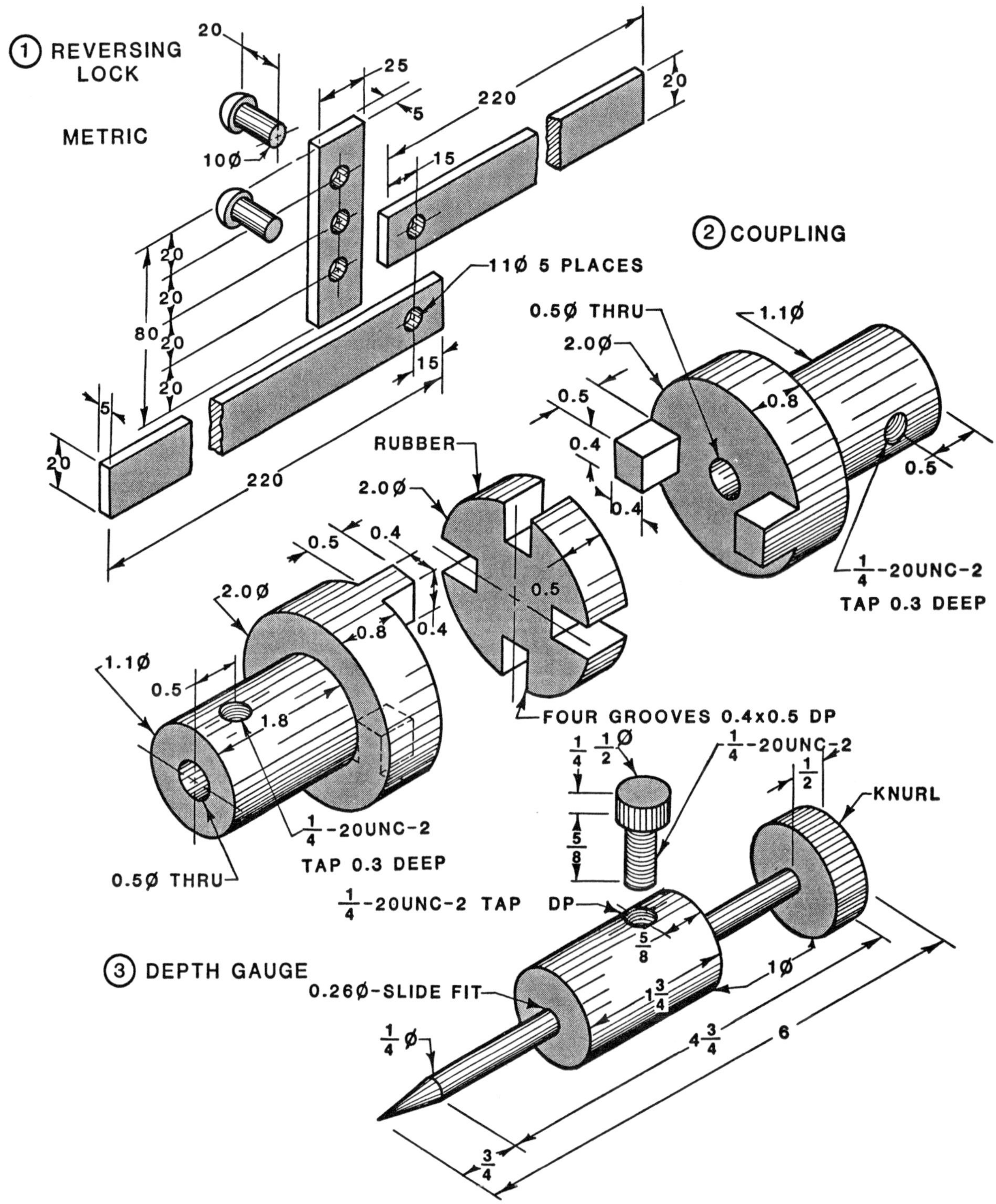

EXPLODED VIEWS

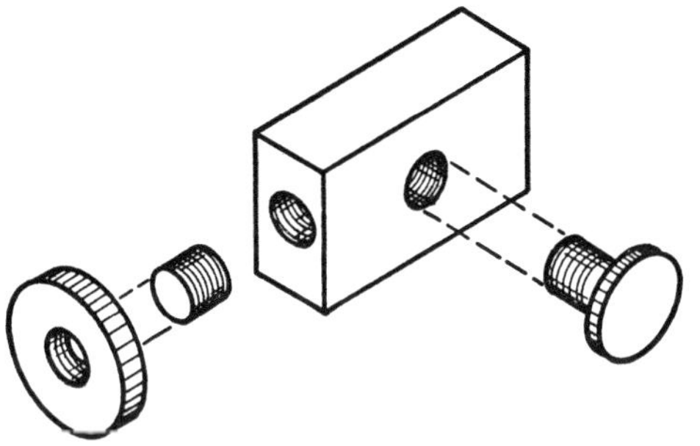

An exploded view is a pictorial drawing that pictures the object as though an explosion had taken place within it.

Exploded views show the adjacent parts in their proper order of assembly.

An exploded view may be considered to be a type of assembly drawing.

Every effort is made to keep one part of the object from overlapping another part. Note that the parts appear to be flying through space.

STUDY THESE EXPLODED VIEWS

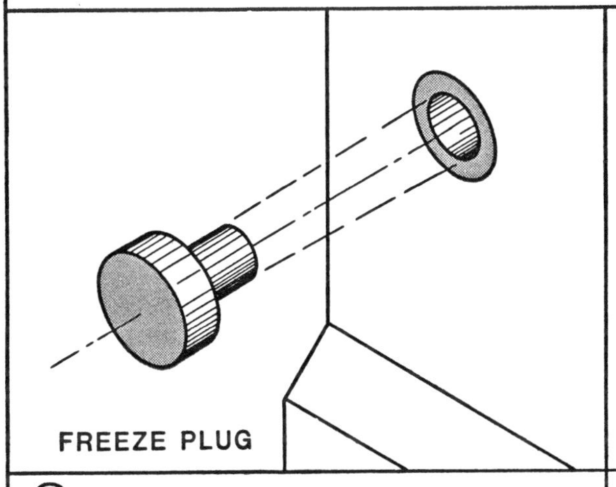

① Note the lines used to denote the normal position of the plug.

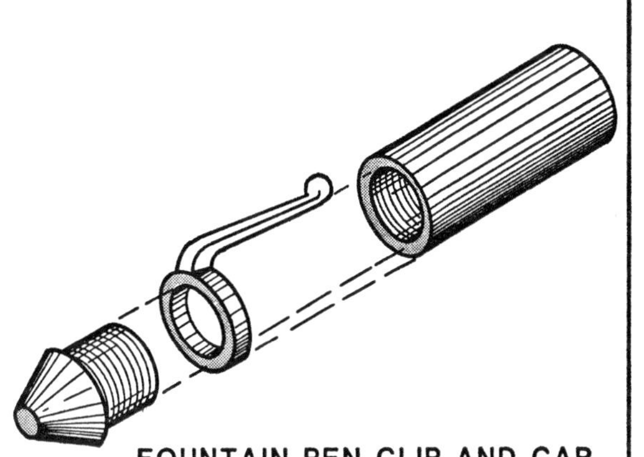

② This drawing could serve as an assembly drawing.

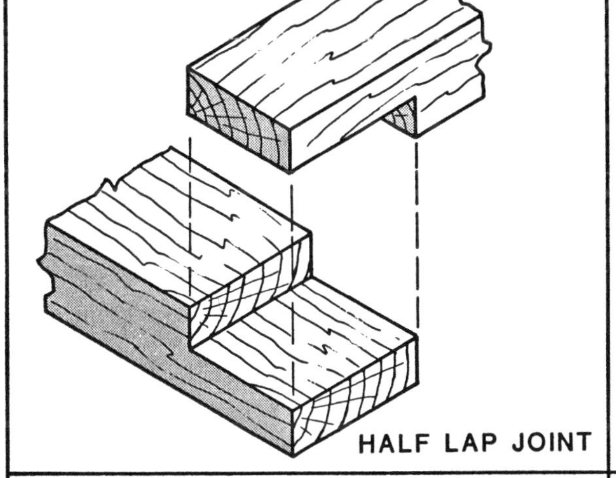

③ Many woodworking drawings may be classed as exploded views.

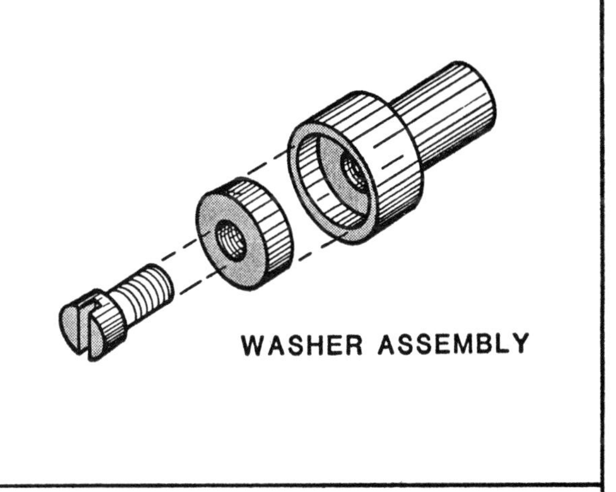

④ "Do-it-yourself" books often use exploded views.

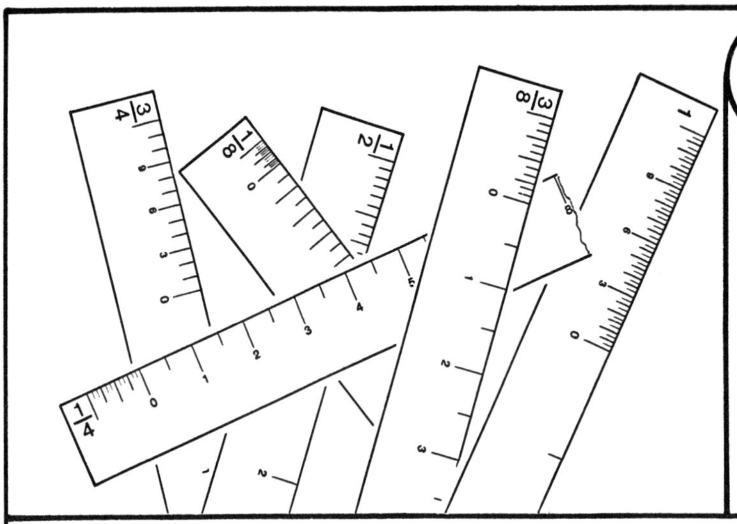

UNIT 16

TESTS

SCALE:		
	FULL	1
	FULL	2
	1/2″=1″	3
	1/2″=1″	4
	1/4″=1″	5
	1/4″=1″	6
	1/4″=1″	7
	1/8″=1″	8
	1/8″=1″	9
	1/8″=1″	10
	3/8″=1″	11
	3/8″=1″	12
	3/8″=1″	13
	3/4″=1″	14
	3/4″=1″	15

MEASURE TO THE CLOSEST 1/16″ WITH A MECHANICAL ENGINEER'S SCALE.

SCALE:		
	FULL	1
	FULL	2
	FULL	3
	1″=10′	4
	1″=10″	5
	1″=20′	6
	1″=20″	7
	1″=30′	8
	1″=30′	9
	1″=40′	10
	1″=40″	11
	1″=50′	12
	1″=50″	13
	1″=50′	14
	1″=60′	15

MEASURE TO THE CLOSEST 1/10″ WITH A CIVIL ENGINEER'S SCALE.

MEASURE EACH FULL-SIZE DIMENSION (SCALE 1:1)

Measure each dimension shown below. List answers on a separate piece of paper.

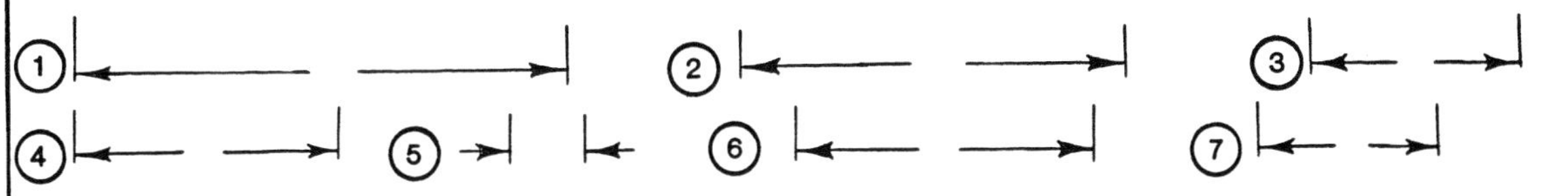

MEASURE EACH HALF-SIZE DIMENSION (SCALE 1/2" = 1")

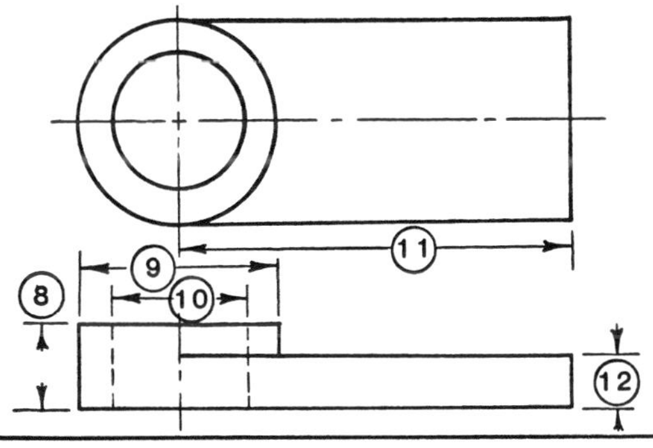

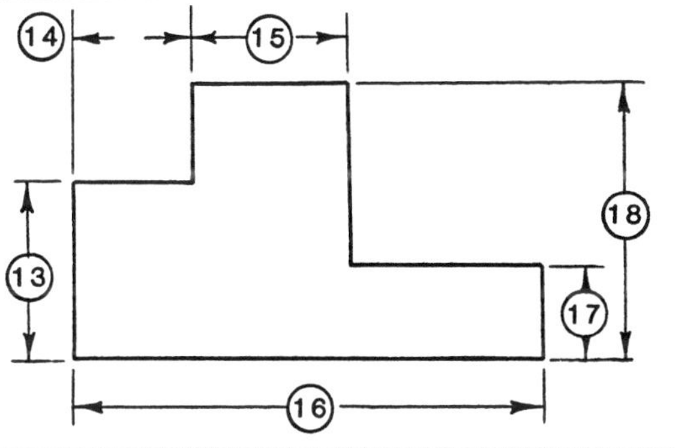

Measure every part of each drawing. And list answers on a separate piece of paper.

READ THE MEASUREMENTS ON THE SCALES AND LINES (SCALE: FULL SIZE)

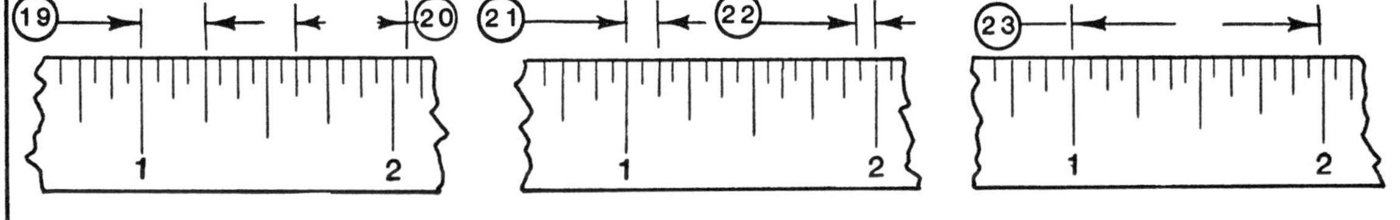

(1) Think of every 1/4" as though it were an inch.

(2) One-half of 1/4" represents 1/2".

(3) One-quarter of 1/4" represents one-quarter of an inch.

(4) Four 1/4" spaces represent 4".

ASSIGNMENT

Measure each dimension with the indicated scales.

SCALE: 1/4" 1"

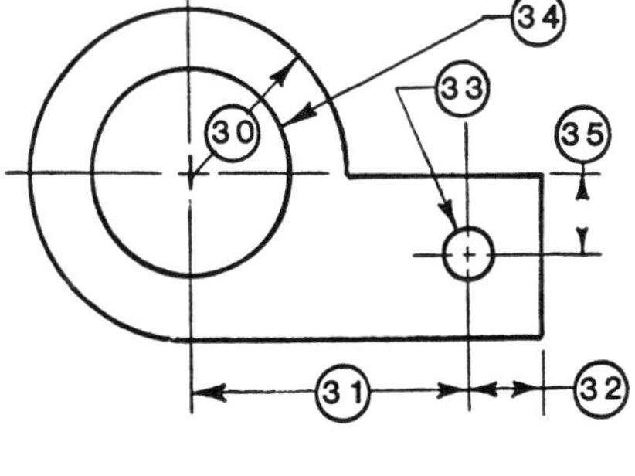

SCALE: 3/8" 1"

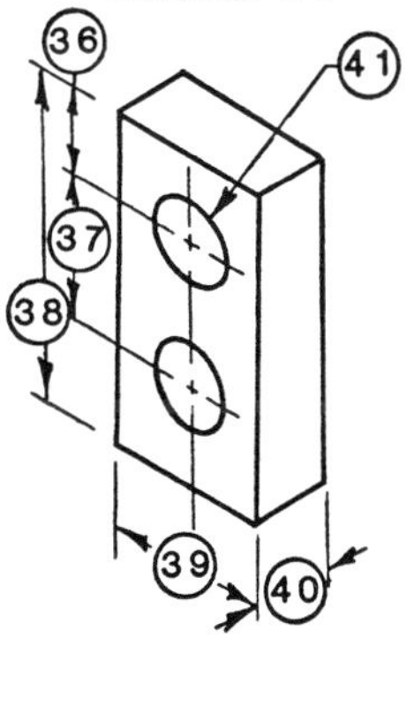

SCALE: 3/4" 1"

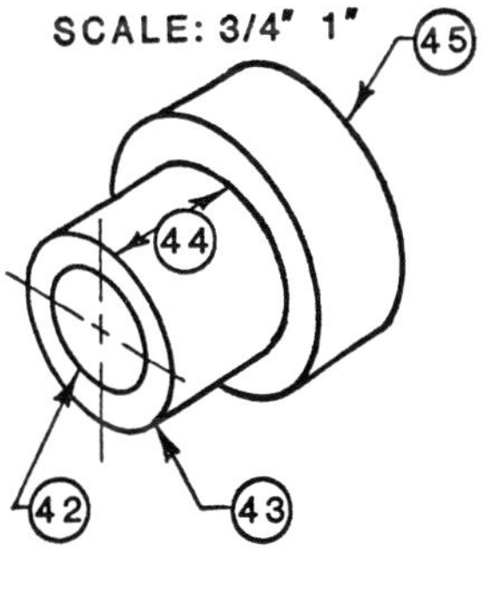

MATCHING VIEW TEST

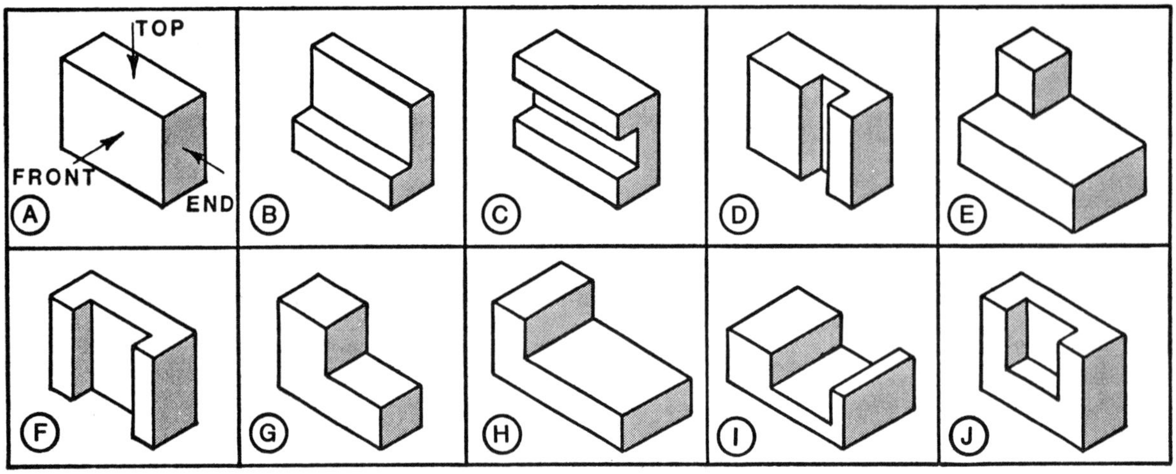

HOW TO TAKE THIS TEST

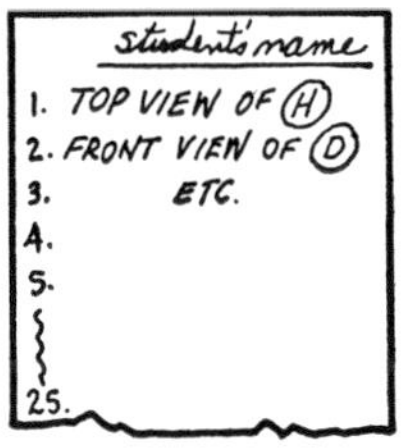

1. Arrange an 8½" x 11" piece of composition paper as shown to the right.

2. Each of the twenty-five views shown below is either a front view, a top view, or an end view of one of the ten objects shown above. Identify each view below in the manner indicated in the example shown at the right.

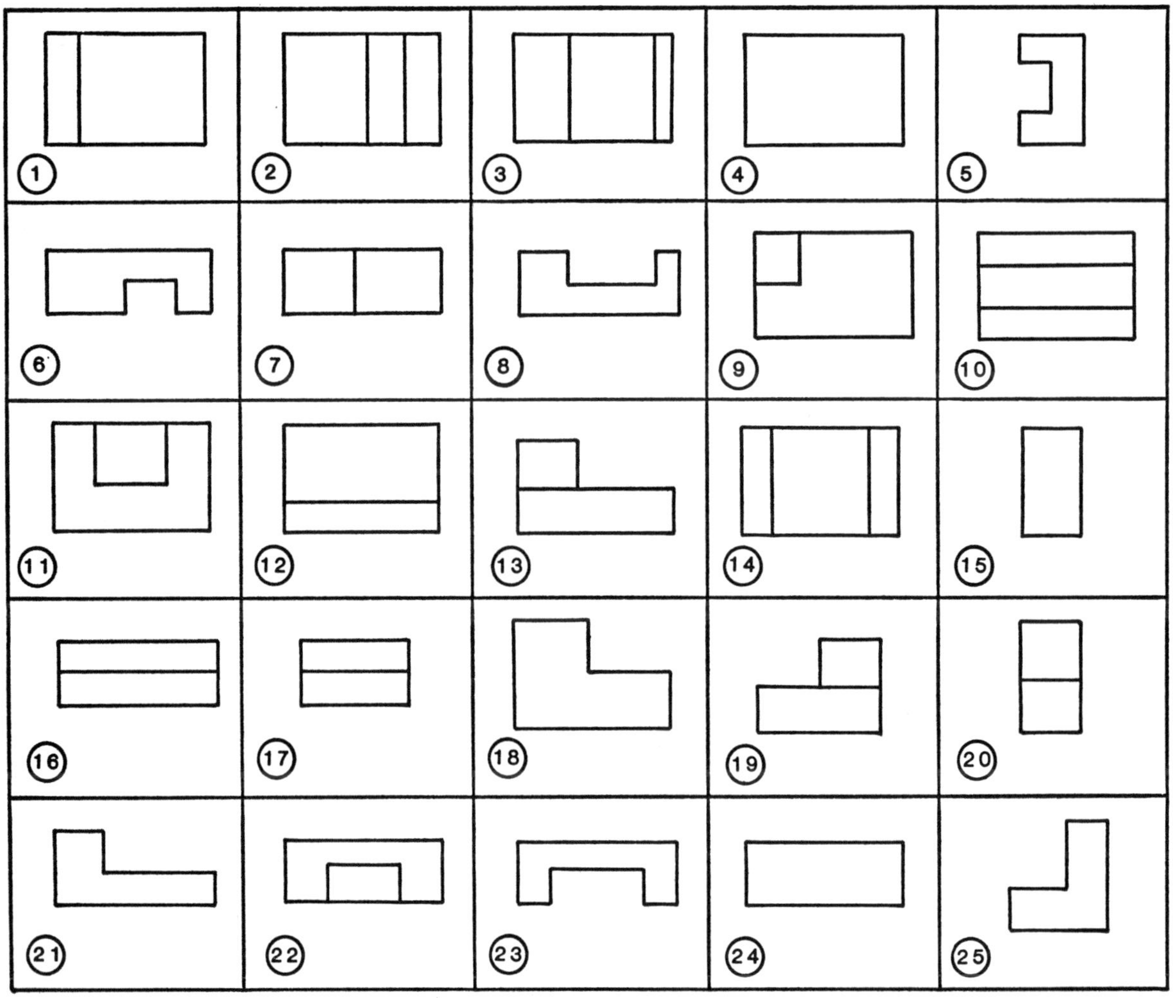

MATCHING TEST

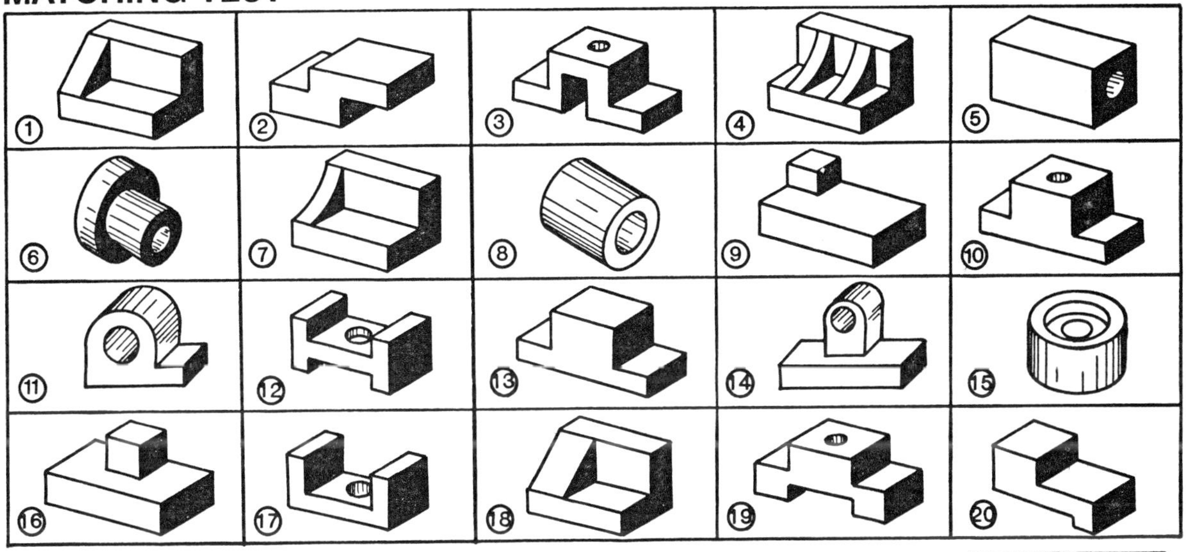

HOW TO TAKE THIS TEST

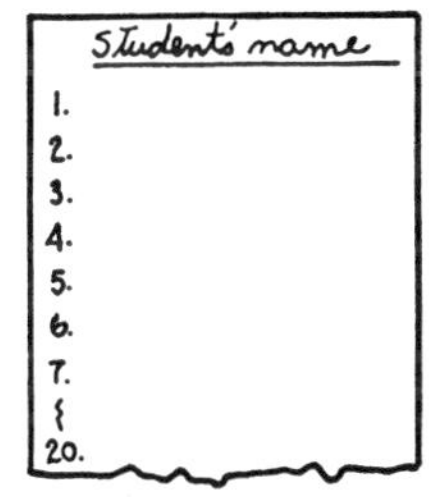

1. Arrange an 8½" x 11" piece of composition paper as shown to the right.
2. Study each drawing below. Try to visualize each object. Locate the drawing of that object in the drawings shown above. Match the letter of the drawing below with the number of the correct picture above. **Example:** Drawing (M) below is an orthographic projection of (1) above.

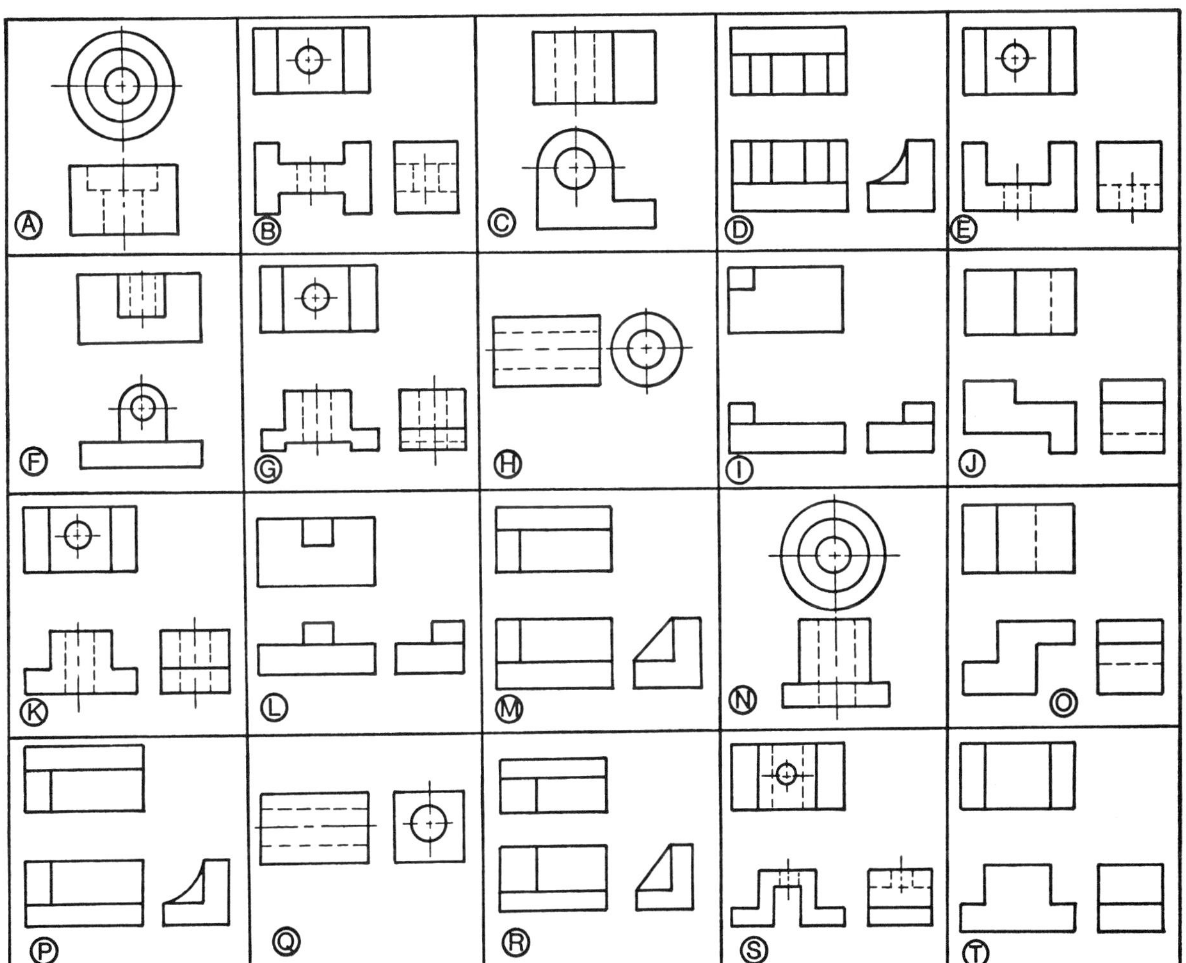

CAN YOU VISUALIZE THESE OBJECTS?

You must now begin to visualize an object from an orthographic projection. This skill will aid you in blueprint reading and in advanced drafting.

VISUALIZE EACH OF THESE OBJECTS FROM THE DRAWINGS BELOW

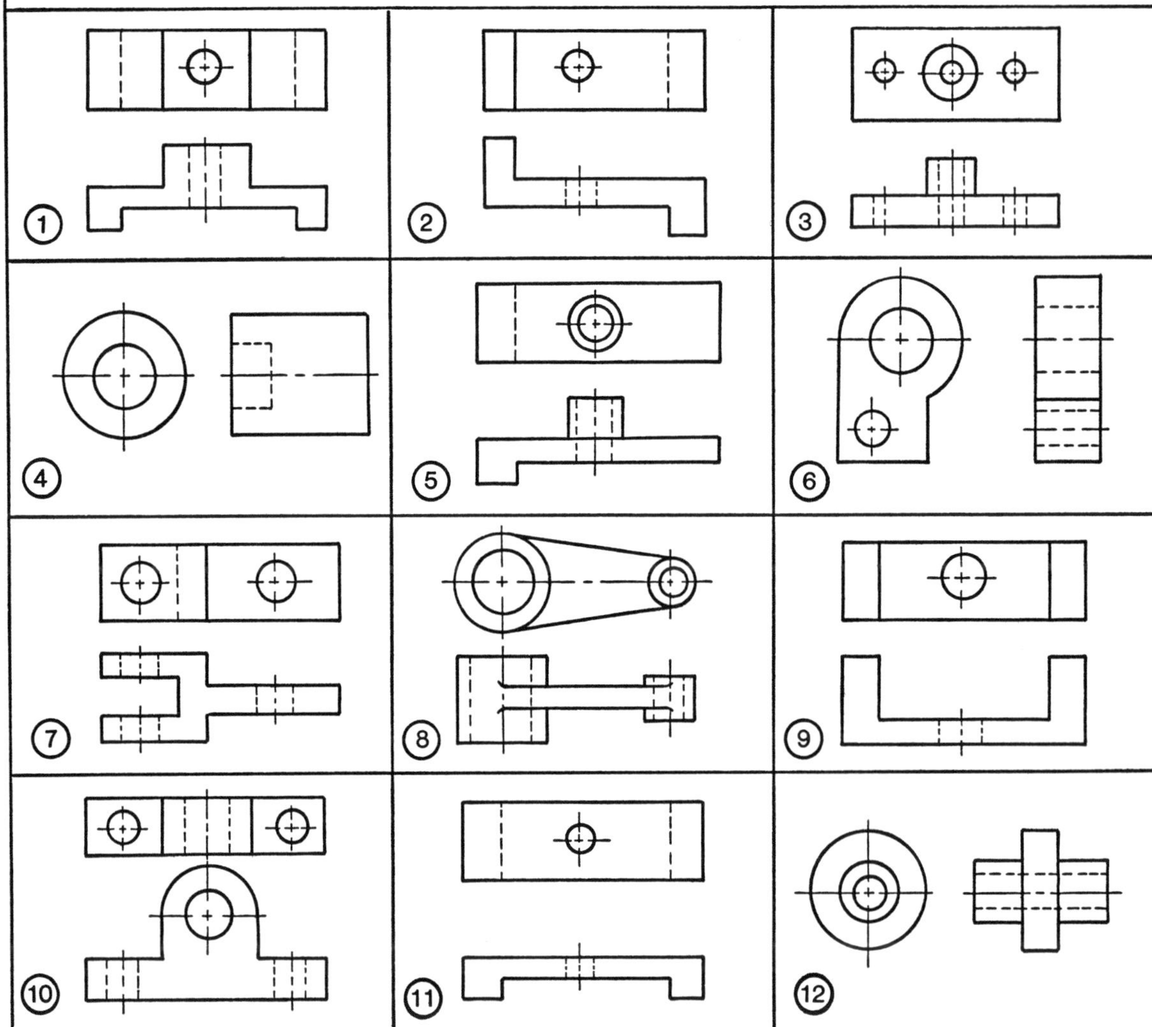

SELECT THE CORRECT PICTORIAL

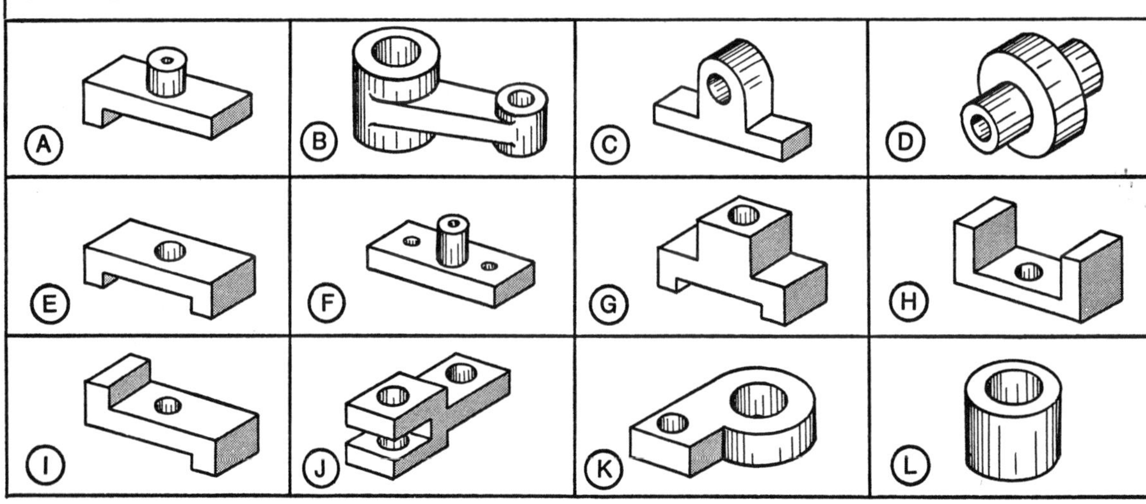

MATCHING TEST

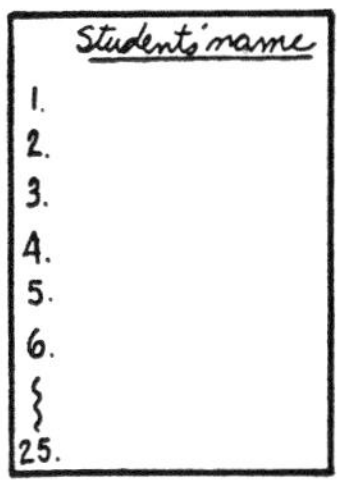

1. Arrange your answer paper as shown to the right.
2. Letter the numbers 1 through 16 in a column.
3. Study each orthographic drawing shown below. Visualize each object. Match the letter of the correct pictorial drawing shown at the bottom of this page with the correct number of the drawings shown directly below.

ORTHOGRAPHIC PROJECTIONS

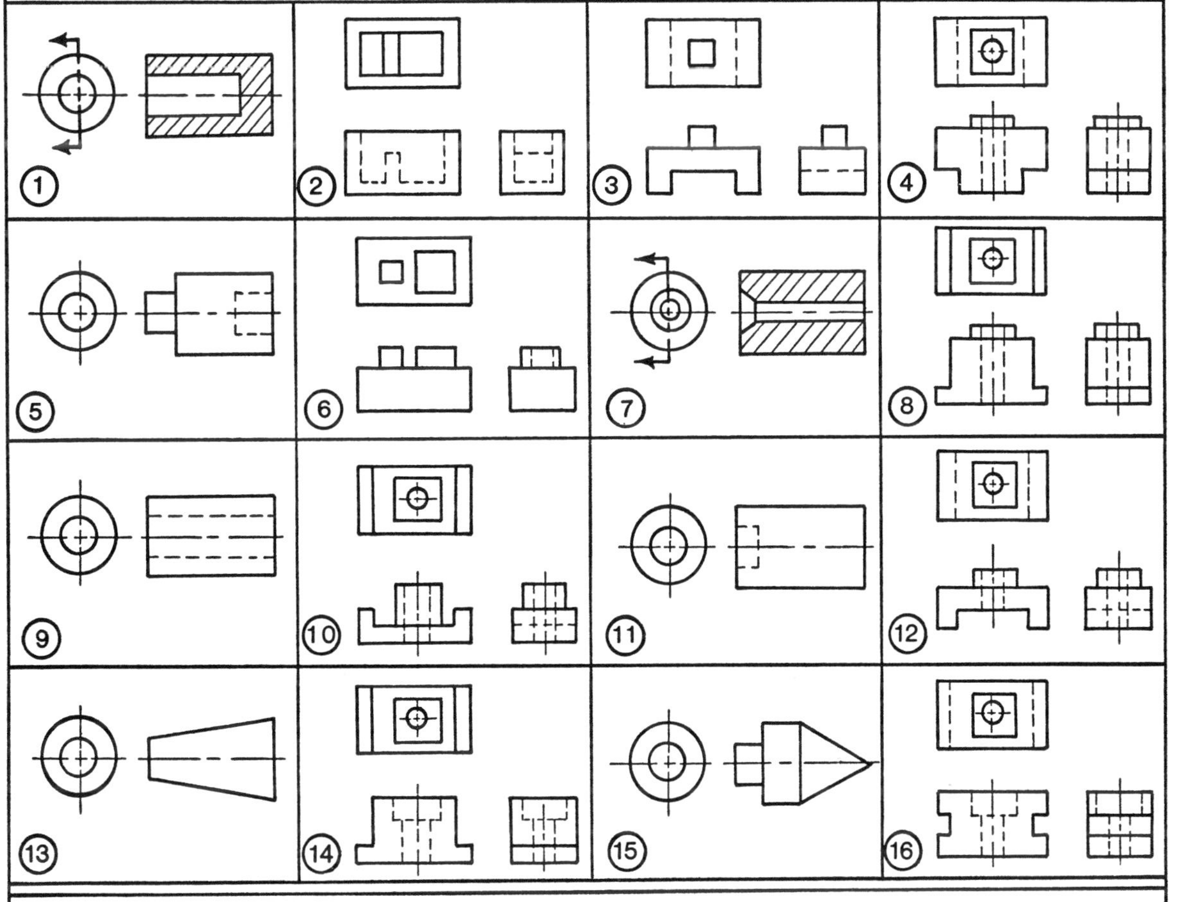

PICTORIAL DRAWINGS

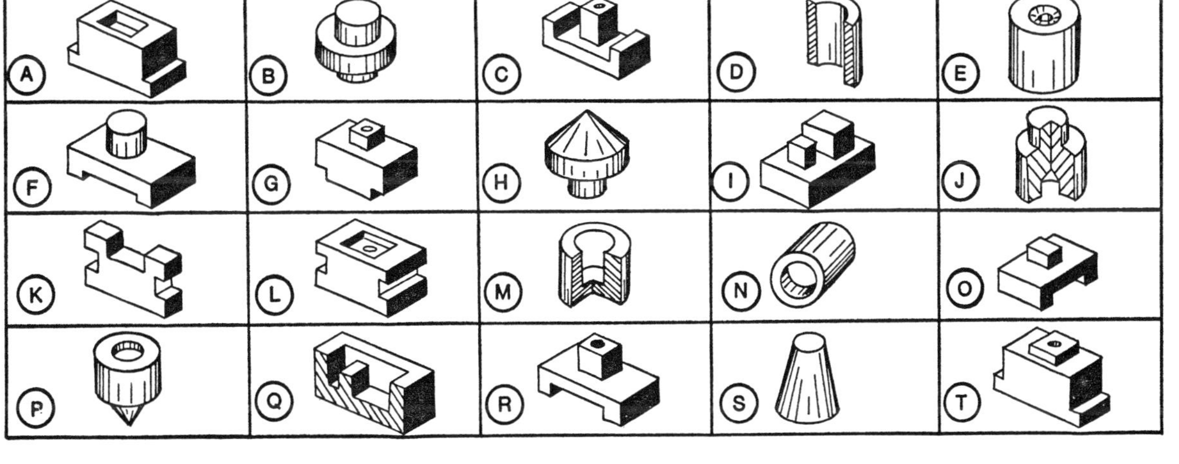

TEST

1. Divide the work area of your paper into four horizontal divisions and five vertical divisions.

2. Number the rectangles 1 to 20 as shown.

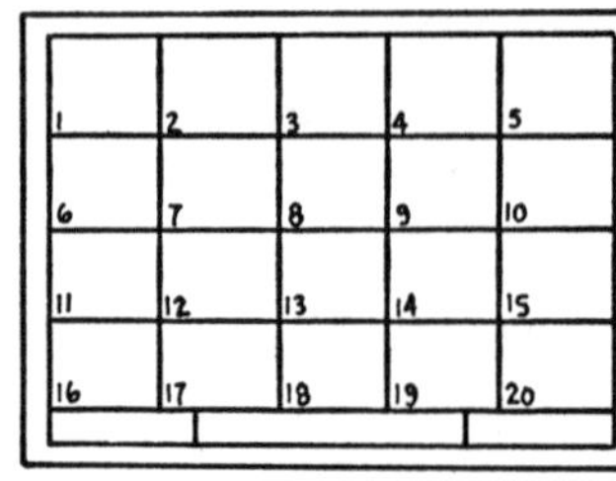

Make a freehand sketch of the orthographic view that you would make first, and show by an arrow in which direction you would project to obtain the other view or views (see example to the right). (Either answer would be correct.)

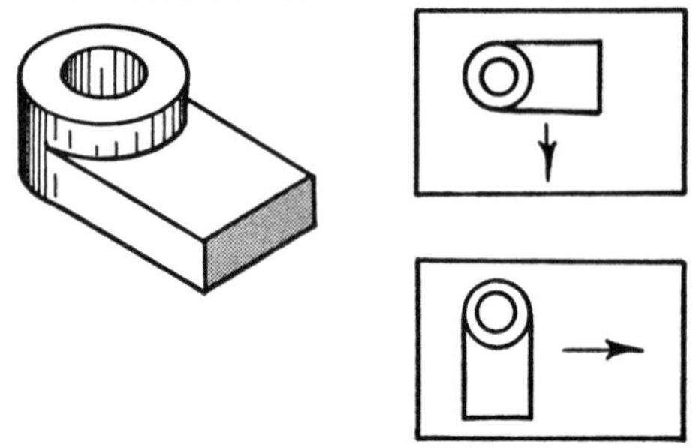

PART ONE

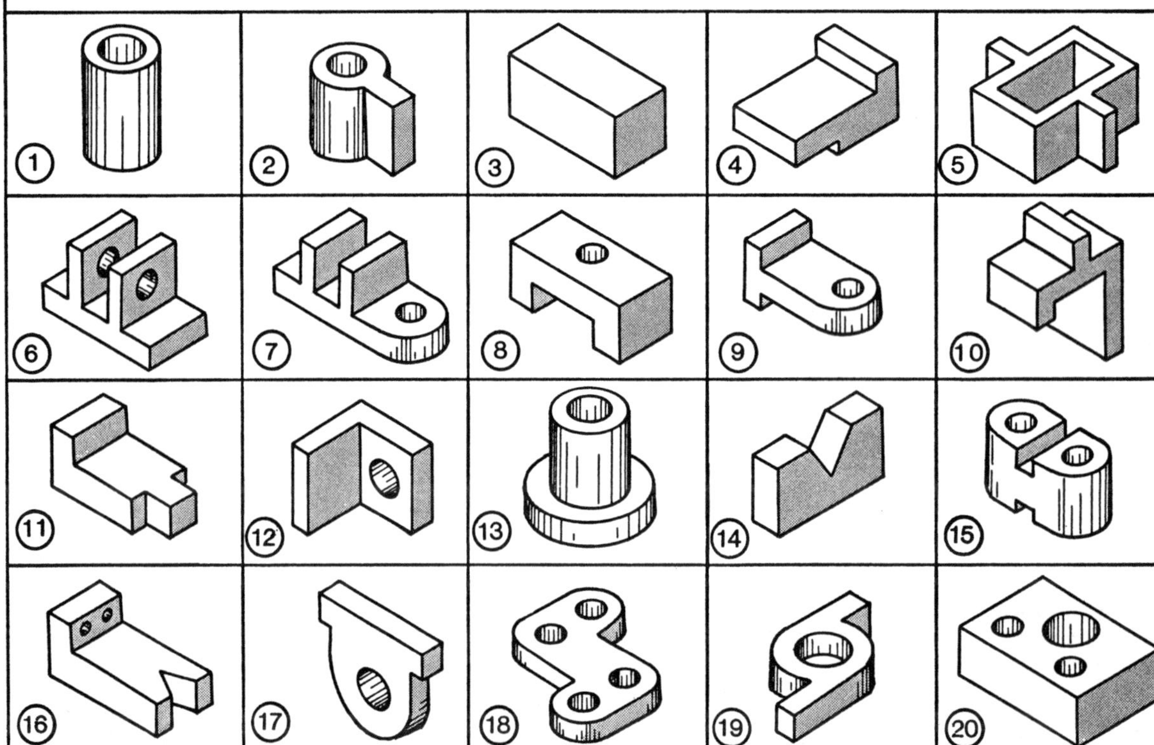

PART TWO

Make a freehand sketch of each. Show how you would dimension each angle shown.

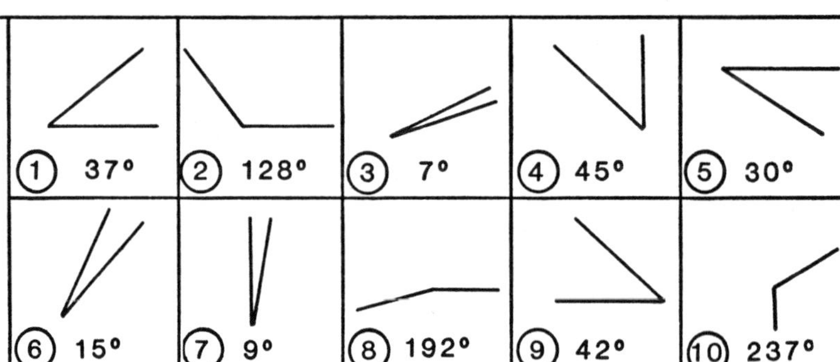

SKETCH THE ORTHOGRAPHIC SOLUTIONS

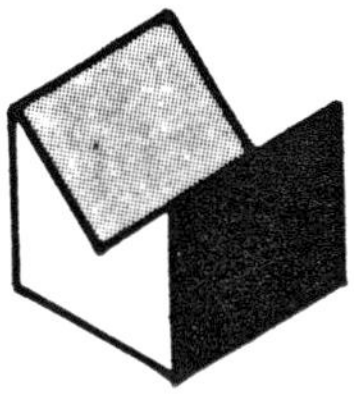

1

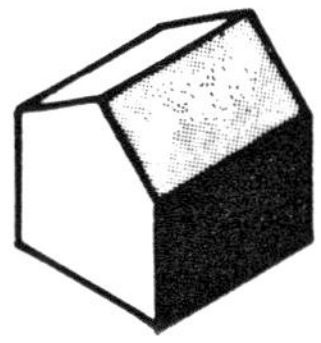

2

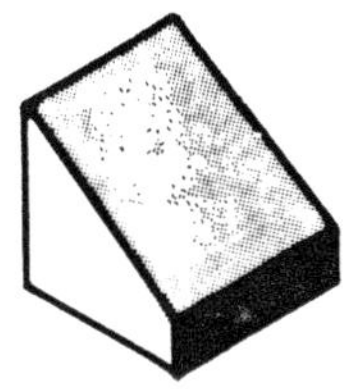

3

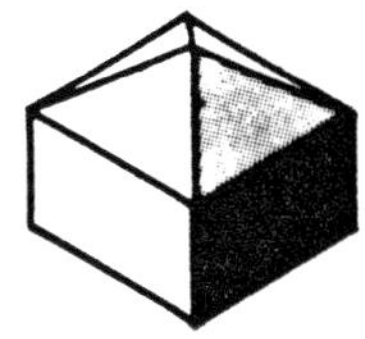

4

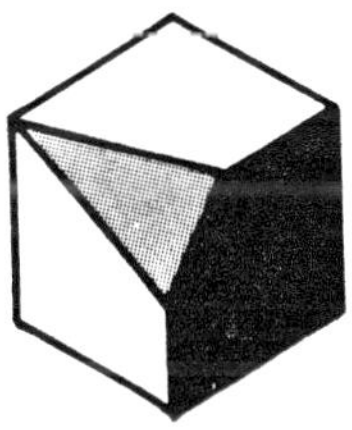

5

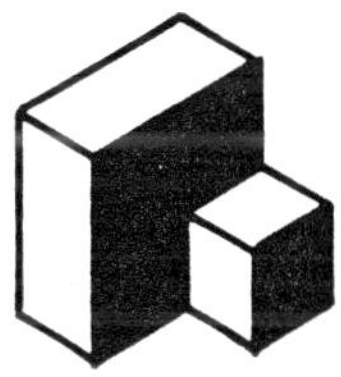

6

7

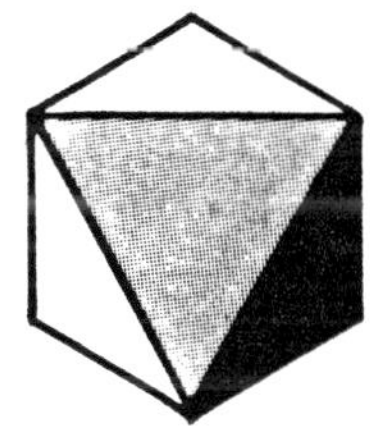

8

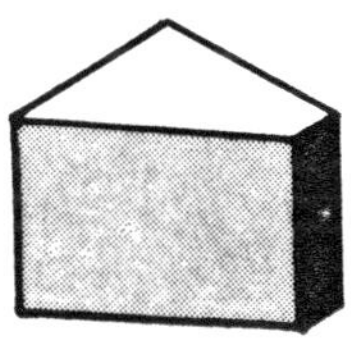

9

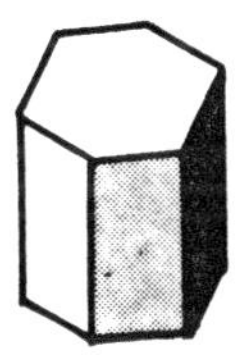

10

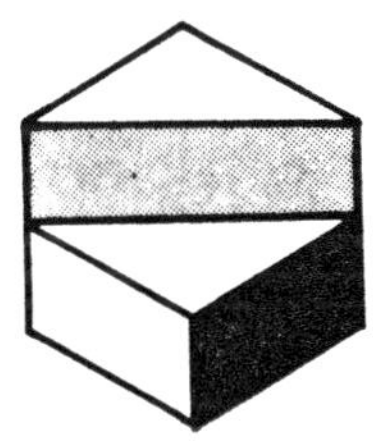

11

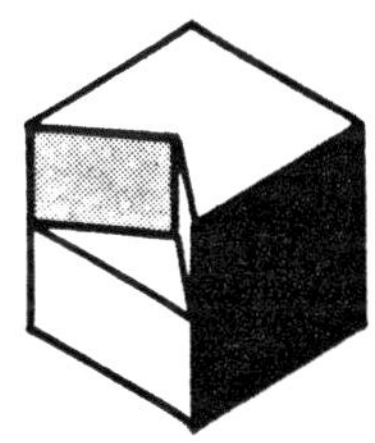

12

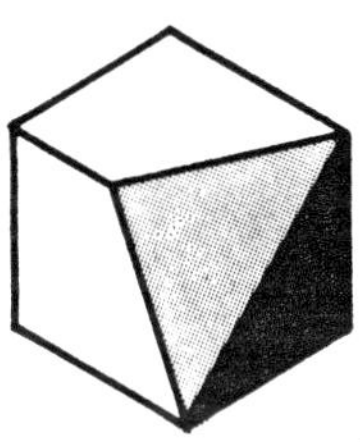

13

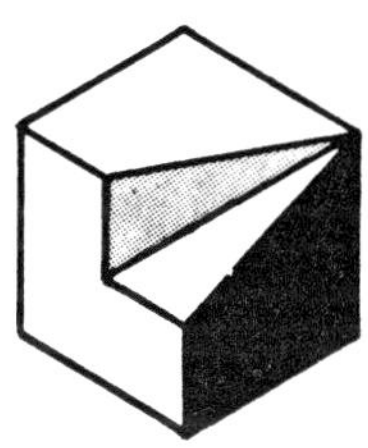

14

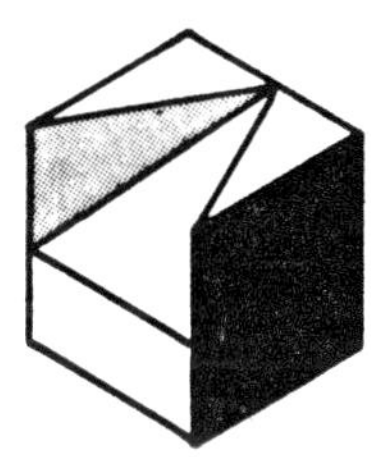

15

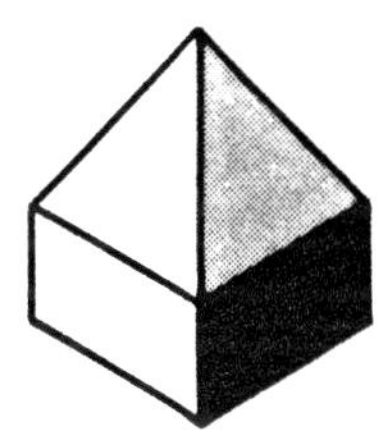

16

17

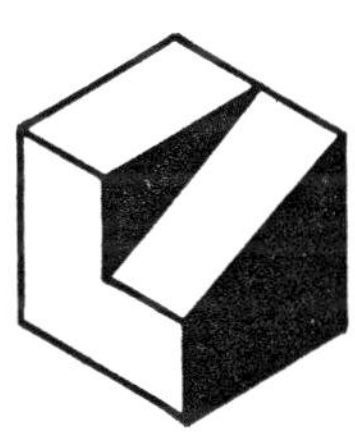

18

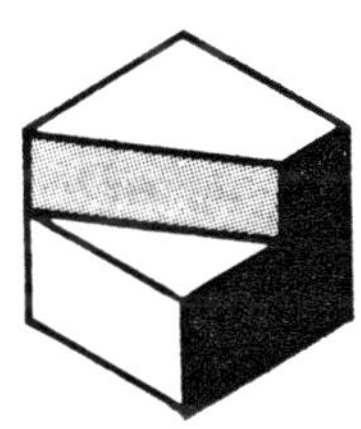

19

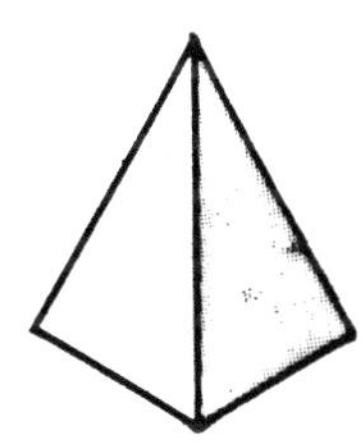

20

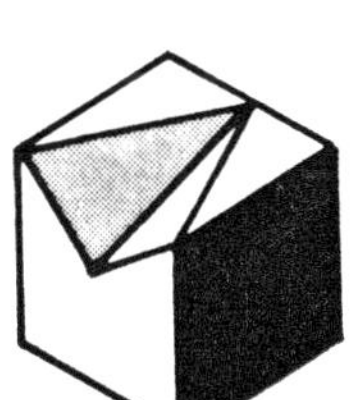

21

22

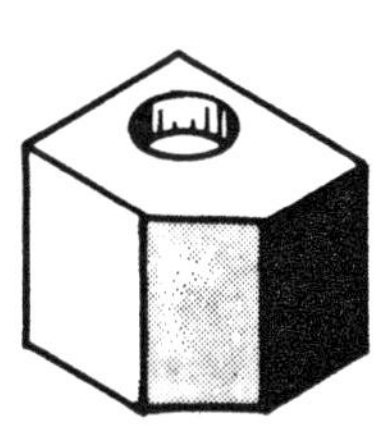

23

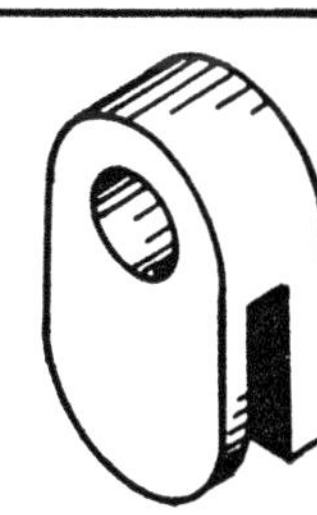

24

SKETCH THE ISOMETRIC SOLUTIONS

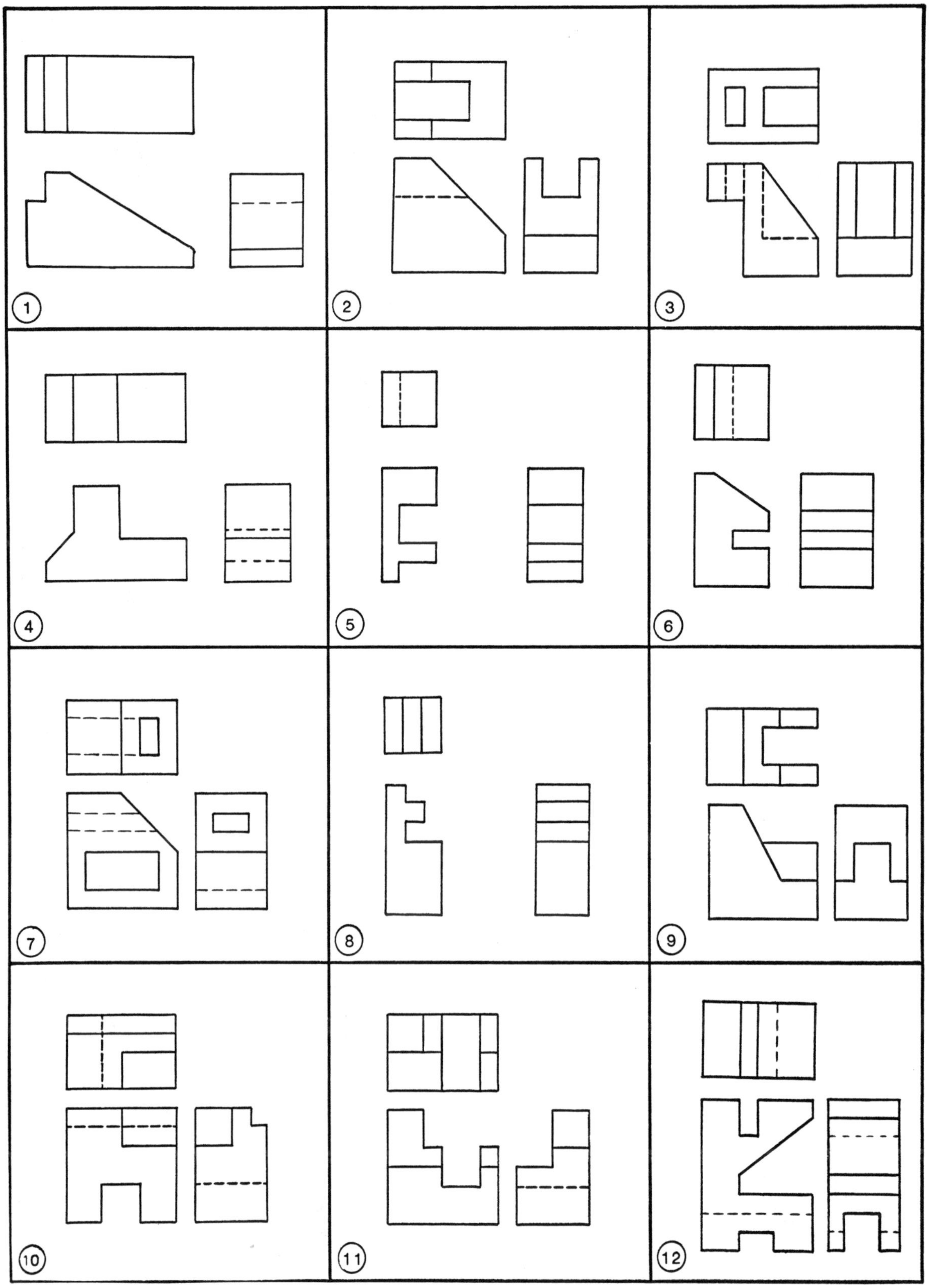

SKETCH THE ISOMETRIC SOLUTIONS

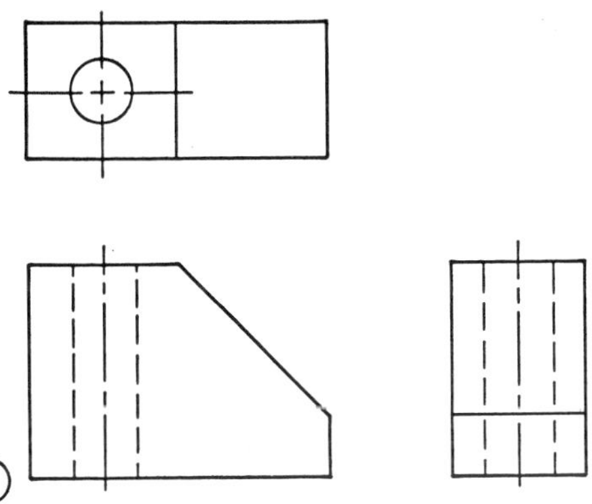

1

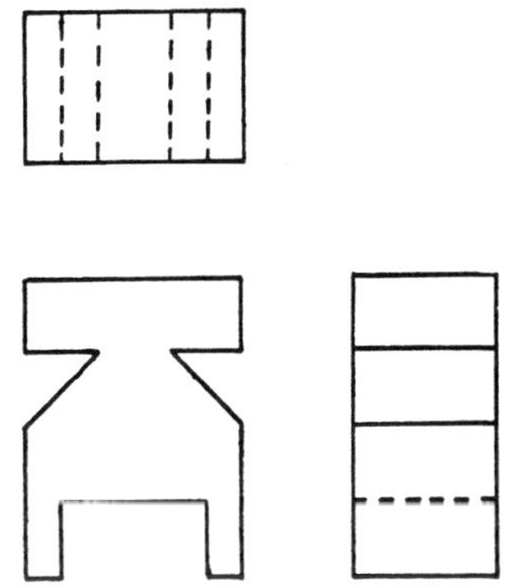

2

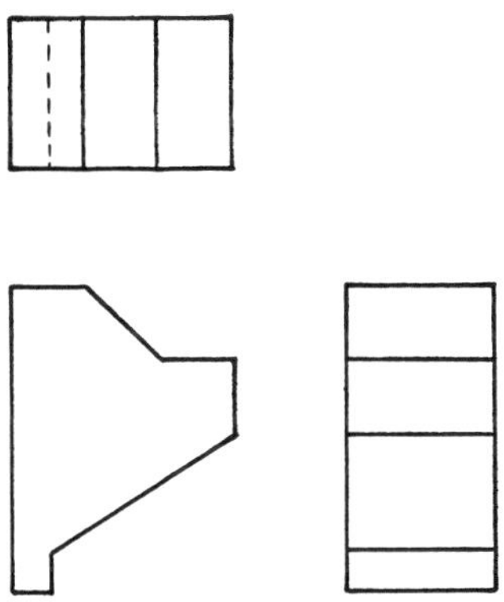

3

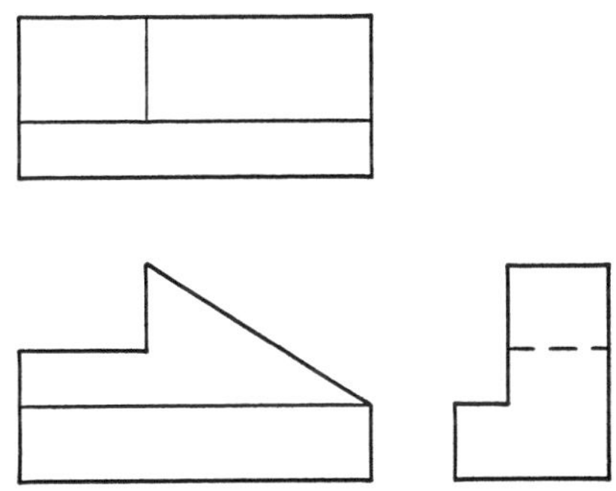

4

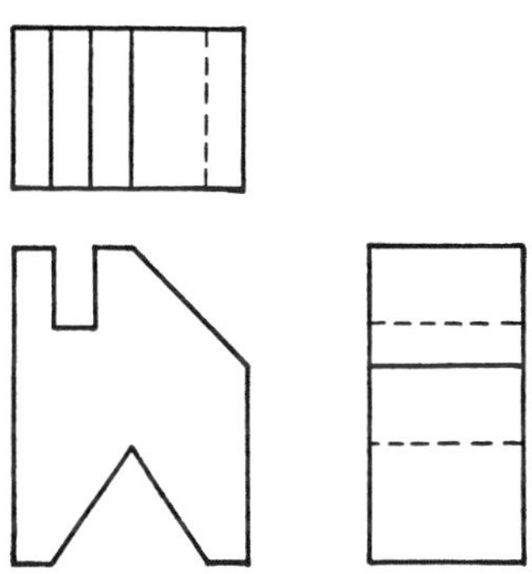

5

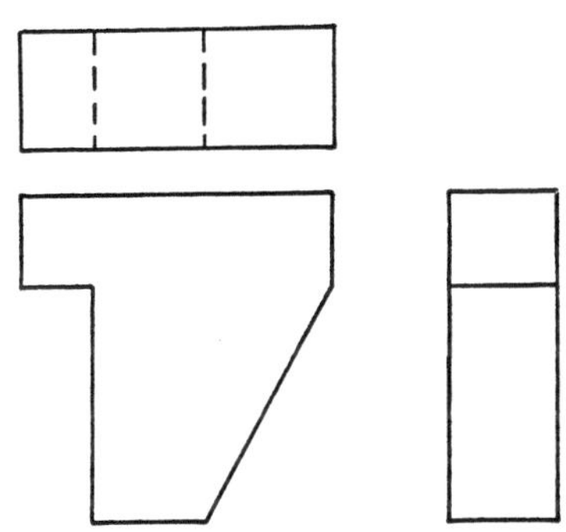

6

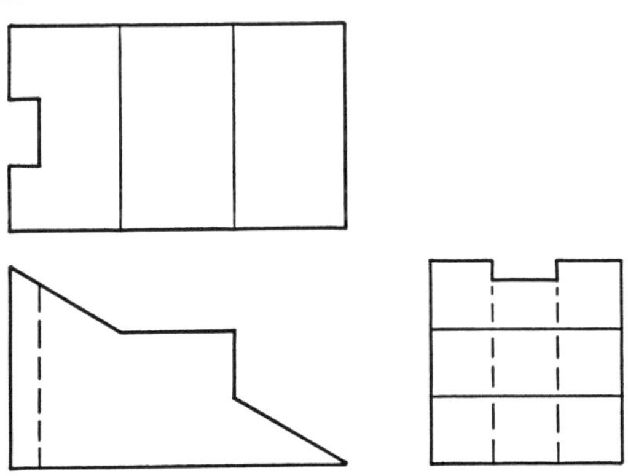

7

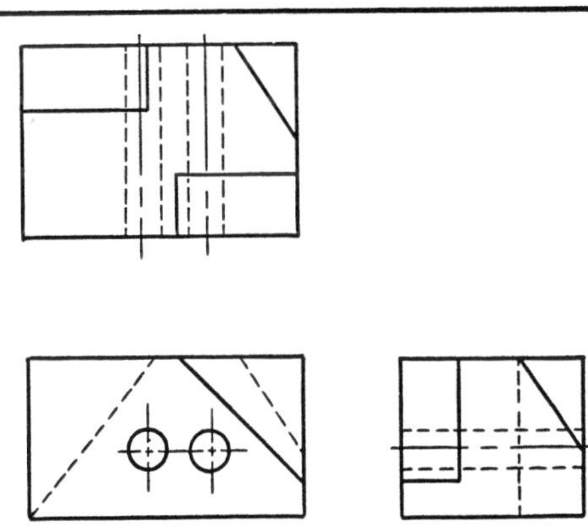

8

UNIT 17

1. **Inch Equivalents (Decimal, Metric)**
2. **Unified and American Screw Threads**
3. **Hexagon Bolts**

TABLES

METRIC-INCH EQUIVALENTS

Inches		Millimetres
Fractions	Decimals	
	.004	.10
	.008	.20
	.012	.30
1/64	.016	.39
	.017	.40
	.020	.50
	.024	.60
	.028	.70
1/32	.031	.79
	.032	.80
	.035	.90
	.039	1.00
3/64	.047	1.19
1/16	.063	1.59
5/64	.078	1.98
	.079	2.00
3/32	.094	2.38
7/64	.109	2.78
	.118	3.00
1/8	.125	3.18
9/64	.141	3.57
5/32	.156	3.97
	.157	4.00
11/64	.172	4.37
3/16	.188	4.76
	.197	5.00
13/64	.203	5.16
7/32	.219	5.56
15/64	.234	5.95
	.236	6.00
1/4	.250	6.35
17/64	.266	6.75
	.276	7.00
9/32	.281	7.14
19/64	.297	7.54
5/16	.313	7.94
	.315	8.00
21/64	.328	8.33
11/32	.344	8.73
	.354	9.00
23/64	.359	9.13
3/8	.375	9.53
25/64	.391	9.92
	.394	10.00

Inches		Millimetres
Fractions	Decimals	
	.472	12.00
31/64	.484	12.30
1/2	.500	12.70
	.512	13.00
33/64	.516	13.10
17/32	.531	13.49
35/64	.547	13.90
	.551	14.00
9/16	.563	14.29
37/64	.578	14.69
	.591	15.00
19/32	.594	15.08
39/64	.609	15.48
5/8	.625	15.88
	.630	16.00
41/64	.641	16.27
21/32	.656	16.67
	.669	17.00
43/64	.672	17.07
11/16	.688	17.46
45/64	.703	17.86
	.709	18.00
23/32	.719	18.26
47/64	.734	18.65
	.748	19.00
3/4	.750	19.05
49/64	.766	19.45
25/32	.781	19.84
	.787	20.00
51/64	.797	20.24
13/16	.813	20.64
	.827	21.00
53/64	.828	21.03
27/32	.844	21.43
55/64	.859	21.83
	.866	22.00
7/8	.875	22.23
57/64	.891	22.62
	.905	23.00
29/32	.906	23.01
59/64	.922	23.42
15/16	.938	23.81
	.945	24.00
61/64	.953	24.21

UNIFIED AND AMERICAN SCREW THREADS

	Coarse Thread Series UNC & NC		Fine Thread Series UNF & NF		Extra Fine Series UNEF & NEF		8-Pitch Thread Series 8 N		12-Pitch Thread Series 12 N		16-Pitch Thread Series 16 N	
Size	Threads Per Inch	Tap Drill	Threads Per Inch	Tap Drill	Threads Per Inch	Tap Drill	Threads Per Inch	Tap Drill	Threads Per Inch	Tap Drill	Threads Per Inch	Tap Drill
.060		80	80	3/64								
1 .073	64	No. 53	72	No. 53								
2 .086	56	No. 50	64	No. 50								
3 .099	48	No. 47	56	No. 45								
4 .112	40	No. 43	48	No. 42								
5 .125	40	No. 38	44	No. 37								
6 .138	32	No. 36	40	No. 33								
8 .164	32	No. 29	36	No. 29								
10 .190	24	No. 25	32	No. 21								
12 .216	24	No. 16	28	No. 14	32	No. 13						
1/4	20	No. 7	28	No. 3	32	7/32						
5/16	18	F	24	I	32	9/32						
3/8	16	5/16	24	Q	32	11/32						
7/16	14	U	20	25/64	28	13/32						
1/2	13	27/64										
1/2	12		20	29/64	28	15/32			12	27/64		
9/16	12	31/64	18	33/64	24	33/64			12	31/64		
5/8	11	17/32	18	37/64	24	37/64			12	35/64		
3/4	10	21/32	16	11/16	20	45/64			12	43/64	16	11/16
7/8	9	49/64	14	13/16	20	53/64			12	51/64	16	13/16
1	8	7/8	12	59/64	20	61/64	8	7/8	12	59/64	16	15/16
1/8	7	63/64	12	1 3/64	18	1 5/64	8	1	12	1 3/64	16	1 1/16
1 1/4	7	1 7/64	12	1 11/64	18	1 3/16	8	1 1/8	12	1 11/64	16	1 3/16
1 3/8	6	1 7/32	12	1 19/64	18	1 5/16	8	1 1/4	12	1 19/64	16	1 5/16
1 1/2	6	1 11/32	12	1 27/64	18	1 7/16	8	1 3/8	12	1 27/64	16	1 7/16
1 3/4	5	1 9/16			16	1 11/16	8	1 5/8	12	1 43/64	16	1 11/16
2	4 1/2	1 25/32			16	1 15/16	8	1 7/8	12	1 51/64	16	1 15/16
2 1/4	4 1/2	2 1/32					8	2 1/8	12	2 11/64	16	2 3/16
2 1/2	4	2 1/4					8	2 3/8	12	2 27/64	16	2 7/16
2 3/4	4	2 1/2					8	2 5/8	12	2 43/64	16	2 11/16
3	4	2 3/4					8	2 7/8	12	2 59/64	16	2 15/16
3 1/4	4	3					8	3 1/8	12	3 11/64	16	3 3/16
3 1/2	4	3 1/4					8	3 3/8	12	3 27/64	16	3 7/16
3 3/4	4	3 1/2					8	3 5/8	12	3 43/64	16	3 11/16
4	4	3 3/4					8	3 7/8	12	3 59/64	16	3 15/16
4 1/4							8	4 1/8	12	4 11/64	16	4 3/16
4 1/4							8	4 3/8	12	4 27/64	16	4 7/16
4 3/4							8	4 5/8	12	4 43/64	16	4 11/16
5							8	4 7/8	12	4 59/64	16	4 15/16
5 1/4							8	5 1/8	12	5 11/64	16	5 3/16
5 1/2							8	5 3/8	12	5 27/64	16	5 7/16
5 3/4							8	5 5/8	12	5 43/64	16	5 11/16
6							8	5 7/8	12	5 59/64	16	5 15/16

HEXAGON BOLTS

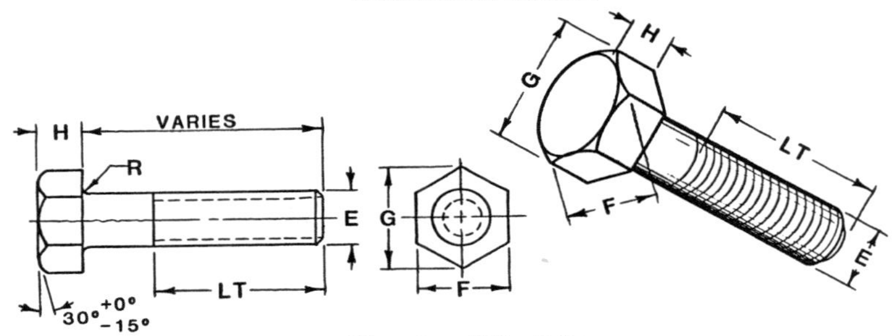

Dimension of Hex Bolts

Nominal Size or Basic Product Diameter		E	F			G		H			R		LT	
		Body Diameter	Width Across Flats			Width Across Corners		Height			Radius of Fillet		Thread Length for Bolt Lengths	
													6 in. and Less	Over 6 in.
		Max	Basic	Max	Min	Max	Min	Basic	Max	Min	Max	Min	Basic	Basic
1/4	0.2500	0.260	7/16	0.438	0.425	0.505	0.484	11/64	0.188	0.150	0.03	0.01	0.750	1.000
5/16	0.3125	0.324	1/2	0.500	0.484	0.577	0.552	7/32	0.235	0.195	0.03	0.01	0.875	1.125
3/8	0.3750	0.388	9/16	0.562	0.544	0.650	0.620	1/4	0.268	0.226	0.03	0.01	1.000	1.250
7/16	0.4375	0.452	5/8	0.625	0.603	0.722	0.687	19/64	0.316	0.272	0.03	0.01	1.125	1.375
1/2	0.5000	0.515	3/4	0.750	0.725	0.866	0.826	11/32	0.364	0.302	0.03	0.01	1.250	1.500
5/8	0.6250	0.642	15/16	0.938	0.906	1.083	1.033	27/64	0.444	0.378	0.06	0.02	1.500	1.750
3/4	0.7500	0.768	1 1/8	0.125	1.088	1.299	1.240	1/2	0.524	0.455	0.06	0.02	1.750	2.000
7/8	0.8750	0.895	1 5/16	1.312	1.269	1.516	1.447	37/64	0.604	0.531	0.06	0.02	2.000	2.250
1	1.0000	1.022	1 1/2	1.500	1.450	1.732	1.653	43/64	0.700	0.591	0.09	0.03	2.250	2.500
1 1/8	1.1250	1.149	1 11/16	1.688	1.631	1.949	1.859	3/4	0.780	0.658	0.09	0.03	2.500	2.750
1 1/4	1.2500	1.277	1 7/8	1.875	1.812	2.165	2.066	27/32	0.876	0.749	0.09	0.03	2.750	3.000
1 3/8	1.3750	1.404	2 1/16	2.062	1.994	2.382	2.273	29/32	0.940	0.810	0.09	0.03	3.000	3.250
1 1/2	1.5000	1.531	2 1/4	2.250	2.175	2.598	2.480	1	1.036	0.902	0.09	0.03	3.250	3.500
1 3/4	1.7500	1.785	2 5/8	2.625	2.538	3.031	2.893	1 5/32	1.196	1.054	0.12	0.04	3.750	4.000
2	2.0000	2.039	3	3.000	2.900	3.464	3.306	1 11/32	1.388	1.175	0.12	0.04	4.250	4.500
2 1/4	2.2500	2.305	3 3/8	3.375	3.262	3.897	3.719	1 1/2	1.548	1.327	0.19	0.06	4.750	5.000
2 1/2	2.5000	2.559	3 3/4	3.750	3.625	4.330	4.133	1 21/32	1.708	1.479	0.19	0.06	5.250	5.500
2 3/4	2.7500	2.827	4 1/8	4.125	3.988	4.763	4.546	1 13/16	1.869	1.632	0.19	0.06	5.750	6.000
3	3.0000	3.081	4 1/2	4.500	4.350	5.196	4.959	2	2.060	1.815	0.19	0.06	6.250	6.500
3 1/4	3.2500	3.335	4 7/8	4.875	4.712	5.629	5.372	2 3/16	2.251	1.936	0.19	0.06	6.750	7.000
3 1/2	3.5000	3.589	5 1/4	5.250	5.075	6.062	5.786	2 5/16	2.380	2.057	0.19	0.06	7.250	7.500
3 3/4	3.7500	3.858	5 5/8	5.625	5.437	6.495	6.198	2 1/2	2.572	2.241	0.19	0.06	7.750	8.000
4	4.0000	4.111	6	6.000	5.800	6.928	6.612	2 11/16	2.764	2.424	0.19	0.06	8.250	8.500